THE LIBRARY
ST. MARY'S COLLEGE OF MARYLAND
ST. MARY'S CITY, MARYLAND 20686

D1565484

Saas-Fee Advanced Course 31

P. Cassen   T. Guillot   A. Quirrenbach

Extrasolar Planets

Saas-Fee Advanced Course 31

Swiss Society for Astrophysics and Astronomy
Edited by D. Queloz, S. Udry, M. Mayor and W. Benz

With 140 Figures, 10 in Color

Patrick Cassen
SETI Institute
515 N. Whisman Rd.
Mountain View CA 94043 USA
pcassen@mail.arc.nasa.gov

Andreas Quirrenbach
Sterrewacht Leiden
PO Box 9513
2300 RA Leiden, The Netherlands
quirrenbach@strw.leidenuniv.nl

Tristan Guillot
Observatoire de la Côte d'Azur
BP 4229
06304 Nice, France
guillot@obs-nice.fr

Volume Editors:
Michel Mayor
Didier Queloz
Stephane Udry
Observatoire de Genève,
Chemin des Maillettes 51,
1290 Sauverny, Switzerland
michel.mayor@obs.uige.ch,
didier.queloz@obs.uige.ch,
stephane.udry@obs.uinge.ch

Willy Benz
Universität Bern,
Physikalisches Institut,
Sidlerstrasse 5,
3012 Bern, Switzerland

This series is edited on behalf of the Swiss Society for Astrophysics and Astronomy:
Société Suisse d'Astrophysique et d'Astronomie
Observatoire de Genève, ch. des Maillettes 51, 1290 Sauverny, Switzerland

Cover picture: Artwork of "51 Pegasus" made by L.R. Cook shortly after the announcement of the discovery of the planet. With permission of the artist.

Library of Congress Control Number: 2005938665

ISBN-10 3-540-29216-0 Springer Berlin Heidelberg New York
ISBN-13 978-3-540-29216-6 Springer Berlin Heidelberg New York

This work is subject to copyright. All rights are reserved, whether the whole or part of the material is concerned, specifically the rights of translation, reprinting, reuse of illustrations, recitation, broadcasting, reproduction on microfilm or in any other way, and storage in data banks. Duplication of this publication or parts thereof is permitted only under the provisions of the German Copyright Law of September 9, 1965, in its current version, and permission for use must always be obtained from Springer. Violations are liable for prosecution under the German Copyright Law.

Springer is a part of Springer Science+Business Media
springer.com
© Springer-Verlag Berlin Heidelberg 2006
Printed in The Netherlands

The use of general descriptive names, registered names, trademarks, etc. in this publication does not imply, even in the absence of a specific statement, that such names are exempt from the relevant protective laws and regulations and therefore free for general use.

Typesetting by authors and SPI Publisher Services using a Springer LATEX macro package
Cover design: *design & production*, Heidelberg

Printed on acid-free paper   SPIN: 11563556   55/SPI   5 4 3 2 1 0

Preface

In April 2001 the Swiss Society of Astrophysics and Astronomy (SSAA) organized its 31st winter "Saas-Fee" course on "Brown Dwarfs and Planets" in a picturesque resort at Grimentz on the Swiss Alps. The range of topics mainly focused on extrasolar planets' science. We entitled these lecture notes "Extrasolar Planets."

Research on extrasolar planets is one of the most exciting fields of activity in astrophysics. In just a decade a huge step has been made from the early speculations on the existence of planets orbiting "other stars" to the first discoveries and the characterization of extrasolar planets. This breakthrough is the result of the growing interest of a large community of researchers as well as the development of a wide range of new observation techniques and facilities. We organized the 31st winter course to cover all relevant aspects of this new field: observation and detection techniques, physics of their interior, and physics of their formation. We were very happy to have three senior lecturers, Andreas Quirrenbach, Tristan Guillot, and Patrick Cassen, cover these three subjects. They provided information to more than 100 participants and also gave updated comprehensive course materials, which is a challenging task considering the rapid development of this field of research. We hope that the level of details and the comprehensive view offered by authors will be appreciated as a comprehensive detailed introduction to this exciting subject.

We would like to warmly thank our three speakers for the high standard of their lectures and notes, as well as their discussion with students. We also thank all participants for their participation, kindness, and enthusiasm in taking part in the events organized. We thank Dominique Briguet of "A La Marena" for his hospitality and his help with the local organization. We would also like to warmly thank Elisabeth Teichamann, our course secretary, who gave us immense support during the preparation of the meeting as well as during the course. This course has been made possible thanks to a grant from the Swiss Academy of Sciences.

Geneva
June 2005

Didier Queloz
Stéphane Udry
Willy Benz
Michel Mayor

Our three lecturers: *(left)* Pat Cassen, *(middle)* Andreas Quirrenbach, *(right)* Tristan Guillot

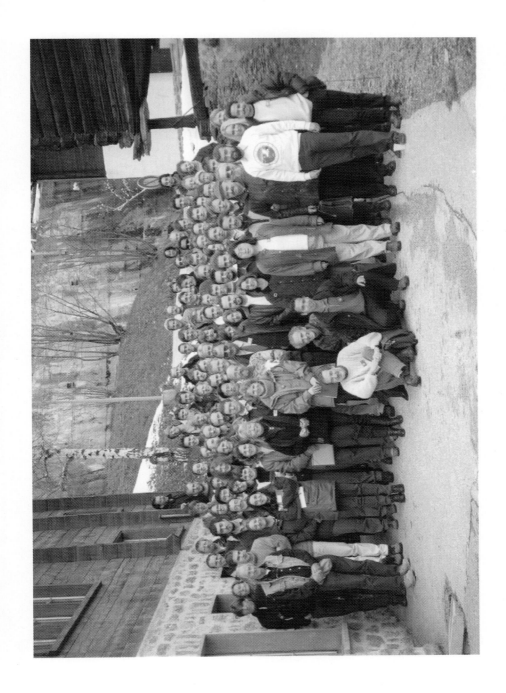

Contents

**Detection and Characterization
of Extrasolar Planets**
A. Quirrenbach 1
1   Methods of Planet Detection................................... 1
    1.1   The Quest for Planets Around Other Stars................. 1
    1.2   What is a Planet? 3
    1.3   Pulsar Planets.. 6
    1.4   Overview of Planet Detection Methods 9
    1.5   "Exotic" Concepts for Planet Detection 11
    1.6   The Search for Extraterrestrial Intelligence................ 15
2   Planet-Forming Disks 21
    2.1   Star Formation: the General Framework 22
    2.2   Observations of Dusty Disks 24
    2.3   Observations of Infalling Material and Evidence
          for Extrasolar Planetesimals............................. 31
3   The Currently Known Extrasolar Planets........................ 35
    3.1   The First Hundred Planets Around Solar-Type Stars 35
    3.2   Distribution of Masses and Orbital Radii 39
    3.3   Orbital Eccentricities.................................... 44
    3.4   Properties of the Parent Stars 47
    3.5   Systems with Multiple Planets 51
    3.6   Interactions Between the Planets in Multiple Systems 58
4   Radial-Velocity Surveys 62
    4.1   The Radial-Velocity Technique 62
    4.2   Limitations of the Radial-Velocity Precision 64
    4.3   Spectrograph Design 71
    4.4   Radial-Velocity Surveys 75
5   Gravitational Microlensing 76
    5.1   Theory of Gravitational Microlensing...................... 77
    5.2   Planetary Systems as Gravitational Lenses 83
    5.3   Microlensing Planet Searches 92

Contents

- 5.4 Astrometric and Interferometric Observations of Microlensing Events 96
- 6 Planetary Transits and Searches for Light Reflected by Planets 101
  - 6.1 The Geometry of Transits 101
  - 6.2 Photometric Error Sources 109
  - 6.3 HD 209458 112
  - 6.4 Photometric Planet Searches 115
  - 6.5 Photometric Space Missions 122
  - 6.6 Searches for the Light Reflected by the Planet 123
- 7 The Effects of Atmospheric Turbulence on Astronomical Observations 130
  - 7.1 The Kolmogorov Turbulence Model 130
  - 7.2 Wave Propagation Through Turbulence 133
  - 7.3 The Effect of Turbulence on Astronomical Images 135
  - 7.4 Fried Parameter and Strehl Ratio 137
  - 7.5 Temporal Evolution of Atmospheric Turbulence 139
  - 7.6 Temporal Structure Function and Power Spectra 139
  - 7.7 Angular Anisoplanatism 140
  - 7.8 Scintillation 141
  - 7.9 Turbulence and Wind Profiles 142
- 8 Introduction to Optical Interferometry 143
  - 8.1 Schematic Design of an Optical Interferometer 144
  - 8.2 Beam Combination Concepts 146
  - 8.3 Source Coherence and Interferometer Response 147
  - 8.4 Bandwidth and Interferometric Field-of-View 148
  - 8.5 Fringe Detection 148
  - 8.6 Phase Decorrelation Mechanisms 153
- 9 Astrometry with Interferometry 160
  - 9.1 Astrometric Signature of Low-Mass Companions 160
  - 9.2 Upper Mass Limits and Astrometric Detection of Gl 876 B 161
  - 9.3 Astrometric Measurements with an Interferometer 162
  - 9.4 Atmospheric Limitations of Astrometry 164
  - 9.5 Dual-Star Interferometry 168
  - 9.6 Interferometric Astrometry from Space 171
  - 9.7 Astrometric Planet Surveys 175
  - 9.8 Astrometric References 176
  - 9.9 The Differential-Phase Method 179
- 10 Nulling Interferometry 181
  - 10.1 Principles of Nulling 181
  - 10.2 Implementation of Achromatic Phase Shifts 189
  - 10.3 Nulling Interferometry from the Ground 195
  - 10.4 Design of Nulling Arrays 201
- 11 Appendix: Useful Definitions and Results from Fourier Theory 210
- References 212

Physics of Substellar Objects Interiors, Atmospheres, Evolution

T. Guillot 243

1. Introduction 243
2. "Our" Giant Planets as a Basis for the Study
   of Substellar Objects 243
   - 2.1 Origins: Role of the Giant Planets for Planet Formation 243
   - 2.2 Gravity Field and Global Properties 244
   - 2.3 Magnetic Fields 246
   - 2.4 Atmospheric Composition 246
   - 2.5 Energy Balance and Atmospheric Temperature Profiles 247
   - 2.6 Spectra 250
   - 2.7 Atmospheric Dynamics: Winds and Weather 251
   - 2.8 Moons and Rings 251
   - 2.9 Oscillations 252
3. Basic Equations, Gravitational Moments and Interior Structures ... 253
   - 3.1 Hydrostatic Equilibrium 253
   - 3.2 Boundary Conditions 253
   - 3.3 Simple Solutions 255
   - 3.4 Mass–Radius Relation 257
   - 3.5 Rotation and the Figures of Planets 260
   - 3.6 Equations of Evolution 265
4. Equations of State 266
   - 4.1 Basic Considerations 266
   - 4.2 Experiments and Theoretical Hydrogen EOSs 272
   - 4.3 Other Elements 277
5. Opacities and Heat Transport 279
   - 5.1 Radiation Absorption – Basic Considerations 280
   - 5.2 Rosseland Opacities 281
   - 5.3 Heat Transport 289
6. Interior Structures of our Giant Planets:
   Numerical Integrations and Results 296
   - 6.1 Basic Principles 296
   - 6.2 Jupiter and Saturn 296
   - 6.3 Uranus and Neptune 304
   - 6.4 Consequences for Formation Models 305
7. Evolution of Giant Planets and Brown Dwarfs 309
   - 7.1 The Virial Theorem 310
   - 7.2 A Semi-Analytical Model 312
   - 7.3 Evolution of Jupiter and Saturn 315
   - 7.4 From Giant Planets to Brown Dwarfs 320
   - 7.5 Extrasolar Giant Planets 326
8. Spectra and Atmospheres 326
   - 8.1 Direct Observations of Substellar Objects 326
   - 8.2 Atmospheric Models: Importance of Condensation 334

9 Pegasi Planets ("51 Peg b-like" Planets) 343
- 9.1 Introduction 343
- 9.2 Evolution of Strongly Irradiated Giant Planets 345
- 9.3 Tidal Effects 352
- 9.4 Atmospheric Dynamics 356

References 361

Protostellar Disks and Planet Formation
P. Cassen 369

1. Introduction 369
2. Observations of Protostellar Disks 370
   - 2.1 T Tauri Stars 370
   - 2.2 Interpretation of T Tauri Spectral Energy Distributions 371
   - 2.3 Accretion Rates of Protostellar Disks 379
   - 2.4 Internal Temperatures of Protostellar Disks 381
3. Theory of Disk Structure and Evolution 383
   - 3.1 Conservation Equations 383
   - 3.2 Turbulence in Disks 390
   - 3.3 Waves in Disks 395
4. Dust-Gas Dynamics 404
   - 4.1 Drift and Settling Velocities in the Absence of Turbulence 404
   - 4.2 Particle Growth and Trajectories in the Absence of Turbulence 407
   - 4.3 The Effect of Turbulence on Particle Settling 409
   - 4.4 The Initial Stages of Accumulation 411
5. Growth of Planetesimals to Planets 414
   - 5.1 Basics of Collisional Planet-building; Runaway Growth 414
   - 5.2 Three-Body Effects on Collision Cross-Section 416
   - 5.3 Evolution of the Velocity Distribution 418
   - 5.4 Calculating Planetesimal Growth 419
   - 5.5 The Final Stage of Accumulation; Rocky Planets in the Terrestrial Planet Region 422
6. The Formation of Gas Giant Planets 425
   - 6.1 Atmospheric Capture or Gravitational Collapse? 425
   - 6.2 Giant Planet Formation by Atmospheric Capture 425
   - 6.3 Giant Planet Formation by Gravitational Collapse 429
7. Planet Migration 431
   - 7.1 Tidal Interaction and Angular Momentum Exchange Between Planet and Disk 431
   - 7.2 Rates of Orbital Evolution 437
   - 7.3 Modeling the Formation of the Solar System 441
   - 7.4 Concluding Comments 444

References 444

Index 449

List of Previous Saas-Fee Advanced Courses

!! 2004 The Sun, Solar Analogs and the Climate
        J.D. Haigh, M. Lockwood, M.S. Giampapa

!! 2003 GravitationalLensing: Strong,Weak and Micro
        P. Schneider, C. Kochanek, J. Wambsganss

!! 2002 The Cold Universe
        A.W. Blain, F. Combes, B.T. Draine

!! 2001 Extrasolar Planets
        T. Guillot, P. Cassen, A. Quirrenbach

!! 2000 High-Energy Spectroscopic Astrophysics
        S.M. Kahn, P. von Ballmoos, R.A. Sunyaev

!! 1999 Physics of Star Formation in Galaxies
        F. Palla, H. Zinnecker

!! 1998 Star Clusters
        B.W. Carney, W.E. Harris

!! 1997 Computational Methods for Astrophysical Fluid Flow
        R.J. LeVeque, D. Mihalas, E.A. Dorfi, E. Müller

!! 1996 Galaxies Interactions and Induced Star Formation
        R.C. Kennicutt, F. Schweizer, J.E. Barnes

!! 1995 Stellar Remnants
        S.D. Kawaler, I. Novikov, G. Srinivasan

* 1994 Plasma Astrophysics
        J.G. Kirk, D.B. Melrose, E.R. Priest

* 1993 The Deep Universe
        A.R. Sandage, R.G. Kron, M.S. Longair

* 1992 Interacting Binaries
        S.N. Shore, M. Livio, E.J.P. van den Heuvel

* 1991 The Galactic Interstellar Medium
        W.B. Burton, B.G. Elmegreen, R. Genzel

* 1990 Active Galactic Nuclei
        R. Blandford, H. Netzer, L. Woltjer

! 1989 The Milky Way as a Galaxy
        G. Gilmore, I. King, P. van der Kruit

! 1988 Radiation in Moving Gaseous Media
        H. Frisch, R.P. Kudritzki, H.W. Yorke

! 1987 Large Scale Structures in the Universe
        A.C. Fabian, M. Geller, A. Szalay

! 1986 Nucleosynthesis and Chemical Evolution
        J. Audouze, C. Chiosi, S.E. Woosley

! 1985 High Resolution in Astronomy
    R.S. Booth, J.W. Brault, A. Labeyrie
! 1984 Planets, Their Origin, Interior and Atmosphere
    D. Gautier, W.B. Hubbard, H. Reeves
! 1983 Astrophysical Processes in Upper Main Sequence Stars
    A.N. Cox, S. Vauclair, J.P. Zahn
* 1982 Morphology and Dynamics of Galaxies
    J. Binney, J. Kormendy, S.D.M. White
! 1981 Activity and Outer Atmospheres of the Sun and Stars
    F. Praderie, D.S. Spicer, G.L. Withbroe
* 1980 Star Formation
    J. Appenzeller, J. Lequeux, J. Silk
* 1979 Extragalactic High Energy Physics
    F. Pacini, C. Ryter, P.A. Strittmatter
* 1978 Observational Cosmology
    J.E. Gunn, M.S. Longair, M.J. Rees
* 1977 Advanced Stages in Stellar Evolution
    I. Iben Jr., A. Renzini, D.N. Schramm
* 1976 Galaxies
    K. Freeman, R.C. Larson, B. Tinsley
* 1975 Atomic and Molecular Processes in Astrophysics
    A. Dalgarno, F. Masnou-Seeuws, R.V.P. McWhirter
* 1974 Magnetohydrodynamics
    L. Mestel, N.O. Weiss
* 1973 Dynamical Structure and Evolution of Stellar Systems
    G. Contopoulos, M. Hénon, D. Lynden-Bell
* 1972 Interstellar Matter
    N.C. Wickramasinghe, F.D. Kahn, P.G. Metzger
* 1971 Theory of the Stellar Atmospheres
    D. Mihalas, B. Pagel, P. Souffrin

---

*   Out of print
!   May be ordered from Geneva Observatory
        Saas-Fee Courses
        Geneva Observatory
        CH-1290 Sauverny
        Switzerland
!! May be ordered from Springer-Verlag

Detection and Characterization of Extrasolar Planets

A. Quirrenbach

1 Methods of Planet Detection

1.1 The Quest for Planets Around Other Stars

The realization that our Sun is just one "average" star amongst billions and billions in the Sky naturally brings with it the question whether some – or perhaps most – of the other stars may also harbor planetary systems like our own. We live in a remarkable epoch, being the first generation that has obtained an affirmative answer to this question, that is undertaking programs to characterize the physical properties of planets outside the Solar System, and that is developing the tools to search for twins of the Earth. For the first time in human history, we are on the verge of being able to address the questions whether there are other habitable worlds, and to search for life elsewhere in the Universe with scientific methods.

The search for extrasolar planets has a long and checkered history (see e.g., Boss 1998a for an easily readable overview). Because of the enormous brightness contrast between planets and their parent stars, the direct detection of planets by taking images of the vicinity of nearby stars would be extremely difficult. Early searches for planets were therefore mostly carried out with the astrometric method, which seeks to detect the motion of the star around the center of mass of the star–planet system (see Sect. 9). First reports on the detection of massive planets ($\sim 10\,M_{\rm jup}$) were published during World War II (Strand 1943; Reuyl and Holmberg 1943), but remained controversial, both with regards to the reality of the results and to the question whether the detected bodies should be called "planets". Much painstaking work over the next few decades lead to the realization that these "detections" were spurious. Continued improvements in the astrometric accuracy finally culminated in the announcement of a planet 1.6 times as massive as Jupiter in a 24-year orbit around Barnard's Star (van de Kamp 1963). A decade earlier Otto Struve had written a remarkable paper, in which he noted the possibility that Jupiter-like planets might exist in orbits as small as 0.02 AU, proposed to search for these

Quirrenbach A (2006), Detection and characterization of extrasolar planets. In: Mayor M, Queloz D, Udry S and Benz W (eds) Extrasolar planets. Saas-Fee Adv Courses vol 31, pp 1–242
DOI 10.1007/3-540-29216-0_1 © Springer-Verlag Berlin Heidelberg 2006

objects with high-precision radial-velocity measurements, and pointed out the feasibility of photometric searches for planets eclipsing their parent stars – all on little more than one journal page (Struve 1952).

By the mid-sixties, the search for extrasolar planets thus appeared to be a thriving field, with eight planetary companions known from astrometric observations (two of them classified as "existence not completely established"), and a number of potentially promising alternative search methods under consideration (O'Leary 1966). By the same time it had also been recognized that brown dwarfs (termed "black dwarfs" at the time) would form a class of their own, with properties intermediate between those of stars and planets. Both astrometric searches for brown-dwarf companions to low-luminosity stars, and attempts at finding them directly with high-resolution imaging techniques, seemed to be successful (Harrington et al. 1983; McCarthy et al. 1985).

Sadly, none of these early claims for detections of planets and brown dwarfs withstood the test of time. It turned out that systematic instrumental errors had been mistaken for the "planetary companion" of Barnard's Star (Gatewood and Eichhorn 1973). What appeared to be the most convincing detection of a brown dwarf, a companion to the star VB 8, could never be confirmed (Perrier and Mariotti 1987; Skrutskie et al. 1987). Other putative planets and brown dwarfs did not fare better. By the mid-nineties, all that remained was a candidate brown dwarf companion of HD 114762, detected with the radial-velocity method (Latham et al. 1989).[1]

This situation changed completely and abruptly with the discovery of 51 Peg b, a Jupiter-like planet in a 4-day orbit (Mayor and Queloz 1995), which has opened a completely new field of astronomy: the study of extrasolar planetary systems. About 150 planets outside our own Solar System are known to date, and new discoveries are announced almost every month. These developments have revolutionized our view of our own place in the Universe. We know now that other planetary systems can have a structure that is completely different from that of the Solar System, and we have set out to explore their properties and diversity.

The following chapters introduce the most important methods that have been employed (or proposed) for the detection of extrasolar planets, and for studies of their physical characteristics. Emphasis is given to observational techniques, their foundations, limitations, and their practical implementation. As far as possible, published results are mentioned in the context of the respective observing techniques, and some outstanding implications for the astrophysics of planets and planetary systems are discussed. This will hopefully elucidate the capabilities, strengths, and weaknesses of the many

---

[1] The radial-velocity technique does not allow measuring the companion mass, but only $m \sin i$, where i is the unknown inclination of the orbit (see Sect. 4). It could therefore not be excluded that HD 114762 B is a low-mass star in a nearly face-on orbit.

complementary observational approaches. It should be kept in mind, however, that the study of extrasolar planets is a rapidly expanding field, in which new and unanticipated results appear almost every month. Technical developments in fields such as adaptive optics, coronography, and interferometry are also occurring at a staggering pace. Nevertheless, the systematic introduction of the fundamental principles and methods attempted in this article will hopefully remain a useful guide for a while to come.

1.2 What is a Planet?

The Definition of "Planet"

Before we can begin to answer the question how planets outside the Solar System can be detected and characterized, we must first agree on an operational definition of the term "planet". The Greek root of the word literally means "unsteady" or "transient"; it was historically applied to the five known "wandering stars" Mercury, Venus, Mars, Jupiter, and Saturn. The Copernican Revolution added the Earth to the list, and the discoveries of Neptune, Uranus, and Pluto completed the census of the large bodies in the Solar System as we know it. The example of Pluto clearly demonstrates the need for a clean definition of the term "planet". With the discovery of a large number of bodies belonging to the Kuiper Belt (Jewitt and Luu 1993; Luu and Jewitt 2002) it has become clear that Pluto is but the largest member of the class of Trans-Neptunian Objects (TNOs). It has therefore be argued that Pluto should be demoted from its rank among the planets. I would side with the majority view, however, that the use of the term "planet" in the Solar System is based on historical developments and should not be changed retroactively.

The history in our own Solar System thus shows that the use of the term "planet" has been expanded from the original five members of this class, to newly discovered objects that shared the most important properties of the established examples. Two of these additions (Neptune and Uranus) were rather undramatic, one was based on the realization that the Earth shared important properties with the planets (it orbits the Sun between Venus and Mars), and one added a physically distinct and different body to the list (Pluto). Progress in our knowledge about the planets has also taught us that our list includes bodies encompassing wide ranges in mass, composition, and other physical characteristics.

When we look outside our Solar System, we should certainly expect to find a variety of objects that share many characteristics with our planets, but that may be different in one or more important ways. It is thus a matter of definition what we call a "planet" and where we draw the boundaries to other classes of objects. From a practical point of view, this definition should be based on properties that are easily verifiable observationally; this favors a definition based on mass over a definition based on the formation history. Nonetheless, we should not expect that we can easily come up with a set of criteria that will in each case allow an unambiguous classification of a newly discovered as a "planet" (or not). For example, if a maximum mass is included among the defining properties, all objects discovered with the radial-velocity technique – and thus with known $m \sin i$, see Sect. 4.1 – could strictly speaking only be called "planet candidates" before additional information on their orbital inclination is secured.

For the purposes of this article, I take a "planet" to be an object that fulfills the following criteria:

- *A planet is an object in orbit around a star or a multiple star system.* This excludes free-floating planet-mass objects. A number of such objects have been detected with direct-imaging surveys in young clusters (e.g., Zapatero Osorio et al. 2000, 2002; Béjar et al. 2001; Lucas et al. 2001).[2] Free-floating objects are not considered further here, although it is possible that some of them originally formed in a circumstellar disk, and were ejected by a collision with another planet (e.g., Bryden 2001).
- *A planet is not in orbit around another planet.* This requirement excludes moons, but one should point out that the distinction between moons and planets is also somewhat fuzzy. For example, the Pluto–Charon systems could be called a double planet rather than a planet with a moon.
- *A planet has a minimum mass of* 10^{22} kg. This distinguishes planets from planetesimals, asteroids, and comets.[3]
- *A planet has a maximum mass of* $13\,M_{\rm jup}$. This sets the boundary between planets and brown dwarfs. The value of $13\,M_{\rm jup}$ has been chosen to roughly coincide with the Deuterium burning limit (e.g., Burrows et al. 1997a). This criterion will often be applied fairly loosely, as objects with $m \sin i < 13\,M_{\rm jup}$ will also be called "planets" even if there is no additional information on i.

As with the word "planet", we will use other terms established in the Solar System and and apply them to analogous bodies and material around other stars; we can thus speak of "moons" and "rings" around extrasolar planets, about "exo-planetesimals", "exo-comets" and "exo-zodiacal dust".

The Thermal Evolution of Giant Planets

A basic understanding of the evolution of planets is an important prerequisite for a discussion of detection methods. The fundamental principle is rather

---

[2] Note that a tentative detection of free-floating planet-mass objects in the cluster M 22 (Sahu et al. 2001) has been retracted (Sahu et al. 2002).

[3] The value of 10^{22} kg is quite arbitrary, of course. I have chosen it because it separates Pluto from the "minor bodies" in the Solar System.

simple: planets are born hot and are initially self-luminous; they cool during their evolution until they reach radiative equilibrium with their parent stars. The age of a planet is therefore an important parameter that determines how difficult it is to detect its thermal emission.

The luminosity evolution of giant planets, alongside with that of brown dwarfs and low-mass stars, is shown quantitatively in Fig. 1; it can be seen from this figure that an old planet is about four orders of magnitude(!) fainter than it was at an age of 1 Myr. Another important conclusion is that luminosity alone is an extremely poor indicator of the mass of substellar objects – information about the age is crucial to distinguish between low-mass stars, brown dwarfs, and planets. (Dynamically determined masses are even better, of course.) We see, however, that at the same age the luminosity of gaseous planets increases strongly with mass. Figure 1 thus provides a very useful first orientation for general considerations about the detectability of giant planets. For more detailed calculations, one has to take into account the planet's temperature, which determines the spectral energy distribution, and modifications to the luminosity evolution due to irradiation by the host star (e.g., Burrows et al. 2003).

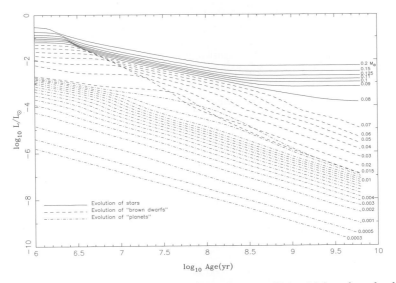

Fig. 1. Evolution of the luminosity (in $L_\odot$) of solar-metallicity M dwarfs and substellar objects vs. time (in yr) after formation. The stars, "brown dwarfs" and "planets" are shown as solid, dashed, and dot-dashed curves, respectively. In this figure, we arbitrarily designate as "brown dwarfs" those objects that burn deuterium, while we designate those that do not as "planets." The masses (in $M_\odot$) label most of the curves, with the lowest three corresponding to the mass of Saturn, half the mass of Jupiter, and the mass of Jupiter. From Burrows et al. (1997a)

1.3 Pulsar Planets

The First Extrasolar Planets

While the considerations of the preceding sections appear to give a solid framework for planet searches, the first firm discovery of objects that fulfill the above definition of a planet came totally unexpected and from a completely different line of research. The extremely stable rotation of pulsars provides a high-precision clock, which can be used for the indirect detection of planets, in a way that is quite similar to the radial-velocity method that will be discussed in detail below (Sect. 4). High-precision monitoring of the time-of-arrival (TOA) of the radio pulses can reveal subtle motions of the pulsar, such as its reflex motion due to the presence of a planetary companion. For a planet with mass m_p in a circular orbit with period P and inclination i, and a "canonical" neutron star mass of $1.35\,M_\odot$, the amplitude of the timing residuals τ is

$$\tau = 1.2\,\mathrm{ms} \left(\frac{m_p}{M_\oplus}\right) \left(\frac{P}{1\,\mathrm{yr}}\right)^{2/3} \sin i \,. \qquad (1)$$

For millisecond pulsars, TOA measurements are possible with a long-term precision of a few µs (e.g., Wolszczan 1994). This implies that planets down to $\sim 0.01\,M_\oplus$ are detectable around pulsars; this limit is far lower than that of any other search method currently contemplated.

After a few false starts (e.g., Bailes et al. 1991; Lyne and Bailes 1992), two planets just a factor of ~ 3 more massive than the Earth were found orbiting the pulsar PSR B1257+12 (Wolszczan and Frail 1992). The two planets are in a 3:2 orbital resonance, which leads to accurately predictable periodic perturbations of the two orbits. The detection of this mutual gravitational attraction between the planets provided the final proof of the reality of the first pulsar planets; the same data set also revealed the presence of a third planet with even lower mass in the same system (Wolszczan 1994). The properties of the planets orbiting PSR B1257+12 are listed in Table 1.

Table 1. Parameters of the PSR B1257+12 planetary system

planet	A	B	C
semi-major axis [light-ms]	0.0035	1.3106	1.4121
eccentricity e	0.0	0.0182	0.0264
orbital period [days]	25.34	66.54	98.22
longitude of periastron	–	249°	106°
planet mass [$M_\oplus$]	$0.015/\sin i_1$	$3.4/\sin i_2$	$2.8/\sin i_3$
distance from pulsar [AU]	0.19	0.36	0.47

After Wolszczan (1999)

The Keplerian timing residuals (1) depend on $m_p \sin i$; this means that the mass of the planet and its orbital inclination cannot be determined independently. In contrast, the strength of the mutual interaction between the planets depends directly on their masses. It has thus become possible to infer the masses and inclinations of planets B and C from modeling of a long series of timing data, which now covers more than a decade (Konacki and Wolszczan 2003). The derived masses are $4.3 \pm 0.2\, M_\oplus$ and $3.9 \pm 0.2\, M_\oplus$, respectively, and the orbital inclinations $53° \pm 4°$ and $47° \pm 3°$ (or $127°$ and $133°$), indicating that the two orbits are nearly co-planar.

Even after taking the three planets and the interaction between planets B and C into account, there remains a long-term systematic variation of the TOA residuals (see the lower two panels of Fig. 2). These residuals could be indicative of the presence of a fourth planet with longer orbital period

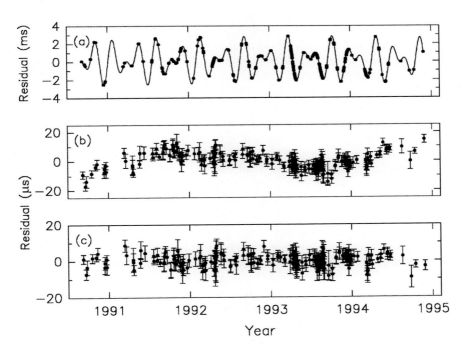

Fig. 2. Timing residuals for PSR B1257+12 at 430 MHz, for three increasingly detailed models. (**a**) Residuals after the fit of the standard timing model without planets. The time-of-arrival variations are dominated by the Keplerian orbital effects from planets B and C. (**b**) Residuals for the model including the Keplerian orbits of planets A, B, and C. Residual variations are determined by gravitational perturbations between planets B and C. (**c**) Residuals for the model including all the standard pulsar parameters, and the Keplerian and non-Keplerian orbital effects. From Konacki and Wolszczan (2003)

(Wolszczan et al. 2000). If the apparent three-year periodicity of the residual signal can be confirmed, this would point to an origin of the disturbance within the pulsar planetary system itself. It will probably be difficult to ascertain the nature of this ionized material – a "coma" ablated from a fourth body or a warped disk are among the possibilities.

Pulsar planets appear to be rare. Only one other pulsar, PSR B1620−26 near the core of the globular cluster M4, has a confirmed planet (Arzoumanian et al. 1996; Joshi and Rasio 1997). The B1620−26 system is rather interesting, too. The planet orbits an inner binary system, which consists of a millisecond pulsar and a white dwarf companion in a half-year orbit. The most likely mass of the planet is $m_p \sin i_p \approx 7\,M_{\rm jup}$, and the semi-major axis and eccentricity of its orbit are $a \approx 60$ AU and $e \approx 0.45$ (Thorsett et al. 1999; Ford et al. 2000).

The Formation of Pulsar Planets

The theories for the formation of pulsar planets can be broadly divided into two classes: (a) scenarios, in which the planets were formed together with a "normal" star, and survived its evolution from the main-sequence to become a red giant and later a rapidly spinning neutron star; and (b) scenarios in which the formation of the neutron star precedes the formation or acquisition of its planets (Podsiadlowski 1993). The first category implies that the planets must be able to survive the formation of the pulsar, which involves a violent transformation in a supernova explosion, and the supernova recoil. This possibility is generally regarded as unlikely, and scenarios of type (b) are favored.

Consequently, most theories of planet formation around millisecond pulsars concern themselves with possible ways to disrupt, evaporate, ablate, or otherwise dismember the companion star, and thus to transform a fraction of the companion's mass into a gaseous disk around the neutron star (Phinney and Hansen 1993). Such a disk could be formed, for example, by an asymmetric supernova explosion in a binary system, which kicks the neutron star into its companion. In this picture, a high-velocity single neutron star with a planetary system is created from the remains of the former binary companion. One could thus speculate that the presence of planets around PSR B1257+12 is related to its unusually high proper motion.

Neutron star disks are clearly very different from those commonly found around pre-main-sequence stars. The disk is exposed to intense radiation and particle flux, close to or even above the Eddington luminosity of the neutron star ($\sim 10^{38}\,{\rm erg\,s^{-1}}$). The metallicity is very high, but initially there are no grains, and the temperature is well above the sublimation temperature of even the most refractory materials. The disk must therefore expand and cool before planets can be formed. Calculations of the evolution of such disks indicate that the formation of "terrestrial" planets such as those of the PSR B1257+12 system may indeed by possible, but the more massive and distant planet around PSR B1620−26 must have a different origin (Phinney and Hansen 1993).

The location of PSR B1620−26 near the core of the globular cluster M4 suggests that the pulsar acquired its planetary companion through an exchange interaction with a cluster star (Sigurdsson 1992, 1995). One plausible formation scenario begins with an old neutron star in a binary system, which interacts with a main-sequence star–planet system (Sigurdsson 1993). The original companion of the neutron star is ejected, while the main-sequence star and its planet are captured. The planet ends up in a wide orbit around the inner binary comprised of the neutron star and the main-sequence star. When the main-sequence star evolves to become a red giant, it transfers mass to the neutron star, spinning it up to become a millisecond pulsar. The chief difficulty of this scenario is the requirement that the age of the millisecond pulsar must be smaller than that of the triple system. However, the expected lifetime of the triple in the dense cluster core is of order $3 \cdot 10^7$ yr, while the estimated age of the binary pulsar is $\gtrsim 10^9$ yr (Ford et al. 2000). This scenario would thus require that the system, currently observed in projection near the edge of the cluster core, is in fact on an orbit that allows it to spend most of its lifetime in the far less dense cluster halo, and thus to escape disruption for a sufficiently long time.

An alternative formation scenario involves a dynamical exchange interaction between a pre-existing binary millisecond pulsar and a wide main-sequence star–planet system, in which the main-sequence star is ejected and the planet left in a wide orbit around the binary pulsar (Ford et al. 2000). Numerical simulations show that the probability of retaining the planet in the encounter is smaller than that of retaining the main-sequence star, but could still be as high as 10%...30%. It is interesting to note that this scenario postulates the formation of a giant planet in a wide orbit around a "normal" star in a globular cluster environment; this is to be contrasted with the apparent absence of "hot Jupiters" in 47 Tucanae (see Sect. 6.4).

1.4 Overview of Planet Detection Methods

The Most Important Detection Techniques

Turning our attention to "normal" stars again, we will now look at the question how we might be able to find planets around them. Many different techniques have been proposed, in spite (or perhaps: because) of the difficulty of the task. The most promising strategies that are used in current detection efforts, or under development for use in the near future, are:

- Direct imaging of the star–planet system.
- Interferometric imaging of the star–planet system.
- Detection of the planetary spectrum in a composite spectrum of star and planet.

- Interferometric detection of the planetary spectrum through the wavelength dependence of the position of the photocenter of the star–planet system ("differential phase method").
- Photometry of planetary transits in front of the star.
- Spectroscopic detection of planetary transits.
- Photometric detection of the light reflected by a planet through its periodic variation with phase angle.
- Astrometric detection of the stellar motion around the star–planet center of mass.
- Radial-velocity measurement of the stellar motion around the star–planet center of mass.
- Imaging of circumstellar disks, which may show signatures of disk–planet interaction.
- Gravitational microlensing.
- Eclipse timing in binaries.

Each one of these techniques has unique strengths and weaknesses, and they vary widely in the information they can provide about the properties of the detected planets. Turning the question the other way around, we can take a list of characteristics that we would like to know about extrasolar planets, and ask which techniques can provide the requested information. Table 2 gives an overview of the most important planetary properties, and how they may be determined. More detailed discussions about the strengths and limitations of individual methods will be given in the subsequent sections. For the moment, the most important observation is that no single approach can give all the desired information; many complementary techniques will be needed to study the different aspects of extrasolar planets and planetary systems.

Typical Order-of-Magnitude Estimates

In order to understand the instrumental and observational requirements for the different planet detection techniques, we need to consider typical values for the potential observables. The large range in the properties of planets obviously implies a large difference in the difficulty to detect them. The Earth and Jupiter provide useful benchmarks (see Table 3), but one should also keep in mind that there are additional classes of planets, e.g., Uranus and Neptune in the Solar System, or the "hot Jupiters" orbiting their central stars at very small orbital radii. The properties of the host star, and the distance from the observer play important roles, too.

The chief difficulty of direct detection methods is the large contrast between the planet and its parent star at a very small angular separation. The reflex motion of the parent star due to the gravitational pull of the planet is very small, so that astrometry and the radial-velocity technique must reach extremely high precision to detect this effect. The photometric signature of transiting planets seems somewhat more easily accessible, at least for giant

Table 2. Important properties of planets, and techniques that can be used to determine them

property	technique	applicability
orbit	astrometry	++
	radial velocity	+
	direct imaging	o
mass	astrometry	++
	radial velocity	+
	microlensing	o
radius	transit photometry	++
radius, albedo	photometry of reflected light	++
radius, temperature	direct detection in mid-IR	++
surface features	photometry of reflected light	+
atmospheric composition	IR or visible spectroscopy	++
	transit spectroscopy	o
presence of moons	transit timing	+
system multiplicity	astrometry	+
	radial velocity	+

The symbols ++, +, and o denote how well the different methods can provide the required information

planets. However, in this case additional complications arise from the small probability that the orientation is such that transits actually occur. It is thus clear from the values listed in Table 3 that there is no "easy" technique for planet detection – this is the reason, of course, why it was not before 1995 that the first planet around a main-sequence star was discovered.

1.5 "Exotic" Concepts for Planet Detection

The subsequent chapters of this review will be devoted to introductions to some of the most promising planet detection techniques. Many more interesting approaches have been proposed, which deserve at least a brief description. It is entirely possible, of course, that one or the other of these "exotic" concepts will turn out to be more fruitful than some of the techniques that are considered "mainstream" today. The variety of physical effects that could in principle be observable should illustrate the diverse opportunities for the immediate and more distant future, and stimulate further ideas about possible ways to obtain more detailed information about planets during various phases of their life cycles, and about their interaction with the host stars.

Table 3. Typical values of observables for Jupiter-like and Earth-like planets

observable	Jupiter	Earth
angular separation	$0\rlap{.}''5$	$0\rlap{.}''1$
brightness contrast at visible $\lambda\lambda$	6×10^{-7}	1.5×10^{-10}
brightness contrast at $10\,\mu\mathrm{m}$	1.5×10^{-7}	1.2×10^{-7}
astrometric amplitude	$500\,\mu\mathrm{as}$	$0.3\,\mu\mathrm{as}$
radial-velocity amplitude	$13\,\mathrm{m\,s^{-1}}$	$0.1\,\mathrm{m\,s^{-1}}$
transit probability	10^{-3}	5×10^{-3}
transit depth	1%	10^{-4}
transit duration	30 h	13 h
timing residuals	2.5 s	1.5 ms

The host star is assumed to be a Sun twin at a distance of 10 pc

Radio Emission from Extrasolar Planets

Five of the planets in the Solar System (Earth, Jupiter, Saturn, Uranus, and Neptune) produce non-thermal cyclotron radio emission, in a process that is thought to be driven by the Solar wind interacting with the planetary magnetospheres. The emission frequency is typically near the electron gyrofrequency in the magnetic field, i.e., of order 30 kHz ... 30 MHz. The emission is very intense (at times, Jupiter is brighter than the Sun at frequencies below 20 MHz), and there exist fairly simple scaling laws that relate the observed radio power to the ram pressure of the Solar wind on the cross-sectional area of the magnetosphere (Zarka et al. 2001). There also exist scaling laws for the magnetic dipole moment of giant planets (Farrell et al. 1999); these scaling laws together can be used to predict the emitted radio power and peak frequency. In a few favorable cases the emission should be observable with current instruments, but no detections have been made so far (e.g., Bastian et al. 2000). This may either be due to the intermittent nature of the cyclotron emission, or to a smaller velocity or density of the stellar wind compared to the Solar wind, or to a smaller magnetic moment of the planet, due perhaps to tidal synchronization. In any case, future low-frequency arrays such as LOFAR or a Square Kilometer Array (Strom et al. 2001) should be able to observe the radio emission from magnetized giant planets.

Interaction-Induced Stellar Activity

Tidal or magnetic interaction between a giant planet in a short-period orbit and its host star might also increase the stellar activity, which could lead to variations in the shape of chromospheric lines in phase with the orbital period. Hints of systematic modulations of the Ca II H and K lines have been found in a few such systems, but they need further confirmation (Cuntz and Shkolnik 2002). It has also been speculated that very strong flares observed in some Solar-type stars could be due to magnetic reconnection between fields of the primary star and a close-in Jupiter-like planet (Rubenstein and Schaefer 2000). A systematic search for unusual flaring stars might thus be a new way of looking for planets, but a better understanding of the planet–star interaction would clearly be needed.

Young Planets Heated by Giant Impacts

The formation of planets proceeds through a phase of giant impacts (e.g., Wetherill 1990). The largest of these impacts may melt all or most of the surface of an Earth-size planetary embryo, and heat it to a temperature of about $1,500\ldots2,500$ K. This would make its thermal emission detectable with a large ground-based interferometer (Stern 1994). Giant impacts on giant planets (such as the event that may have tipped the rotation axis of Uranus) will likely heat them to similar temperatures, making them even more easily detectable. The cooling times are of order a few hundred to several thousand years; this should make the number of impact-heated objects at any given time large enough to expect one detection per every few hundred pre-main-sequence stars surveyed. One would then still have to establish the planetary nature of the detected object, of course, and distinguish it from more massive companions or other possible interlopers.

Planets Swallowed by Giant Stars

As a Solar-mass star evolves off the main sequence, it expands and ascends the red giant branch of the Hertzsprung–Russell diagram. During that evolutionary phase, the star develops a large convective envelope with a radius of up to $\sim 100\,R_\odot$. Planets within that radius will be accreted by the star, and thus deposit energy, angular momentum, and elements such as Lithium in the stellar envelope. It has therefore been argued that the infrared excess (due to a substantial expansion of the star and ejection of a shell) and high Li abundance observed in $\approx 5\%$ of the G and K giants could be caused by the accretion of a giant planet or a brown dwarf (Gratton and D'Antona 1989; Siess and Livio 1999).

In an even later evolutionary stage, when the star becomes an asymptotic giant branch (AGB) star, it swells to an even larger size and develops a strong wind. Planets with even larger orbital radii can then get engulfed in the extended atmosphere, or interact with the wind flow. Episodic accretion of wind material on the planet may give rise to optical flashes and affect SiO maser emission (Struck et al. 2002). The details of these interactions are quite complex and poorly understood at the moment; this limits their potential use as a diagnostic tool and indicator for the presence of planets.

Planets Around White Dwarfs

It should be clear from the preceding paragraph that the orbits of planets can change drastically during the star's post-main-sequence evolution. Low-mass companions will spiral into the star due to the viscous and tidal forces exerted by the bloated atmosphere during the giant phase, but it might also be possible that some of them may be left in an orbit of radius $a \lesssim 1\,\mathrm{AU}$ around the ensuing white dwarf. This would be a very favorable situation for detecting the planet, because its radius would be ~ 10 times larger than that of the parent star! In the Rayleigh–Jeans portion of the combined spectrum (i.e., for observations at wavelengths much longer than that corresponding to the peak of the Planck function), the ratio of the total emission to that of just the white dwarf is given by

$$\frac{I_\mathrm{tot}}{I_\mathrm{WD}} = 1 + \frac{R_p^2 T_p}{R_\mathrm{WD}^2 T_\mathrm{WD}} \approx 1 + 100\,\frac{T_p}{T_\mathrm{WD}}\,. \qquad (2)$$

For example, a planet with $T_p = 200\,\mathrm{K}$ orbiting a white dwarf with $T_\mathrm{WD} = 10,000\,\mathrm{K}$ would dominate the total emission of the system at long wavelengths.

Several groups have conducted near-infrared searches for substellar companions of white dwarfs, and some low-mass companions have been reported, but no planet has been discovered (e.g., Zuckerman and Becklin 1992). The above argument suggests, however, that searches should be conducted in the mid-infrared, where the planet can produce a strong excess over the white dwarf spectrum (Ignace 2001). The Spitzer (formerly SIRTF) infrared mission should have sufficient sensitivity to detect such planets out to a distance of $\sim 10\,\mathrm{pc}$.

Occultations by the Moon or Artificial Satellites

The planet detection schemes discussed in the previous few paragraphs intend to take advantage of special situations in which the the signature of the planet is not swamped by the nearby bright host star. In the general case, one may try to address the contrast problem by blocking the light from the star. This could either be achieved by using the dark limb of the Moon as an occulting edge (Elliot 1978), or by building a spacecraft carrying an occulting screen (Schultz et al. 1999, 2000; Copi and Starkman 2000). In either case, the observations would be carried out with a space telescope, which has to maintain alignment with the occulter to a precision of a fraction of an arcsecond (the typical angular separation of the planet from its parent star). The main obstacles for Lunar occultations are the rather large brightness of even the dark side of the Moon, and the difficulties of maneuvering the telescope. While it is possible to find orbits that give rather long ($\sim 1\,\mathrm{h}$) occultations of arbitrary stars, an enormous amount of propellant would have to be used to change targets.

Artificial occulters face similar problems. The diameter of the occulting screen clearly has to be larger than the telescope aperture, i.e., at least $\sim 10\,\mathrm{m}$.

To subtend an angle of no more than 0″.1, it must therefore be placed at a separation of at least 20,000 km from the telescope. Furthermore, the intensity of the starlight in the shadow of the occulter is not zero; it must be computed with Fresnel's diffraction theory (e.g., Born and Wolf 1997).[4] Application of Babinet's Principle gives the approximation (Schultz et al. 1999)

$$\frac{I}{I_0} \approx \frac{16}{\pi^2} \cdot \frac{\lambda a}{D^2} = \frac{16}{\pi^2} \cdot \frac{\lambda}{\varphi D} , \qquad (3)$$

where I and I_0 are the intensity in the presence and in absence of the occulting disk, λ the observing wavelength, a the distance between the occulter and the telescope, D the diameter of the occulter, and φ the angle subtended by the occulter as seen by the telescope. Equation (3) is valid for $D^2 \gg \lambda a$. For the above numbers ($D = 10$ m, $a = 20{,}000$ km) and $\lambda = 500$ nm, the on-axis intensity is still 16% of the value in the absence of an occulter. This shows that diffraction at the edge is a serious problem. With λ and φ fixed, and clear limits on the potential to increase D and a, the only viable way of obtaining a better starlight suppression is the use of a tapered occulter, i.e., a screen which is not completely opaque but has a transmission that continuously increases from 0 at the center to 1 at the edge (Copi and Starkman 2000). Manufacturing such a screen with precisely prescribed transmission function is a considerable technological challenge. This, together with the requirement of maneuvering the screen and telescope very precisely, has so far prevented serious consideration of this approach for a planet-detection mission.

A variation of the occultation idea is the use of a coronograph, which includes an occulting spot in the focal plane of the telescope. Compared to an external occulter, a coronograph has the disadvantage that the starlight is blocked only after passage through the telescope optics. The telescope therefore has to be built to very stringent specifications on wavefront quality and light scattering level. Nevertheless, this approach is currently regarded more promising than that of an external occulting screen.

1.6 The Search for Extraterrestrial Intelligence

The Drake Equation

Speculations about the possibility of life, of conscious beings, and of civilizations elsewhere in the Universe have a long history (Dick 1982, 1998). The search for extraterrestrial intelligence (SETI) as a scientific endeavor was born with the realization that our own technology had advanced to the point that radio signals could be transmitted and detected over interstellar distances

---

[4] One may recall Poisson's famous bright spot. Using Fresnel's theory, Poisson – who was very critical of that theory – predicted the seemingly absurd appearance of a bright spot behind a circular obstruction. This spot was almost immediately found experimentally by Arago, a great triumph of the wave theory of light.

(Cocconi and Morrison 1959). Soon the question was raised how many civilizations in the Galaxy might be engaged in attempts at communicating with each other, leading to the formulation of the famous *Drake Equation* (Drake 1962)

$$N = R_* \cdot f_p \cdot n_h \cdot f_l \cdot f_i \cdot f_c \cdot L \,. \qquad (4)$$

The individual factors in this equation have the following meanings:

- N: the number of communicating civilizations in Galaxy;
- R_*: the rate of star formation in the Galaxy (expressed in stars per year);
- f_p: the fraction of stars that harbor planetary systems;
- n_h: the average number of planets (or moons) with conditions that are suitable for the genesis of life;
- f_l: the fraction of habitable planets on which life actually develops;
- f_i: the probability that evolution produces intelligent life;
- f_c: the fraction of intelligent civilizations that try to communicate over interstellar distances;
- L: the length of the communication phase (in years).

Unlike the other equations in this book, which (hopefully!) quantify our insights and knowledge, it is the main purpose of the Drake equation to organize our ignorance. We know that $R_* \approx 1\,\mathrm{yr}^{-1}$ (Trimble 1999), and we can now state fairly confidently that $f_p \geq 0.01$.[5] The determination of n_h is one of the great observational challenges for the coming ten to twenty years, as described extensively in this overview. With some luck, we might even be able to obtain an estimate for f_l from astronomical observations. This factor may also be amenable to biochemical experimentation in the tradition of the famous Miller–Urey experiments (Miller 1953) and modern attempts to generate synthetic life forms (Szostak et al. 2001). At present, in the absence of any evidence for extraterrestrial life, we have to admit that f_l could be anywhere between 10^{-9} and 1.

The next factor, f_i, is equally uncertain. Biologists are deeply divided about the question whether life necessarily evolves towards intelligence once it gets going. On the one hand, one may point out that a staggeringly improbable series of events has lead to the emergence of intelligent life on Earth (Gould 1989); on the other hand, one can argue that convergence is a ubiquitous property of life, which makes it likely that particular biological properties and features will sooner or later manifest themselves as part of the evolutionary process (Conway Morris 1998). In addition, we do not understand the biological basis of intelligence at all. What is the "quantum leap" that separates *homo sapiens* from *pan troglodytes*, the chimpanzee? Would *homo neanderthalensis* have become capable of constructing radio telescopes, if he

---

[5] This estimate is based on the number of planets detected in the Solar neighborhood (Sect. 3.1), with a "safety factor" applied for the possibility that the efficiency of planet formation may vary with the Galactic environment.

hadn't been displaced by a more advanced species? Finding an answer to these questions seems to be a key step towards a better estimate of f_i.

The factors f_c and L fall into the realm of sociology. It is tempting to speculate that $f_c \approx 1$, given the human drive for exploration, but we do not know with certainty that this extrapolation from our anthropocentric view of the world is really justified. The value of L depends on external factors such as global epidemics and giant impacts by comets and asteroids, and on internal factors that could lead to a quick end of a "semi-intelligent" civilization – wars or the exhaustion of natural resources. It appears possible that our own species and our offspring may populate the Earth at least for the remainder of the main-sequence lifetime of the Sun ($L \approx 5 \cdot 10^9$ yr), but if we are not careful, we may not live to see $L_{\text{homo}} = 100$ yr.

We may thus characterize the emergent fields of exo-planetary astronomy and astrobiology as attempts to systematically explore the individual factors of the Drake Equation, from the left to the right. In contrast, SETI (which should perhaps better be called "Search for Extraterrestrial Technology" or "Search for Interstellar Communication") is an attempt to bypass this painstaking process by going directly for the grand prize. The chances of success are very uncertain, as the above arguments are consistent with estimates that range from an average distance between "neighbors" of ≈ 30 pc, to a Galaxy that is void of life save that on a lonely, solitary Earth.

The Fermi Paradox

If civilizations are common in the Galaxy, one may ask the question why we have not found any incontrovertible evidence for their existence yet. More to the point, it has been argued that the absence of extraterrestrials from the Solar System implies that we are alone in the Galaxy, and that any searches for extraterrestrial civilizations are futile. The chain of arguments, which is known as *Fermi's Paradox*, goes as follows:

1. Lets's assume that our civilization is not the only one in our Galaxy that has developed technology.
2. Then our civilization must be "typical". This means that it is not the most advanced of all, and that other civilizations share our desire to explore.
3. Space travel is not too difficult for civilizations "slightly" more advanced than ours.
4. The time scale to colonize the whole Galaxy is $\lesssim 10^8$ yrs, i.e., small compared to the age of the Galaxy.
5. Then one must conclude that the Solar System should have been colonized a long time ago. But this is not the case.

So it appears that we have encountered a logical difficulty if we believe in the ubiquity of life. However, each step in this chain has potential loopholes, some of them more severe, others less.

The assumption in step (1.) can certainly be questioned. As explained in the previous section, we know very little about f_i, the likelihood for intelligence to emerge through evolution. If this factor is small, we may indeed be alone in the Galaxy.

Step (2.) seems to be quite plausible. The development of intelligence on Earth may have been a singular event, but if we assume that it has occurred in *one* other place, it very likely occurred in *many* other places. Then it is very unlikely that we are the most advanced civilization, given that there are many Solar-type stars (i.e., stars with comparable mass and metallicity) that are several Gyrs older. Life on a terrestrial planet around any of these stars would have a several-Gyr head start compared to the Earth. And assuming that *none* of these earlier civilizations would be interested in exploring the Galaxy (or that *all* of them would refrain from doing so for ethical reasons) seems extremely unlikely, too.

To justify step (3.), we can invoke some physical considerations. Several methods of attaining speeds necessary for interstellar travel ($v \gtrsim 0.1c$) have been suggested, including pulsed fusion and antimatter-powered rockets, light sails pushed by lasers, and interstellar ram jets (Crawford 1990). The biggest hurdle to overcome for interstellar travel are the enormous energy requirements; accelerating a spaceship to a substantial fraction of the speed of light in a reasonable time would require a few times the current global power production. This is a staggering power requirement, but it is plausible that it could be met by humanity very soon. If our power production grows at an average rate of only $\sim 1\% \, \text{yr}^{-1}$, it will take less than 1,000 years, and the power requirement for interstellar travel will only be a fraction of a per cent of the global power consumption.[6] For comparison, a Saturn V rocket during lift-off consumed $\sim 0.5\%$ of the global power production. It thus seems likely that a civilization that is only slightly more advanced than ours will have the technical means to travel through interstellar distances.

The argument in step (4.) is based on the assumption that civilizations will establish colonies, and that each colony will again establish sub-colonies once it gets firmly established. With reasonable assumptions about the mean distance between these colonies, and about the time it takes for a colony to establish itself and to spawn a new settlement, one can estimate that it takes only a few Myrs to reach every habitable planet in the Galaxy (Crawford 2000).

Step (5.) also contains an assertion that can be questioned. While we have no scientifically valid evidence for the presence of other life forms in the Solar System, we do not have strong evidence for their absence, either. Observational limits on artificial probes within the Solar System are very weak, and small probes may even be hiding among us (Tough 2000).

---

[6] One should be somewhat careful with arguments based on sustained exponential growth, of course. At the same growth rate, humanity would need to generate more than $1 \, L_\odot$ in less than 10,000 years.

In summary, all possibilities are still open. Other intelligent forms of life may be here (within the Solar System), they may be there (within the Galaxy), or they may be nowhere.

SETI Strategies

The fundamental task of SETI is finding artificially generated signals beamed towards the Earth, and distinguishing them from the astrophysical and instrumental backgrounds. A good strategy is looking for signals whose time-bandwidth product approaches the limiting value $B\tau \approx 1$ set by the uncertainty principle (Tarter 2001). Most SETI experiments have been searches for narrow-band signals in the radio regime, where stars are comparatively weak emitters. (FM and TV transmitters are factors 10^6 to 10^9 brighter than the quiet Sun.) The main difficulty of this approach is the enormous search space that needs to be covered (position, frequency, frequency drift due to relative acceleration of emitter and receiver, pulse format). One either has to concentrate on "magic frequencies", assuming that the transmitter will choose e.g., a frequency close to the 1,420 MHz hydrogen line, or construct a spectrometer with many millions of frequency channels, and provide sufficient computer power for the data processing.

An alternative approach are optical searches for very short laser pulses. The flux from a Solar-type star at a distance of 300 pc is $\sim 4 \cdot 10^5$ phot m^{-2} s^{-1} in a broad-band optical filter. The arrival of two or more photons with in sub-microsecond time window is thus exponentially suppressed by Poisson statistics; such events could therefore be attributed to a laser beacon. The operational optical SETI experiments use two or more fast avalanche photodiodes in coincidence to detect nanosecond pulses.

A further choice has to be made regarding the targets of the search. One can either point the (optical or radio) telescope at selected nearby stars, thereby maximizing the sensitivity and therefore the chances to detect a nearby, relatively weak transmitter, or conduct a sky survey, which optimizes detectability of more distant, very strong emitters. Sometimes an intermediate solution is chosen, by piggy-backing a SETI receiver to a telescope during observations taken for a different purpose. This circumvents the difficulty of obtaining a large amount of dedicated telescope time, but makes the search somewhat less efficient, depending on the nature of the primary observing program.

Instead of searching for signals beamed explicitly towards us, one might also envisage looking for electromagnetic signals generated for the internal communication needs of extraterrestrial civilizations, leaking more or less isotropically from their home planet into space. Since we have been using radio transmitters for the better part of a century, intelligent life on Earth is now detectable out to a distance of ~ 25 pc, but the signal is relatively weak – current targeted SETI programs can detect the equivalent power of strong TV transmitters out to ~ 0.3 pc, not a very useful distance.

The Scientific Impact of SETI

SETI projects have always been justified on the basis of the large impact that a positive result would have. But a well-designed scientific experiment should also satisfy the criterion that even a null-result constitutes significant progress, and provides new insight. So far, the failure of all SETI efforts to pick up a signal does not tell us very much, because the search space that has been covered is still quite small. To quantify this statement, we can use the SETI figure of merit (Dreher and Cullers 1997)

$$\mathcal{S}(P_T) \equiv N_{\text{stars}}(P_T) \cdot \ln(\nu_h/\nu_l) \cdot \eta_{\text{pol}} \cdot N_{\text{looks}}. \tag{5}$$

Here P_T is the effective isotropically radiated power (EIRP) of the transmitter[7], $N_{\text{stars}}(P_T)$ the number of stars observed by a SETI project for which a transmitter of strength P_T is detectable, ν_h and ν_l the upper and lower limits of the frequency range covered, $\eta_{\text{pol}} \in \{0.5, 1\}$ a polarization efficiency factor, and N_{looks} the number of observations for each target star. In Fig. 3, $\mathcal{S}(P_T)$ is plotted as a function of P_T for the most important ongoing surveys. Taking for simplicity all factors on the right-hand side of (5) except the first to be of order unity,[8] we can draw some simple conclusions from this figure. First of all, there are few (if any) civilizations in the Galaxy transmitting in excess of 10^{23} W equivalent power to us at the frequencies covered by the radio surveys (mostly near the 1,420 MHz line). Second, we can define a "reasonable" equivalent transmitter power P_T, which we think an advanced civilization could afford, by multiplying the actual power with the antenna gain. The curves in Fig. 3 then tell us how many stars have been surveyed to that depth. Conversely, we can wage a guess the rate of incidence of civilizations (one per 10^x stars, where x is your guess), look up that number on the y-axis, and infer the equivalent power that these civilizations would have to use to be detected in our surveys.

In the coming decades, new radio arrays with large collecting area (the Allen Telescope Array, and perhaps later a Square Kilometer Array) will enable much more sensitive all-sky surveys than currently possible. The detection of artificial signals in these surveys is largely a computational problem; the extrapolation of Moore's Law therefore predicts a dramatic increase in the search capabilities in a relatively short time. By the middle of the 21st century, it should be possible to search a significant fraction of all stars in the Galaxy to an interesting limit, comparable to the strongest current man-made signals. A null result from such a survey would indeed place strong constraints

---

[7] The actual power needed for a directed transmission is much smaller, of course. Using an optical telescope with a resolving power of 0″.1 to send a beacon, for example, reduces the power requirement by a factor 10^{14} compared to an isotropic emitter. This factor is called *antenna gain*.

[8] The logarithmic frequency factor is actually as small as 10^{-3} for some of the surveys shown (Tarter 2001).

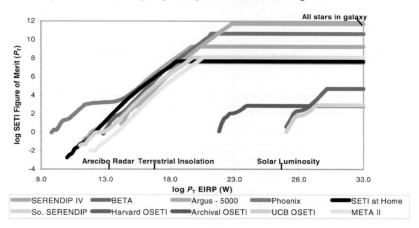

Fig. 3. SETI Figure of Merit as defined in (5) for current searches. The power axis represents the average effective isotropically radiated power for narrow-band continuous wave searches, but is the peak transmitter power for short optical pulses. From Tarter (2001)

on the number of civilizations in the Galaxy. On the other hand, we might be lucky and receive transmissions from intelligent beings even before we can launch the planned space missions TPF/DARWIN which will be capable of identifying the chemical signature of primitive life through spectroscopy of the atmospheres of nearby habitable planets.

2 Planet-Forming Disks

The origin of planetary systems is intimately linked to the formation of their parent stars. The theory of star formation and observations of young stellar objects are large fields of astronomy; we will only summarize briefly those aspects that are relevant for planet detection methods. For excellent introductions into star formation see e.g. Shu et al. (1987), van Dishoeck and Blake (1998), and Dutrey (1999). It is now generally accepted that circumstellar disks play an important role during the pre-main-sequence phase, and that planets are formed in these relatively massive disks. The disk hypothesis for the formation of the Solar System actually dates back to Kant (1755) and Laplace (1796); it provides a framework in which the following salient features of the Solar System can be explained:

- The planetary orbits are nearly circular and coplanar.
- The orbital motions and rotations of the Sun, planets, and moons are predominantly in one sense.

- The Solar System is differentiated, with systematic trends in the planetary properties with distance from the Sun.
- The Solar System contains a large number of small bodies (asteroids and comets), again with properties that vary systematically with distance from the Sun.

These properties are not prescribed by Kepler's Laws or other fundamental laws of physics, but they are a direct consequence of the way the Solar System formed (e.g., Encrenaz 2001). We may therefore expect that extrasolar planetary systems may share these general characteristics with ours because they were formed by the same processes and in a similar environment. It should be pointed out, however, that mechanisms such as orbital migration and two-body scattering might in many cases have played a more prominent role than in the Solar System. The detections of massive planets in orbits with small radii, and of planets in highly eccentric orbits (see Sect. 3) are certainly an indication that a large range of outcomes is possible from the processes that shape nascent planetary systems.

2.1 Star Formation: the General Framework

"Bimodal" Star Formation

Stars form in molecular clouds, the coldest (typically $10\ldots50\,\mathrm{K}$) and densest ($n \approx 10^3\,\mathrm{cm}^{-3}$) phase of the interstellar medium. Because of the low temperatures and extremely high opacity, the first phases of star formation can only be observed at radio, millimeter, and mid-infrared wavelengths. These wavelength ranges contain rich spectral information, which can be used to diagnose the physical state of the gas, to trace large-scale motions, and to pinpoint the sites at which stellar embryos are forming. Rotational and vibrational transitions of molecules such as H_2, CO, CS, NH_3, H_2O, CH, OH, and many others, provide information on gas temperature, density and kinematics, the far-infrared continuum emission probes the distribution and temperature of dust, near-infrared images reveal young embedded stars, and infrared fine structure lines and radio free-free emission trace hot gas that has been ionized by newborn massive stars (e.g. Genzel 1992).

One can broadly distinguish between two distinct star-forming environments: cold dark clouds, which form only low-mass stars, and giant molecular complexes, which are associated with high-mass star formation (Evans 1999). In giant complexes, of which the Orion A cloud and NGC 3603 are prominent examples, high-mass and low-mass stars form in close proximity to each other (e.g., Eisenhauer et al. 1998); the stellar density in the Orion region is $\sim 2 \cdot 10^4\,\mathrm{pc}^{-3}$. In contrast, low-mass star forming regions are much less dense, and form only stars with masses $\lesssim 2\,M_\odot$. There are also differences between individual regions; whereas the Taurus–Auriga clouds have a filamentary structure with isolated star formation, the ρ Ophiuchi cloud is a cluster with a higher density of newly born stars.

The Solar neighborhood is a mix of stars born in different environments. Roughly 80% of all Solar-mass stars may have originated in dense regions like Orion, but for any individual old star it is practically impossible to ascertain whether it formed in this way or in relative isolation. Consequently, we are looking at planetary systems (or stars without planetary companions!) that were exposed to widely different environmental influences during their youth. It is quite possible that the ionizing flux of young massive stars modifies or even destroys the circumstellar disks in the surrounding cluster before they can form planets (Armitage 2000); this may indeed be happening near the Orion Trapezium (Johnstone et al. 1998). In this context one should keep in mind that all current information on extrasolar planets comes from observations of low-mass stars (up to $\approx 1.5\,M_\odot$). It is certainly possible that all known planets are associated with stars that were born in relative isolation. Massive stars, which form exclusively in dense clusters, may therefore conceivably have planetary systems with vastly different properties, or no planets at all. This interesting question can for example be addressed with astrometric surveys of high-mass stars and of pre-main-sequence stars (see Sect. 9.7).

Molecular Cloud Collapse, Fragmentation, and Low-Mass Star Formation

Molecular clouds typically have a clumpy structure; the densest clumps are associated with star formation. According to the famous *Jeans criterion*, a parcel of gas with temperature T, density ρ, and mean molecular mass $\overline{\mu}$ will collapse under its own gravitational attraction if its mass is above the critical value

$$m_J \equiv \left(\frac{\pi k T}{4 G \overline{\mu} m_H}\right)^{3/2} \rho^{-1/2}$$

$$\approx 5 \cdot 10^4 \left(\frac{T}{100\,\mathrm{K}}\right)^{3/2} \left(\frac{\rho}{\mathrm{cm}^{-3}}\right)^{-1/2} M_\odot \,. \tag{6}$$

This expression shows that only clumps with masses large compared to individual stars can start collapsing. (Even for gas as cold as $10\,\mathrm{K}$, a density of $2.5 \cdot 10^6\,\mathrm{cm}^{-3}$, far in excess of typical values for quiescent molecular clouds, would be required for a $1\,M_\odot$ clump to collapse.) After the collapse has been initiated, fragmentation will produce sub-clumps, which will then proceed to form individual stars. Conservation of angular momentum naturally leads to an increasingly flattened morphology, and finally to the formation of a disk.

Starting with these processes, the formation of a low-mass star can be described by four main phases (e.g. Shu et al. 1987):

1. A slowly rotating pre-stellar core forms by losing its magnetic and turbulent support.

2. The core collapses "from the inside out". A central protostar and disk are formed, which are deeply embedded within an infalling envelope of dust and gas. The luminosity of the protostar is derived from accretion.
3. Deuterium ignites in the central region, which becomes convective. A stellar wind develops, primarily perpendicular to the disk. This is the bipolar outflow phase.
4. Accreting material falls on the disk rather than the star. The opening angle of the wind widens; it blows away the surrounding gas. This is the T Tauri phase, in which the pre-main-sequence star is surrounded by a remnant circumstellar disk.

The later two of these phases correspond to four observationally distinct classes of objects, described in Fig. 4 (André 1994). During the pre-main-sequence evolution, there is a general trend for the peak of the spectral energy distribution to move towards shorter wavelengths, partly because of increasing temperatures, partly because of decreasing opacity towards the central source. During the same time, the mass of the envelope/disk decreases from nearly a Solar mass to a small fraction of that value. The formation of planetary systems must coincide with the phases that are accompanied by a sufficiently massive disk; the time available for planet formation is therefore limited by the time scale for disk dispersal.

Observational data on the lifetime of pre-main-sequence disks can most easily be obtained from a census of the circumstellar disk fraction in young clusters, spanning a significant range in age (Haisch et al. 2001b). In the youngest clusters the disk fraction is very high ($\gtrsim 80\%$); it decreases rapidly with cluster age. About half of the stars lose their disks within ~ 3 Myr; at and age of ~ 6 Myr essentially all disks have disappeared. Strictly speaking these data refer to the hot inner disk that gives rise to the near-infrared excess, but the lifetime of the outer disk, in which planets can form, seems to be closely related (Haisch et al. 2001a). We can thus infer that planets must form within a few Myrs.

2.2 Observations of Dusty Disks

Pre-Main-Sequence Disks

As explained in Sect. 2.1, T Tauri stars are new-born Solar-type stars with ages of 10^5 to 10^7 yrs, which have emerged relatively unobscured from their natal molecular clouds (Bertout 1989). The detection of an infrared excess around many of these stars, and the subsequent modeling of their spectral energy distribution, led to the wide acceptance of a circumstellar disk interpretation (Beckwith and Sargent 1993). A key argument in favor of this model was the realization that a spherical geometry is prohibited, since the dust in the system would then completely obscure the central star; an axial symmetry

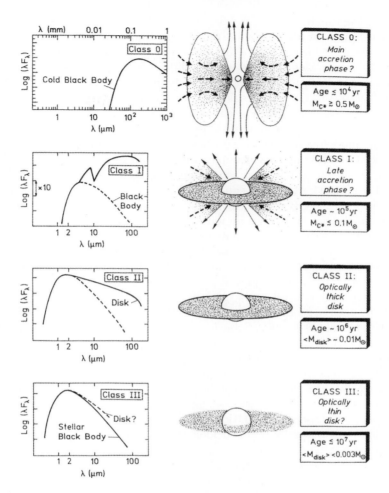

Fig. 4. The main stages of star formation (from *top* to *bottom*). The *left* column shows schematic spectral energy distributions. The overall geometry of the protostar and its immediate environment is sketched in the *middle* column. The *right* column gives approximate ages and disk masses. For details see the text. From André (1994)

is also supported by observations of bipolar optical jets and molecular outflows.

Observations of a number of pre-main-sequence disks in the millimeter/submillimeter continuum can be fitted with a λ^{-1} emissivity law for the dust, which implies that grain growth is occurring in them (Dutrey 1999). This is in general agreement with the ideas about grain growth discussed above. Since the emission is mainly optically thin, it allows measuring the disk mass;

values in the range $10^{-3} \ldots 10^{-1}\,M_\odot$ have been derived.[9] The star HL Tau is a good example for a very young ($\sim 0.1\,$Myr) object of mass $\sim 1\,M_\odot$. Its spectral energy distribution can be fitted by a toroidal circumstellar envelope of mass $\sim 0.11\,M_\odot$ (Men'shchikov et al. 1999). This model requires the presence of very large ($\gtrsim 100\,\mu$m) dust grains in the inner 100 AU, and much smaller grains ($\lesssim 1\,\mu$m) in the outer regions, again in agreement with expectations.

The first resolved images of circumstellar dust disks were obtained in the mid-eighties with interferometric observations of the millimeter continuum emission, and a few years later with optical and infrared high-resolution imaging (Koerner 1997). Famous examples are the compact objects seen silhouetted against the bright H II region associated with the Orion Trapezium Cluster (O'Dell et al. 1993; McCaughrean and O'Dell 1996; Bally et al. 2000). The objects closest to the OB stars have a cometary morphology, with a tail pointing away from the hot stars, and an ionization front towards them. Objects further from the OB stars have a more elliptical appearance with a star visible in the center, as expected for a sample of pre-main-sequence objects surrounded by disks with diameters of a few hundred AU.

Many other pre-main-sequence disks have been imaged in scattered light at infrared and visible wavelengths over the past few years, both from the ground and from space (e.g., Roddier et al. 1996; Potter et al. 2000; Stapelfeldt et al. 1998; Krist et al. 2000, 2002; Schneider et al. 2003; McCaughrean et al. 2000). The object HH-30 is a particularly interesting case (Burrows et al. 2006). The dust disk is seen nearly edge-on, and appears as a dark lane with a diameter of $\sim 500\,$AU. The central star is completely obscured, so that the surface of the flaring disk can clearly be seen in scattered light. A highly collimated jet is aligned with the rotation axis of the disk; clumps of gas are ejected along the jet axis at a speed of $\sim 200\,\mathrm{km\,s^{-1}}$. The fortuitous orientation of HH-30 thus allows an unusually detailed look at the geometry of the disk and jet, demonstrating that the collimation must take place within $\sim 30\,$AU of the star.

The steeply descending radial density and temperature profiles make the detection of the outer disks (beyond $\sim 200\,$AU) in the mm continuum very difficult. This region can be probed through its CO emission, however. Disks around T Tauri stars such as GM Aur are found to have outer radii of $100 \ldots 800\,$AU; velocity gradients along the major axes of the disks clearly demonstrate Keplerian rotation, as expected (Koerner et al. 1993; Guilloteau and Dutrey 1998; Dutrey et al. 1998). Many other molecules have been detected in pre-main-sequence disks and provide detailed diagnostics of the physics and chemistry of the gas. Some molecules (HCN, H_2CO) are depleted by factors ~ 100 due to condensation onto grains (Aikawa and Herbst 1999; Dutrey et al. 1997). While large CO disks are common around stars at an age

---

[9] Note that gas masses derived from observations of dust or CO are based on assumptions about the dust/gas and CO/H_2 ratios, respectively. In circumstellar disks these ratios can be substantially different from the "standard" ISM values. For details see Dutrey (1999).

of $\sim 10^6$ yrs, they have not been found around older stars, at $10^7 \ldots 10^8$ yrs (Liseau and Artymowicz 1998; Greaves et al. 2000).

Debris Disks

Dusty circumstellar disks around main-sequence stars are quite different in nature from pre-main-sequence disks, as their dust mass is at least 10 times higher than their gas mass. The Infra-Red Astronomical Satellite (IRAS) discovered that some bright nearby stars (among them Vega and Fomalhaut) emit much more strongly at wavelengths between 25 µm and 100 µm than expected (Aumann et al. 1984; Backman and Paresce 1993; Lagrange et al. 2000). Coronographic observations of β Pictoris, the star with the strongest infrared excess, soon showed that this "Vega phenomenon" is due to a circumstellar dust disk (Smith and Terrile 1984). A disk- or ring-like morphology has also been found in several other cases (including Vega, Fomalhaut, ε Eri, and HR 4796 A, Zuckerman 2001), but some more distant Vega-excess stars are associated with reflection nebulosities reminiscent of the Pleiades, indicating that their far-IR excess may also be caused by local heating of interstellar dust (Kalas et al. 2002).

The properties of a few prominent and well-studied main-sequence dust disks are listed in Table 4. They are all much more luminous than the dust disk of the Solar System, and they are associated with relatively young stars. Indeed, there is a clear relation between the fractional dust luminosity $f_d \equiv L_{\rm dust}/L_*$ with stellar age τ_*; it can be represented by a power law $f_d \propto \tau_*^{-1.76}$ (Spangler et al. 2001). There is a substantial gap in our knowledge of dust

Table 4. Properties of debris disks

star	d [pc]	r_c [AU]	$T_{\rm dust}$ [K]	$s_{\rm dust}$ [µm]	$L_{\rm dust}/L_*$	$\tau_{\rm PR}$ [yr]	$\tau_{\rm coll}$ [yr]	τ_* [yr]
α Lyr	8	150	85	$10 \ldots 100$	$2 \cdot 10^{-5}$	10^7	10^7	$4 \cdot 10^8$
α PsA	7	150	60	≈ 10	$8 \cdot 10^{-5}$	10^7	10^6	$2 \cdot 10^8$
β Pic	19	75	110	≈ 1	$2 \cdot 10^{-3}$	10^6	10^4	$1.2 \cdot 10^7$
ε Eri	3	39	50	≈ 10	$7 \cdot 10^{-5}$	10^8	10^6	$8 \cdot 10^8$
Zodiacal dust			280	$1 \ldots 100$	$1 \cdot 10^{-7}$			
Kuiper Belt			< 40	> 1	$< 10^{-7}$			

Listed are the distance d, characteristic radii r_c (derived from the color temperature), dust color temperature $T_{\rm dust}$, estimated grain sizes $s_{\rm dust}$, disk luminosity compared to the star $L_{\rm dust}/L_*$, Poynting–Robertson lifetime of typical grains $\tau_{\rm PR}$, collisional time scale $\tau_{\rm coll}$ at r_c, and estimated age of the star τ_*. Properties of Solar System dust (Zodiacal dust and dust in the Kuiper Belt) are included for comparison. Adopted from Mann (2001), with the age of β Pic taken from Zuckerman et al. (2001)

disks between relatively young stars and the Sun, due to the limited sensitivity of the IRAS and Infrared Space Observatory (ISO) missions. This gap will hopefully be closed soon by the Space Infra-Red Telescope Facility (SIRTF), which should be sensitive down to the mass in small grains inferred for our present-day Kuiper Belt ($6 \cdot 10^{22}$ g) surrounding a Solar-type star at 30 pc (Meyer et al. 2001).

Dust grains spiral into their parent stars due to Poynting–Robertson drag, and they are destroyed by mutual collisions (Dermott et al. 2001). Typical time scales for these processes are listed in Table 4 together with the ages of the stars.[10] We see that in all Vega-excess disks the life time of dust particles is much smaller than the age of the star. This leads to the important conclusion that the dust disk cannot be a remnant from the star formation process, but must be replenished by erosion of planetesimals or cometary bodies; hence the expression *debris disk* for these objects. The general trend of decreasing disk mass with age is then consistent with the evolution of the Solar System; the prominent Vega-excess disks correspond to the time of heavy bombardment in the early Solar System. If the dust disk of β Pic is generated by collisional destruction of 1-km planetesimals, a very large mass ($\sim 125\,M_\oplus$) must reside in planetesimals (Artymowicz 1997). An alternative scenario, based on the evaporation of cometary bodies, will be discussed in Sect. 2.3.

The amount and nature of gas in the disks of Vega-excess stars has been investigated with emission and absorption spectroscopy; the largest body of data is (not surprisingly) available for β Pic. Radio observations give an upper limit of $\sim 1.5\,M_\oplus$ to the atomic hydrogen content from a non-detection of the 1,420 MHz line (Freundling et al. 1995). The data on H_2 seem to be contradictory. Tentative detections of the S(0) and S(1) rotational transitions with ISO imply an H_2 mass of $\sim 0.2\,M_{\rm Jup}$ (Thi et al. 2001), whereas the lack of H_2 absorption against the stellar disk in the ultraviolet places an upper limit of $\sim 3 \cdot 10^{-4}\,M_{\rm Jup}$ on that same quantity (Lecavelier des Etangs et al. 2001). This discrepancy could in principle be resolved by a clumpy structure of the molecular hydrogen – the clumps would show up in emission, but the line-of-sight towards the star may not be covered. It seems difficult, however, to hide such a large mass of H_2 in this way; more data are clearly required to resolve this issue. In contrast to H_2, many ions and neutral species, as well as CO, have been found in absorption against β Pic (Vidal-Madjar et al. 1994; Lagrange et al. 1998; Roberge et al. 2000); the relative abundances of the refractory elements appear to be close to Solar. The amount of neural gas can be estimated from Na I emission (Olofsson et al. 2001) and from a tentative detection of the 157.7 μm [C II] fine structure transition with ISO

---

[10] Note that the Poynting–Robertson and collisional time scales depend strongly on the particle size and position in the disk. The tabulated values therefore give only rough indications on the correct orders-of-magnitude. For more details see Artymowicz (1997) and Dermott et al. (2001).

(Kamp et al. 2003); the latter authors conclude that a gas mass in the range $0.2\ldots4\,M_\oplus$ and a gas-to-dust ratio of $0.5\ldots9$ is consistent with all observations.

Evidence for Planets in Disks?

As we have seen above, the existence of debris disks in itself provides a strong argument for the existence of solid bodies in these systems, as there must be a mechanism to replenish the dust such as the erosion of km-sized planetesimals. This does not tell us anything about the existence of planets, however; for some reason the later stages of their formation process may simply have failed. It might be possible, however, to detect planets indirectly by their influence on the morphology of circumstellar disks. It has been pointed out earlier that sufficiently massive planets open gaps in disks, which may be observable directly or indirectly (cf. Fig. 5). After most of the disk material has been

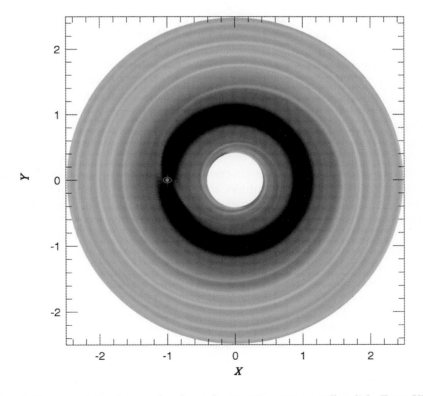

Fig. 5. Numerical simulation of a planet forming in a circumstellar disk. From Kley et al. (2001)

cleared, dust rings may remain an provide signposts of recent planet formation (Kenyon and Bromley 2002).

The temperature of grains at any given radius in the disk is determined by radiative equilibrium with the star; the spectral energy density (SED) of the disk is therefore the superposition of the Planck functions appropriate for grains at the proper temperatures (neglecting spectral features). If a gap is opened in the disk, there are no grains in the corresponding temperature range, which should result in a dip in the SED. Unfortunately, this dip is not very deep and extremely broad, extending over three orders of magnitude in wavelength. Without further information, it is impossible to identify such a feature, and to distinguish it from variations in the overall disk and dust parameters (Steinacker and Henning 2003). Interferometric imaging in the mid-infrared or at mm/sub-mm wavelengths is a better technique for the detection of gaps in disks (Wolf et al. 2002). An obvious caveat is that for observations at $\lambda \sim 10\,\mu m$ the emission is dominated by dust at $\sim 1\,AU$, where the temperature is $\sim 300\,K$. This means that gaps at larger radii have to be observed at longer wavelengths. The Atacama Large Millimeter Array (ALMA) will be very well suited for this task.

Most observational data on gaps in disks come from somewhat older objects, already close to or in the debris disk phase. A good example is HD 4796 A, which is surrounded by a prominent ring-like structure that has been observed at near- and mid-infrared wavelengths (Schneider et al. 1999; Telesco et al. 2000). It is tempting to attribute the relatively sharp truncation at the inner and outer edges of the annulus to the dynamical action of one or more "shepherd planets". The inner edge of the bright ring appears to be slightly asymmetric, which has been interpreted as an indication of a planet in an eccentric orbit (Wyatt et al. 1999). A similar disk with and apparent gap at a radius of $\sim 250\,AU$ has been found around HD 141569 A, a $\sim 5\,Myr$-old intermediate-mass star (Weinberger et al. 1999, 2000). However, improved coronographic imaging of this object with the ACS instrument on HST shows that the previously observed structure in the disk is not a ring but rather tightly wound spiral, which could be due to tidal interaction with the nearby binary HD 141569 BC (Clampin et al. 2003). As this example shows, it is quite difficult to proof beyond doubt that an observed structure in a disk cannot be caused by anything but a planet.

Millimeter and sub-millimeter images of the most prominent Vega-excess stars (α Lyr, α PsA, β Pic, and ε Eri), which probe relatively cool dust on scales $\gtrsim 50\,AU$, have consistently revealed surprising morphologies with strong clumping, quite different from simple smooth disks (Greaves et al. 1998; Holland et al. 1998, 2003; Liseau et al. 2003). Perhaps most intriguing is the dust disk of ε Eri, whose morphology has been modeled with a resonant patter due to a mean-motion resonance with a planet (Ozernoy et al. 2000; Quillen and Thorndike 2002). These models predict changes of the observed structure with time. It should simply revolve around the star if the planetary orbit is circular, or change its shape with orbital phase if the orbit is eccentric. It will thus be possible to distinguish between different models on the orbital time scale of the presumed planet (≈ 140 yrs).

The disk of β Pic is a special case again, because its brightness has enabled so much high-quality information to be gathered, and because its edge-on orientation allows observations of out-of-plane distortions. Large-scale asymmetries and warps have been reported on all scales accessible to observations (10...400 AU), both in reflected light and in thermal infrared emission (Heap et al. 2000; Pantin et al. 1997). The warps could be caused by a planet in an orbit that is slightly inclined with respect to the disk plane; a fairly large range of planetary masses and orbital radii would be compatible with the observations (Mouillet et al. 1997). Substructure in the outer disk, somewhat reminiscent of Saturn's rings, has been detected by comparing details in space-based and ground-based images (Kalas et al. 2000). They might have been generated by a close stellar encounter in the last $\sim 100,000$ yrs, which could then also have triggered the outer warps. The planetary hypothesis remains the best explanation of the recently detected small-scale warp within the inner ~ 20 AU, which is even stronger than the outer distortions (Weinberger et al. 2003; Wahhaj et al. 2003). It is also possible that two planets in orbits that are inclined with respect to each other and with respect to the disk cause the complicated warped structure. An unambiguous resolution of this question will probably have to await observations with even higher spatial resolution.

While imaging observations probe the structure of circumstellar disks on scales of many AU, photometry is a better tool to diagnose inhomogeneities in edge-on disks at much smaller radii. Deep periodic occultations of the T Tauri star KH 15D have indeed been attributed to eclipses by dust, indicative of either a clump or warp in the circumstellar disk at a radius of ~ 0.2 AU (Hamilton et al. 2001; Herbst et al. 2002). It is tempting to speculate that this distortion is caused by the gravitational action of a planet, but many alternative explanations are still viable. In any case, continued monitoring of this star will yield interesting information on the structure of circumstellar material close to a young star.

2.3 Observations of Infalling Material and Evidence for Extrasolar Planetesimals

The scenario of planetary system formation outlined above implies that a large number of planetesimals remain in the circumstellar disk during the phase when large planet-mass bodies are already there. Scattering of these small planetesimals by the planets leads to drastic changes of their orbits. Some of them are thrown out of the system or into bound orbits with very long periods, others impact one of the planets, and yet others get so close to the star that they evaporate. These processes correspond to the formation of the Oort Cloud in the Solar System, and the epoch of "heavy bombardment" of the terrestrial planets that is still evident in the cratering record of the Moon.

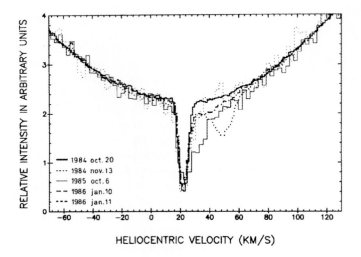

Fig. 6. Spectra of β Pic in the region of the Ca II K line. They have been normalized such that the "continua" (i.e., the stellar line broadened by rotation) coincide. Remarkably, the variations of the line profile all arise redshifted relative to the stellar radial velocity. From Ferlet et al. (1987)

There are strong indications that this phase has now been observed spectroscopically in several young planetary systems (Grinin 1999; Grady et al. 2000b). Intermittent narrow absorption features have been detected in the red wings of the UV and visible lines of β Pictoris (Ferlet et al. 1987, see Fig. 6). These absorption events have time scales from a few hours to many days. The only explanation consistent with most of the observational material is that of swarms of evaporating planetesimals; the dense absorbing cloud is essentially a large cometary coma. The required evaporated mass per event (if abundances appropriate for planetesimals are adopted) usually corresponds to km-size planetesimals. The infall episodes occur up to 200 times per year, but the event rate varies substantially from year to year. This, together with the large predominance of redshifted over blueshifted features, suggests that the observed infalling bodies belong to an orbital family (perhaps fragments of a disintegrated large comet)[11] with the mean orbit oriented in such a way that they approach the star when they are seen against it (Artymowicz 1997, see Fig. 7). The evaporation of comets could also be the dominant process responsible for replenishing the gas disk of β Pictoris (Lecavelier des Etangs 1998).

Redshifted absorption features have also been detected in the spectra of Herbig Ae/Be stars, which are pre-main-sequence stars of intermediate mass and thus the progenitors of objects like β Pictoris. In fact, variations of the

---

[11] Analogous groups of Sun-grazing comets exist also in the Solar System (Sekanina 2002 and references therein).

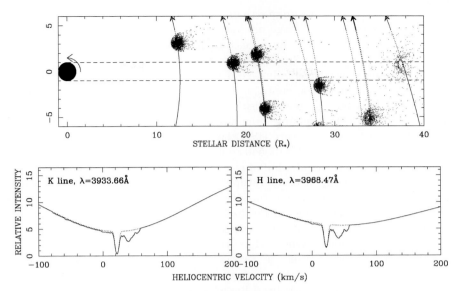

Fig. 7. Top panel: Simulated view of falling evaporating bodies (planetesimals or comets) passing in front of the star (*black circle* at the origin; the observer is to the right). Each comet evaporates creating a coma filled with dense gas containing singly ionized calcium, occulting a fraction of the stellar disk. *Bottom panels*: The cometary comae from the top panel cause multiple, redshifted, variable absorption features superimposed on both the stable narrow circumstellar and the broad stellar K and H lines of Ca II. From Artymowicz (1997)

line profiles appear to be ubiquitous, in particular in stars that are believed to be surrounded by an edge-on disk (Grady et al. 1996). The interpretation of the spectral variability is more difficult for these objects, however. They are still surrounded by a fairly massive gaseous disk, which makes it more difficult to distinguish between accretion of relatively unprocessed disk material and cometary events. Detailed multi-line analyses of UX Orionis have concluded that for this star the infalling (and outflowing) gas cannot be heavily hydrogen-depleted, as would be expected if it originated from the evaporation of solid bodies (Natta et al. 2000; Mora et al. 2002). Accretion of material with Solar-like chemical composition appears to be the dominant process for most very young stars, giving way to more episodic infall of metal-rich gas at an age of ~ 10 Myr (Grady et al. 1997, 2000a; Mora et al. 2003). The best cases of comet evaporation associated with younger stars are 51 Oph (Roberge et al. 2002) and WW Vul (Mora et al. 2003).

Our current understanding of the relation between the occurrence of planetesimals/comets and stellar age is limited by the relatively small number of observed stars, and by the low duty cycle with which they have been monitored. High-precision photometry from space (see Sect. 6.5) is a much more efficient observing technique than high-resolution spectroscopy, and may lead to the detection of a large number of extrasolar comets (Lecavelier des Etangs et al. 1999). It will be necessary, however, to discriminate between eclipses caused by comets, by planets, or by dust clumps.

The information from infalling comets, combined with mid-infrared spectra of Herbig Ae/Be stars, contains important clues about the ways dust is processed in pre-main-sequence disks. Silicates contained in interstellar dust are predominantly amorphous, but observations from the ground and with the Infrared Space Observatory (ISO) have shown that the disks surrounding β Pic and the Herbig star HD 100546 contain significant amounts of crystalline silicates (Knacke et al. 1993; Malfait et al. 1998). The production of these crystals requires processing at temperatures near 1,000 K, which are only reached in the innermost disk near the star (Hill et al. 2001). Since the disks of these stars are most likely being replenished by evaporating comets that formed at radii beyond the snow line, the crystalline material must have been incorporated in these comets. One is thus compelled to conclude that there must have been a large-scale transport process connecting the different regions of the disk at the epoch of comet formation (see Fig. 8). Similar indicators of crystalline silicates have also been found in some Solar System comets, and lead to the suggestions that large-scale mixing has been important here, too (Hanner et al. 1994). Observations of extrasolar comets can thus help us understand the chemical and mineralogical evolution of grains and particles from which planets can form.

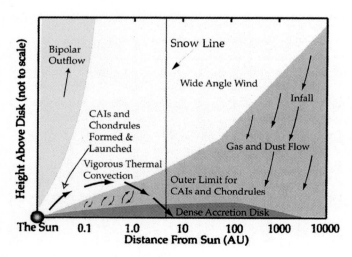

Fig. 8. Schematic drawing of a protostellar nebula (distance on a logarithmic scale) centered on the proto-Sun. The region of dust processing is shown (< 1 AU together with the snow line (≈ 5 AU), infalling envelope, bipolar outflow, and less collimated wind. Comet formation is expected to occur from ≈ 5 AU to beyond 40 AU. From Hill et al. (2001)

The evaporation of comets may be observable not only at a very young age, but also near the end of the stellar life time. When the star evolves off the main sequence, it expands and becomes much more luminous; this leads to the evaporation of icy bodies at increasing radii, up to several hundred AU (Stern et al. 1990). During the late phase of stellar evolution, convection of material from the core can enhance the outer layers with carbon. The oxygen in the winds from such carbon stars will be almost completely locked in CO, with a predicted H_2O abundance $\lesssim 10^{-11}$ (Willacy and Cherchneff 1998). The discovery of water vapor around the carbon star IRC+10216 with an implied abundance $\sim 10^{-7}$ can therefore only be explained by an external addition of H_2O to the wind, most plausibly through the evaporation of the Kuiper Belt around this star (Melnick et al. 2001). The observed rate of H_2O mass loss requires the evaporation of 10 $M_\oplus$ of ice over the life time of the strong stellar wind; this is comparable to the original mass of water ice in the Solar System's Kuiper Belt. More sensitive future observations with ESA's Herschel telescope will be able to probe the presence and mass of Kuiper Belts around a much larger sample of carbon-rich stars.

3 The Currently Known Extrasolar Planets

The discovery of a planet orbiting the star 51 Peg (Mayor and Queloz 1995, see Fig. 9) has opened a new field of observational astrophysics: the systematic study of planetary systems, their dynamical properties, formation, and evolution. About 150 extrasolar planets have now been detected, giving us a first glimpse at the diversity of these objects, and allowing the first statistical inferences. Unexpected discoveries have stimulated interesting new theoretical developments. Not the least of the big surprises in this field was the discovery of 51 Peg b itself, a giant planet with an orbital period of only 4 days! It probably formed at a much larger distance from its parent star, and migrated subsequently to its current position. In this chapter we will take a look at this and other phenomena, including orbital eccentricities and dynamical interaction between planets in multiple systems.

3.1 The First Hundred Planets Around Solar-Type Stars

The Presently Known Planets

The radial-velocity surveys (see in Sect. 4.4 for more details) have discovered about one hundred planets over the past seven years. One can thus rightfully speak about extrasolar planets as a new branch of observational astrophysics. The first discoveries showed that our Sun is not unique in having planetary companions. The field has progressed very quickly from this exploratory stage to a phase where it has become possible to perform the first meaningful statistical analyses of the properties of giant planets, of the distribution of their

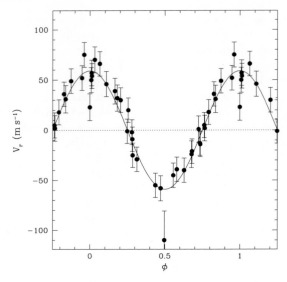

Fig. 9. Discovery plot of 51 Peg b, showing radial velocity (corrected for a slow variation of γ) as a function of orbital phase. The solid line represent an orbital fit in which the eccentricity was fixed at $e = 0$. From Mayor and Queloz (1995)

orbits, and of the characteristics of their parent stars. It is certainly necessary to keep selection effects related to the precision of the Doppler surveys and the limited time covered so far by systematic monitoring campaigns in mind, but most main-sequence stars of spectral type F or later within 50 pc and brighter than $V \approx 8$ have now been observed for a few years. This puts statistical inferences on a fairly firm footing, with the cautionary remark that our knowledge about planets around M dwarfs is still quite incomplete because of the intrinsic faintness of these stars.

The number of publications announcing new planet detections (and sometimes calling previous announcements into question) is growing rapidly; it is therefore quite a challenge to keep an authoritative list of the known extrasolar planets. Several groups maintain web pages with such lists, among them Geoff Marcy and co-workers, (http://exoplanets.org), Michel Mayor and colleagues (http://obswww.unige.ch/exoplanets), and Jean Schneider (http://www.obspm.fr/planets). The inclusion of a specific claimed planet detection in these lists is sometimes a matter of personal judgment, and the underlying philosophies are somewhat different. The International Astronomical Union has established a Working Group on Extrasolar Planets (Boss et al. 2003), which is about to publish a list with fairly conservative selection criteria (including publication in a refereed journal) at http://www.ciw.edu/boss/IAU/div3/wgesp. These compilations can serve as useful starting points for synoptic studies, and as guides to the original literature on planet detections.

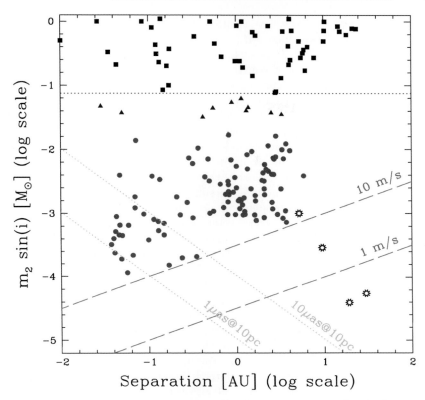

Fig. 10. Minimum mass $m_p \sin i$ versus orbital semi-major axis for the known extrasolar planets (*circles*) and low-mass stellar companions (*squares*). Hipparcos data have shown that most of the objects plotted as triangles are low-mass stars in nearly face-on orbits, i.e., their true mass is above the hydrogen burning limit (0.075 $M_\odot$), shown as a dashed line (Halbwachs et al. 2000). The four giant planets in the Solar System are shown as stars. The detection limits of radial-velocity and astrometric surveys are also shown. Courtesy Stéphane Udry

The Parameter Space Covered by Planet Surveys

The Doppler technique has a strong detection bias in favor of massive planets in short-period orbits, because the velocity amplitude scales with $K_* \propto m_p \cdot P^{-1/3}$ (19), or equivalently $K_* \propto m_p \cdot a^{-1/2}$. With a single-measurement precision of $3\,\mathrm{m\,s^{-1}}$, planets giving rise to wobbles with $K_* \gtrsim 10\,\mathrm{m\,s^{-1}}$ can reliably be detected (Marcy et al. 2000a); this limit is shown in the mass-separation plane in Fig. 10.[12] The lowest-mass planet known, HD 49674 B.[13]

---

[12] With a sufficient number of observations it should also be possible to get orbits for planets with K_* close to the measurement precision (Cumming et al. 2002).

[13] Editor note added in proof: The lowest mass planet know at this time is HD 160691 (Santos et al. 2004)

During the revision of the text (Butler et al. 2003), does indeed occupy the lower left corner of this diagram, just above the $10\,\mathrm{m\,s^{-1}}$ line. Its parameters are $m_p \sin i = 0.12\,M_{\mathrm{jup}}$, $P = 4.948\,\mathrm{days}$, $a = 0.057\,\mathrm{AU}$, and $K_* = 14\,\mathrm{m\,s^{-1}}$. The best sensitivity can of course only be reached for stars that are photospherically quiet, as discussed in Sect. 4.2. Planets have also been found around young, more active stars, but more data points are then required for a reliable orbital solution even for relatively large K_* (Kürster et al. 2000).

Another limiting factor in the parameter space covered so far is the limited time over which high-accuracy radial-velocity monitoring has been performed. It is obvious that the time needed from the first measurement to a secure detection is of the order of the orbital period; at high signal-to-noise perhaps half or a quarter of that time may be sufficient (Eisner and Kulkarni 2001a). Among all currently known planets, the one with the longest period, $P = 5,360\,\mathrm{days}$, and the largest orbital semi-major axis, $a = 5.9\,\mathrm{AU}$ is 55 Cancri d (Marcy et al. 2002). It occupies an orbit with moderate ellipticity, $e = 0.16$, and weighs in at $m_p \sin i = 4.05\,M_{\mathrm{jup}}$, thus giving rise to a stellar reflex motion with amplitude $K_* = 49.3\,\mathrm{m\,s^{-1}}$. As its designation implies, 55 Cnc d is a member of a multiple system (see Sect. 3.5). It is apparent from Fig. 10 that 55 Cnc d is the only known extrasolar planet with an orbital semi-major axis larger than that of Jupiter; this is consistent with the ~ 15 years over which observations with $\lesssim 10\,\mathrm{m\,s^{-1}}$ have been performed.

The cutoffs at the bottom and to the right in Fig. 10 can thus easily be understood as selection effects due to the precision and time coverage limitations of the present radial-velocity data. It is important to point out that no such limitations exist towards the left and top in this diagram. The paucity of planets in these areas is a true astrophysical phenomenon, not an observational artifact; it will be further discussed in Sect. 3.2. One should further add that there is no very strong detection bias with regard to the other orbital elements, with the exception of the inclination. (Nearly face-on orbits are strongly disfavored, of course, because $K_* \propto \sin i$.) It is therefore a reasonable approximation to assume that the observed eccentricity distribution is representative of the intrinsic one. In fact, a large range of eccentricities have been observed, from circular orbits up to $e = 0.927$ (Naef et al. 2001a, see also Sect. 3.3).

It is certainly noteworthy that a large fraction of the accessible mass – separation – eccentricity parameter space is actually populated with giant planets (Figs. 10 and 13). This was not at all anticipated ten years ago, when it was thought that the orbits of gas giants should be similar to those of Jupiter and Saturn, with periods of several years and low eccentricities. The great diversity of giant extrasolar planets is the first big surprise in this field, which has stimulated a wealth of new ideas about the formation and evolution of planetary systems.

On the other hand, it is also clear from Figs. 10 and 13 that one cannot argue that our Solar System is special in any way. A few planets have actually been detected that are somewhat similar to Jupiter, which is located near the corner of the currently accessible parameter space; Saturn analogs are simply out of reach at the present Doppler accuracy and monitoring time. It is therefore quite possible that many Jupiter/Saturn analogs will be discovered by future higher-accuracy Doppler surveys or other detection methods.

The Frequency of Giant Planets Around Solar-Type Stars

About 150 planet detections in a little less than 90 distinct systems, among $\sim 2{,}000$ stars surveyed, gives a lower limit of $\sim 5\%$ of all nearby Solar-type stars that have planets in the detectable parameter range, i.e., with radial velocity variations $\gtrsim 10\,\mathrm{m\,s^{-1}}$ and separations $\lesssim 3\,\mathrm{AU}$. The best-studied sample comprises 51 stars that have been monitored at Lick Observatory over the past 15 years. Eight planetary systems (two of them triple, one double, and five single) have been detected among these 51 stars, corresponding to a $\sim 15\%$ detection rate (Fischer et al. 2003a). This number will probably rise somewhat over the next few years, as more planets close to the detection limits will be found. On the other hand, we also now that the majority of stars do not have companions in the currently observable range; about 60% of the stars observed by the Lick/Keck/AAT team have a radial velocity r.m.s. of $5\,\mathrm{m\,s^{-1}}$ or less (G. Marcy, priv. comm.).

Somewhat stronger statements can be made in those parts of the parameter space where the observational data give an essentially complete picture. First of all, brown dwarf companions in orbits of less than 3 AU are very rare, with an incidence well below 1% (Fischer et al. 2002b; Vogt et al. 2002). Second, it is relatively easy to identify "hot Jupiters", i.e., planets with orbital periods $\lesssim 5$ days. 11 such planets are known to date, which means that $\sim 0.5\%$ of all stars in the Solar neighborhood are orbited by these objects.

3.2 Distribution of Masses and Orbital Radii

Planetary Masses

Figure 11 shows the histogram of the minimum mass $m \sin i$ for all currently known companions to Solar-type stars. Perhaps the most important conclusion that can immediately be drawn from this figure is that this distribution is bimodal, i.e., that there are two distinct populations of companion objects. This establishes "planets" as a physically distinct class of their own; they are not just the low-mass tail of the stellar binary population. The bimodal nature of the minimum mass distribution provides supporting evidence for the view that planets and stellar companions form through different mechanisms.

It has already been pointed out that the regimes of planets and stellar companions are separated by the "brown dwarf desert" $10\,M_{\mathrm{jup}} \lesssim m \sin i \lesssim 80\,M_{\mathrm{jup}}$. Only very few objects have been detected in this mass range, although they could easily be found by the radial-velocity surveys. In fact, most of the objects with $m \sin i$ in this mass range are actually stellar companions in nearly

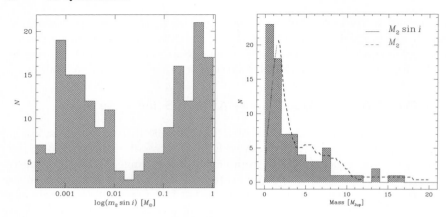

Fig. 11. *Left*: Distribution of minimum masses from the currently known low-mass companions to Solar-type stars. Although the radial-velocity method has a higher sensitivity to higher-mass companions, the observed distribution rises very steeply towards the low-mass domain. From $\sim 0.01\,M_\odot$ up to the stellar regime, only a few objects have been detected; this region is frequently called the "brown dwarf desert". This gap in the mass distribution of low-mass companions to Solar-type stars supports the view that there are two distinct populations (*planets and stars*), with different formation mechanisms. From Santos et al. (2002). *Right*: Same figure but with linear mass scale. The dashed line indicates a statistical estimate of the "true" planetary mass distribution. Updated from Jorissen et al. (2001)

face-on orbits; they produce an astrometric wobble sufficiently large to be detected with the Hipparcos satellite (Halbwachs et al. 2000). These objects are marked with triangles in Fig. 10, because their actual mass lies above the hydrogen burning limit. Conversely, the planet candidates with $m \sin i \lesssim 10\,M_{\rm jup}$ have usually not been detected with Hipparcos, which confirms that they are not stellar companions seen face-on (Zucker and Mazeh 2001a). The precision of Hipparcos is insufficient to rule out brown dwarf companions in these cases, but the a priori distribution of $\sin i$ (see Sect. 4.1) implies that the vast majority of the planet candidates really have masses below $10\,M_{\rm jup}$. The bottom line is that the brown dwarf desert is even less populated than it might appear from an $m \sin i$ histogram.

The small number of planets in the two lowest mass bins in Fig. 11 is clearly due to incompleteness close to the detection limit. Aside from this incompleteness, the number of planets rises somewhat towards lower masses when a logarithmic x-axis is chosen; this corresponds to a steep increase towards lower masses on a linear scale.

With a sufficiently large sample of detected planets, it is possible to disentangle the $\sin i$ projection effects from the observed $m \sin i$ histogram on a statistical basis, and to derive the underlying true mass distribution (Jorissen et al. 2001; Zucker and Mazeh 2001b). This is shown as a dashed line in Fig. 11. The sharp drop-off around $10\,M_{\rm jup}$, the low-level tail extending to

$\sim 20\,M_{\rm jup}$, and the steep rise towards lower masses are again immediately apparent. Zucker and Mazeh (2001b) derive a flat distribution for $dN/d\log m_p$, which corresponds to $dN/dm_p \propto m_p^{-1}$; Marcy et al. (2003a) give a similar scaling, $dN/dm_p \propto m_p^{-0.7}$, for $m_p < 8\,M_{\rm jup}$. It appears tempting to extrapolate these power laws to lower masses, in order to predict the numbers of Neptune-mass or even smaller planets. This could be misleading, however, because the formation of massive gas planets may be governed by physical processes that are different from those determining the frequency of lower-mass planet formation. Any extrapolation should therefore await a better understanding of the physics of planet formation.

Orbital Semi-Major Axes

With the first few detections of extrasolar planets it became rapidly apparent that gas giants occur with a large diversity of orbital separations, ranging from 0.05 AU (51 Peg b, Mayor and Queloz 1995), over 0.23 AU (ρ CrB B, Noyes et al. 1997), 0.43 AU (70 Vir B, Marcy and Butler 1996), 2.1 AU (47 UMa b, Butler and Marcy 1996) to 5.2 AU (Jupiter) and 9.6 AU (Saturn). The current statistical basis of ~ 150 extrasolar planets allows a much more detailed view of the distribution of orbits up to $a = 3$ AU. In Fig. 12 the minimum mass of the known planets is plotted as a function of their orbital semi-major axes, with a linear mass scale, which shows some trends and features more clearly than the logarithmic scale of Fig. 10:

- There is a remarkable "pile-up" of planets in orbits with $a \approx 0.05$ AU, i.e., with periods $P \approx 4$ days.

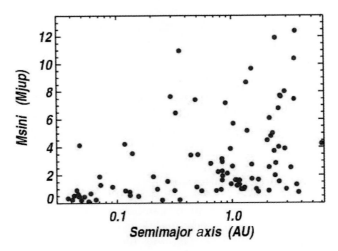

Fig. 12. Minimum mass $m_p \sin i$ versus semi-major axis for the known extrasolar planets, with a linear mass scale. From Marcy et al. (2003b)

- Orbits with $a \approx 0.3\,\mathrm{AU}$ are slightly less common than smaller and larger orbits.
- There are no planets with $m \sin i \geq 4.1\,M_\mathrm{jup}$ at $a \leq 0.3\,\mathrm{AU}$, whereas one third of all known planets at $a > 1\,\mathrm{AU}$ has a mass above this value.

As pointed out in Sect. 3.1, these findings cannot be due to an observational bias; they clearly tell us something about the formation and/or orbital evolution of giant planets.

Orbital Migration

Perhaps the most striking result from the Doppler surveys is the discovery that there are giant planets in orbits with $a < 3\,\mathrm{AU}$ at all. In fact, because our understanding of the Solar System implies that the gas giants were formed at $a > 5\,\mathrm{AU}$ and stayed there over the past 5 Gyrs, it was generally expected that this would also be the case in extrasolar systems. So could planets like 51 Peg b have formed much closer to their parent stars, at the location where they are found now? Given our general knowledge about star and planet formation (Sect. 2), there are several arguments why this appears exceedingly unlikely (Lin et al. 1996):

- At 0.05 AU the temperature in the pre-planetary disk is about 2,000 K, too hot for refractory materials to condense. Therefore planetary cores cannot form there.
- The surface density within $\sim 0.5\,\mathrm{AU}$ is too small for $\sim 10\,M_\oplus$ planetary cores to form.
- Even if a core is present (or if a core is not needed to form a gas giant), there is likely not enough gas to form a $\sim 1\,M_\mathrm{jup}$ planet.
- During their formation phase, young planets have radii up to ~ 10 times larger than their present values (Bodenheimer and Pollack 1986). Combined with the high temperature implied by small orbital distances, this gives very low escape speeds. The planet will thus be susceptible to evaporation, and to ablation by the stellar wind.

These arguments do not complete rule out the possibility of in situ formation at small radii (Bodenheimer et al. 2001), but the much more plausible conclusion is that the planets now found at small orbital radii must have formed at much larger distances, and subsequently migrated to their present locations.

Planet Survival and "Hot Jupiters"

The realization that orbital migration is an important mechanism for shaping planetary systems also poses a new question: why are the observed planets found in their present locations, i.e., why did their migration stop at the observed semi-major axes? Why did they not spiral all the way into the stellar photosphere? One possible mechanism involves the transfer of angular momentum from the spinning star to the planet through tidal interaction. Young stars rotate fairly rapidly, so that the Keplerian period at $a \gtrsim 0.05\,\mathrm{AU}$ is longer than the stellar rotation period. In that case the tides raised on the star exert an outward torque on the planet; the time scale of the consequent orbital evolution is (Goldreich and Soter 1966; Lin et al. 1996)

$$\tau_a \equiv -\frac{a}{\dot{a}} = \frac{P}{9\pi}\left(\frac{a}{R_*}\right)^5 \frac{m_*}{m_p} Q_* \,, \qquad (7)$$

where $Q_* \approx 1.5\cdot 10^5$ is a parameter, which describes how efficiently the tidal energy is dissipated within the star. It is apparent from (7) that the tidal torque depends very sensitively on a. The tidal interaction therefore sets in very suddenly during the inward migration; this could be a possible reason for the "pile-up" of the observed orbits at $\sim 0.05\,\mathrm{AU}$. An alternative explanation might be the truncation of the disk by the stellar magnetosphere, which could also occur at a similar radius (Shu et al. 1994).

Determining the fate of close-in planets is further complicated by the concurrent evolution of the star and the disk. Due to the spin-down of the star through torquing by the wind, the rotation period will eventually become longer than the orbital period of hot Jupiters. This reverses the sign of the orbital evolution, and may cause the planet to spiral into the star. The time scale for this process is $\propto m_p^{-1}$ (7), which could explain the absence of massive planets in short-period orbits (Pätzold and Rauer 2002; Jiang et al. 2003). In a computation of the orbital and structural evolution of planets that took into account Roche lobe overflow and consequent mass loss of the planet, three classes of objects could be identified: the planets with the lowest initial masses were completely destroyed, intermediate-mass planets lost some of their mass and ended up in stable orbits at $\sim 0.04\,\mathrm{AU}$, and the most massive planets did not migrate very far (Trilling et al. 1998). A combination of migration, mass loss, and the opening of gaps in the disk is thus likely needed to explain the observed minimum mass – semi-major axis distribution (Zucker and Mazeh 2002; Udry et al. 2003b).

Finally, one can pose the question what consequence the migration of giant planets has for the formation of terrestrial planets. A migrating gas giant would certainly destroy already existing small planets, and it would probably also suppress the subsequent formation of terrestrial planets at $\sim 1\,\mathrm{AU}$ (Armitage 2003). This suppression occurs because the gas that flows into the inner disk behind the migrating planet is depleted of dust as a result of having already formed planetesimals at larger radii. It is thus likely that the systems with hot Jupiters were not able to form terrestrial planets, even if there are dynamically stable orbits for habitable planets in those systems.

Planets with Long Periods

Although the radial-velocity technique selects against planets with a long orbital period ($K_* \propto P^{-1/3}$), the sensitivity of the current surveys is clearly good enough to detect Jupiter analogs (see Fig. 10). Nonetheless, no such object has been found with certainty yet, and it is hardly possible to make statistical inferences about the occurrence of planets with periods $\gtrsim 5\,\mathrm{yr}$. This situation will likely change dramatically within a few years, when the surveys covering a large number of stars with high precision, which were initiated or enlarged after the detection of 51 Peg b, will reach a sufficient time baseline. The typical orbital period of the newly announced extrasolar planets is already shifting towards longer periods, reflecting the increasing time baseline of the ongoing Doppler surveys. But most of the confirmed long-period planets that are known today still have relatively high masses (Santos et al. 2001b; Fischer et al. 2002b; Marcy et al. 2002), with the exception of 47 UMa c, the outer member of a system with two planets (Fischer et al. 2002a).

Planets with long periods first tend to show up as linear trends in radial-velocity data; deviations from linearity become apparent with continued monitoring, until finally the survey covers a full Keplerian period (e.g., Naef et al. 2001b). Most planet search teams will wait until roughly this time before announcing a planet detection and publishing an orbital solution. Important statistical information can be gleaned earlier, however. For example, radial-velocity variations indicative of distant planets appear to be significantly more common for stars that are already known to harbor an inner planet than for single stars (Fischer et al. 2001). Again, only time will tell how the properties of inner and outer planets are related to each other.

3.3 Orbital Eccentricities

The second big surprise (after the discovery of the "hot Jupiter" 51 Peg b) was the highly eccentric orbit of 70 Vir b, with $e = 0.40$ (Marcy and Butler 1996). In contrast, Jupiter and Saturn have $e \approx 0.05$, and even Mercury's ($e = 0.21$) and Pluto's ($e = 0.25$) orbital eccentricities are modest in comparison. It had generally been expected that extrasolar planets would also be found in nearly circular orbits, because they form in a circumstellar disk, and dissipation in that disk should generally lead to the circularization of the orbits. Even more extreme examples have subsequently been found, such as HD 89744 b with $e = 0.7$ (Korzennik et al. 2000) and HD 80606 b with an astonishing $e = 0.927$ (Naef et al. 2001a).

The radial-velocity curves of eccentric orbits deviate strongly from the sinusoidal variations associated with circular orbits. This characteristic shape makes it easy to distinguish planetary companions from other sources of radial-velocity jitter, as in the case of ι Dra (Frink et al. 2002). This may introduce a slight bias towards higher eccentricities in planet samples, but the effect is probably rather insignificant.

A plot of orbital eccentricity versus period for the known extrasolar planets is shown in Fig. 13, together with the same quantities for five of the Solar-System planets and stellar binaries. A few properties of the distribution of orbital parameters are worth noting:

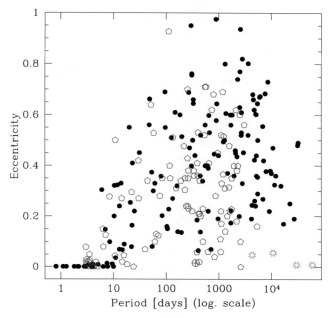

Fig. 13. Eccentricity versus orbital period for the known extrasolar planets (*pentagons*), stellar binaries (*filled dots*), giant planets in the Solar System (*stars*), and the Earth. From Santos et al. (2003b)

- All planets with short-period orbits have small eccentricities ($e \lesssim 0.1$ for $p \lesssim 10$ days).
- For longer periods, larger eccentricities are fairly common.
- The eccentricity–period diagrams for planets and stellar binaries look remarkably similar.
- There is also a class of long-period planets with nearly circular orbits (see also Vogt et al. 2002).
- The giant planets in the Solar System have small eccentricities, but they are not unusual.

To interpret these observations correctly, we have to consider not only the original orbit, but also the evolution of the eccentricities after the formation of the planet. This is a new insight that could only come from data on extrasolar planets and not from the Solar System, quite analogous to the case of the evolution of the major axes due to migration.

Tidal Circularization

The small eccentricities of the short-period orbits are generally attributed to *tidal circularization*, due to dissipation within the planet. The tidal bulges raised on a planet in an eccentric orbit lead to a decay of the eccentricity on a time scale (Goldreich and Soter 1966; Bodenheimer et al. 2001)

$$\tau_e \equiv -\frac{e}{\dot{e}} = \frac{4Q_p}{63n}\frac{m_p}{m_*}\left(\frac{a}{R_p}\right)^5$$

$$= \frac{Q_p}{10^6}\frac{m_p}{M_{\text{jup}}}\left(\frac{M_\odot}{m_*}\right)^{3/2}\left(\frac{a}{0.05\,\text{AU}}\right)^{13/2}\left(\frac{R_{\text{jup}}}{R_p}\right)^5 \text{Gyr}, \quad (8)$$

where Q_p is the tidal dissipation parameter for the planet (analogous to Q_* defined above), and $n = 2\pi/P$ the mean motion of the planet. Very little is known about plausible values for Q_p, but indirect arguments indicate that $10^5 \lesssim Q_p \lesssim 10^6$ for Jupiter (Goldreich and Soter 1966). Assuming that extrasolar gas giant planets possess similar Q_p values (which is a bit of a wild guess, see Marcy et al. 1997), the orbits of old hot Jupiters should be circularized according to (8). This equation also shows that the circularization time scale depends steeply on the orbital distance, both directly ($\tau_e \propto a^{13/2}$) and indirectly, because strongly irradiated planets are somewhat bloated, which leads to a further shortening of τ_e at small a. The tidal circularization scenario provides thus a fairly natural explanation of steep decrease in the upper envelope for e at $P \lesssim 10$ days.

Origin of Eccentric Orbits

The poor efficiency of tidal circularization for longer-period orbits is a necessary, but clearly not a sufficient condition for the existence of planets with high-eccentricity orbits. A rather common mechanism (or even several mechanisms) that can induce large values of e either during the formation of planets or thereafter through interactions is obviously needed to explain the observations (Fig. 13). A special case are planets orbiting a component of a wide stellar binary, such as the companion of 16 Cyg B (Cochran et al. 1997). Such systems can oscillate between high- and low-eccentricity states, if the inclination of the orbital plane of the planet with respect to that of the stellar binary is appreciable (Kozai 1962).

For those planets that are not found in wide stellar pairs (which is the large majority), gravitational interaction of planets in multiple systems is the most plausible way to generate a large value of e. If several giant planets form in a massive disk, their mutual perturbations induces a gradual increase in their orbital eccentricities (Lin and Ida 1997). The orbits may eventually become unstable and cross each other, so that several planets can merge and form a very massive planet, which tends to end up in an orbit with high eccentricity ($0.2 \lesssim e \lesssim 0.9$) and relatively small semi-major axis ($0.5\,\text{AU} \lesssim a \lesssim 1\,\text{AU}$). Alternatively, if a large number of massive planets form nearly simultaneously through fragmentation in a disk or protostellar envelope, dynamical relaxation leads to the ejection of most of the planets, while the remaining ones end up in highly eccentric orbits (Papaloizou and Terquem 2001). The distribution of eccentricities (and of the orbital inclination with respect to the stellar equator, and of the mutual inclinations of orbits in multiple systems) thus encapsulate information on the dynamical history of planetary systems, which are in turn directly related to the formation mechanism.

3.4 Properties of the Parent Stars

In addition to the properties of the known planets themselves, we may ask the questions whether there is a direct relation between these properties and the characteristics of the host stars, and whether the population of stars harboring planets is in any way different from their parent population of stars in the Solar neighborhood.

Stellar Metallicities

The most striking observation related to the properties of planet host stars is that they are more metal-rich on average than comparison stars without known planets in the Solar neighborhood (see Fig. 14). Another way of stating this is saying that the probability of finding a giant planet increases with metallicity, as seen clearly in the right panel of Fig. 14. This is an interesting physical relationship between planets and their parent stars.

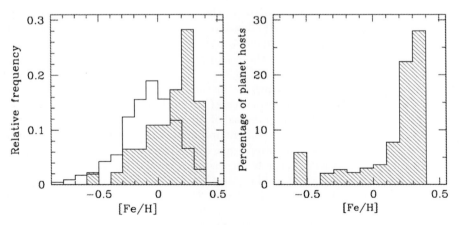

Fig. 14. *Left panel*: Metallicity (i.e., [Fe/H]) distributions for stars with planets (*shaded histogram*) compared with the same distribution for field dwarfs in the Solar neighborhood (*open histogram*). In this panel, both distributions are normalized by the respective number of data points. Most planet hosts are more metal rich than our Sun. *Right panel*: The percentage of stars that have been found to harbor a planet, for each metallicity bin. This plot shows clearly that the probability of finding a giant planet increases with metallicity. Updated from Santos et al. (2001a)

In establishing the metallicity–planet correlation, one has to be a bit careful about selection biases (Butler et al. 2000). In this context it is important that the targets of the large radial-velocity surveys have not been pre-selected on the basis of metallicity; stellar type (FGKM dwarf) and apparent brightness have indeed been the main selection criteria. Nonetheless, the rate of planet detections might be modified by some more subtle metallicity-dependent effects. Metal-rich stars have deeper absorption lines, which improves the attainable Doppler precision. Furthermore, metal-rich stars are brighter than metal-poor stars of the same spectral type, which leads to a Malmquist-type bias that could enrich magnitude-limited samples with metal-rich stars. However, while present to a certain degree, these selection effects are too small to explain the observed metallicity enhancement among planet host stars. A number of studies with somewhat different approaches therefore all come to the conclusion that the metallicity–planet correlation is a well-established physical fact (Gonzalez et al. 2001b; Santos et al. 2001a, 2003a; Reid 2002).

Why are Planets Found Preferentially around Metal-Rich Stars?

Two main hypotheses have been advanced to explain the enhanced metallicity of planet host stars: either planets form preferentially in the disks of metal-rich stars, or the atmospheres of planet-bearing stars have been polluted with high-Z material, perhaps from planets that have migrated all the way into the star, or from planetesimals scattered into star-impacting orbits by migrating planets (Quillen and Holman 2000). In the first scenario, the higher metallicity is the cause of the enhanced occurrence of planets, in the latter scenario, the presence of planets causes an enhanced metallicity. Several tests have been devised to distinguish between these two hypotheses, with somewhat mixed results.

One test consists of comparing the metallicity enhancement separately for stars with different kinematic properties (Barbieri and Gratton 2002). If high metallicity is the cause for the presence of planets, there should be no correlation between the occurrence of planets and galactocentric distance. It is only the overall metallicity that is important, and within each metallicity bin the distribution of stars with perigalactic distance should be the same for stars with and without planets. If on the other hand the presence of planets causes enhanced metallicities, one should expect that at any galactocentric distance planet host stars should be more metal-rich than average, and in each metallicity bin, stars with planets should tend to have smaller perigalactic distances. Barbieri and Gratton (2002) find precisely this effect from a reconstruction of Galactic orbits of planet hosts and comparison stars without planets, and conclude that scenarios in which the presence of planets is the cause of higher metallicities are strongly favored.

A different class of tests is based on more direct attempts to determine the effect of planet engulfment on the atmospheric composition of the parent star. Hydrodynamic simulations show that Jupiter-like planets spiraling into stars with $1.0\,M_\odot \leq m_* \leq 1.3\,M_\odot$ are partially or totally dissolved within the convection zone, and can thus indeed enhance the metallicity significantly (Sandquist et al. 1998). This would in particular deposit the isotope ^{6}Li in the atmosphere, which is normally destroyed during the early phases of stellar evolution.[14] The detection of ^{6}Li in the planet host star HD 82943 has therefore been interpreted as evidence for planet engulfment (Israelian et al. 2001).

The strongest argument for the alternative explanation, i.e., that planet engulfment does not play a dominant role, comes from an analysis of metallicity with stellar temperature. More massive stars (within the range of interest) have much shallower convective envelopes (Murray et al. 2001); adding a given amount of planetary material should thus have a much stronger effect on their surface composition. No such trend of metallicity with mass of the convection zone is observed, which means that a "primordial" source of the metallicity excess is much more likely (Santos et al. 2001a, 2003a). The "standard" model of giant planet formation through planetesimal formation and runaway gas accretion actually predicts that low-metallicity systems are much less likely to form planets than their high-metallicity counterparts, because the surface density of solid material in the pre-planetary disk plays a critical role for planet formation (Youdin and Shu 2002). One could thus even argue that the enhanced metallicities of planet hosts provide an empirical argument for this planet formation scenario over the disk instability model, in which no strong dependence on metallicity is expected (Boss 2002).

With arguments for either interpretation, the question about the cause of the planet–metallicity correlation has not been completely settled yet. It is certainly possible, that accretion of iron-rich material, high primordial metallicity, and selection effects all play a certain role (Murray and Chaboyer 2002).

Stellar Rotation Rates

Among the effects of tidal interaction between the star and planet in a small orbit is a spin-up of the star, which ultimately leads to synchronization of the stellar rotation rate with the orbital motion of the planet. The time scale of this spin-up is given by (Goldreich and Soter 1966; Trilling 2000)

$$\tau_s \equiv -\frac{\omega}{\dot\omega} = Q_* \omega \left(\frac{R_*^3}{Gm_*}\right) \left(\frac{m_*}{m_p}\right)^2 \left(\frac{a}{R_*}\right)^6$$

$$= 13 \, \frac{Q_*}{10^5} \cdot \frac{\omega}{2\cdot 10^{-6}\,\mathrm{s}^{-1}} \cdot \frac{m_*}{1.05 M_\odot} \times \qquad (9)$$

$$\times \left(\frac{1.2 R_\odot}{R_*}\right)^3 \left(\frac{0.45 M_\mathrm{jup}}{m_p}\right)^2 \left(\frac{a}{0.051\,\mathrm{AU}}\right)^6 \,\mathrm{Gyr}\,,$$

---

[14] ^{6}Li is destroyed at relatively low temperatures ($\sim 1.6\cdot 10^6$ K) through (p,α) reactions. During the early evolutionary phases, the proto-star is completely convective, so that the cool surface material is mixed with the hot stellar interior, which leads to the destruction of all ^{6}Li.

where ω is the angular velocity of the stellar rotation, and Q_* the tidal dissipation parameter of the star defined already in the context of (7). The scaling parameters in the second line of (10) have been chosen to match estimates for the 51 Peg system; the time scale for rotational synchronization is in that case somewhat longer (by a factor of a few) than the age of the system. Among the known "hot Jupiters" τ Boo has by far the shortest synchronization time ($\tau_s \approx 0.8\,\mathrm{Gyr}$, Trilling 2000). This star is the also the only one in a sample of planet hosts for which the rotation period is almost identical to the orbital period of the planet (Barnes 2001). This strongly suggests that tidal spin-up has indeed occurred in the case of τ Boo.

Planets in Multiple Stellar Systems

To explore the factors that influence the formation and evolution of planetary systems, one should clearly look for planets in diverse environments, i.e., around as large a variety of parent stars as possible. In this respect, binaries and multiple stellar systems provide rich opportunities, because in such systems the stability of pre-planetary disks and of planet orbits is restricted to limited distance ranges. In a binary system with component separation a_B, stable regions exist at small radii around each component (up to $\lesssim 0.3 a_B$, depending on the mass ratio), and around the binary system at radii $\gtrsim 2 a_B$. The statistics of planets in binaries could therefore hold important clues to issues related to the formation and migration process. Furthermore, the relative orientation of the stellar rotation axes, the binary orbit, and the planetary orbits could help us understand the processes that govern the distribution of angular momentum during the early phases of star and planet formation. While the mass of the Solar System is dominated by the Sun, most of its angular momentum resides in the orbital angular momenta of the planets – it would certainly be interesting to investigate the angular momentum distribution in more complicated systems.

The first planet orbiting a component of a wide binary was found around 16 Cyg B; it has already been mentioned that in this case the stellar companion might be responsible for pumping up the eccentricity of planetary orbit (Cochran et al. 1997). Meanwhile it has been discovered that 16 Cyg A is a binary with separation $3\rlap{.}''4$ itself (Patience et al. 2002), and another planet has been found in a similar hierarchical triple system (HD 178911, Zucker et al. 2002). The statistics have been improved both by searching for planets in known stellar binaries, and by searching for stellar companions to known planet hosts with speckle and adaptive optics techniques (Patience et al. 2002; Luhman and Jayawardhana 2002). Progress has also been made on the theoretical front; for example, numerical simulations indicate that terrestrial planets can actually form in systems like α Cen (Barbieri et al. 2002; Quintana et al. 2002). Much needs still to be done, however, before we can link the potential information content of planets in binary systems to the pressing current questions about the physical processes that shape planetary systems in general.

3.5 Systems with Multiple Planets

The Zoo of Planetary Systems

The story of extrasolar planetary systems – in the sense of multiple planets orbiting one and the same star – began with the discovery of a second and third planet around v And (Butler et al. 1999). Because of the analogy with the Solar System we might expect that different types of planets (Earth-like, gas, and ice giants) might form and evolve together. We can study the general architecture of multiple systems and ask questions such as: What is the spread in the planetary masses? Do the masses increase from the inside out, or the other way round? Are the orbits (nearly) coplanar? In addition, the gravitational interaction between the planets can be so strong that orbital resonances play a dominant role for their dynamical behavior (see Sect. 3.6), providing a rich new field of investigation.

The most important parameters of the ten known multiple systems are summarized in Table 5. (Note that the star HD 83443 was also believed to harbor two planets, but recent measurements demonstrated that the existence of the outer planet is not firmly established (Butler et al. 2002) Two stars (v And and 55 Cnc) are known to harbor three planets each; pairs of two planets have been detected in the other 8 systems. The notation (b, c, d, ...) follows the sequence in which the planets have been discovered; there is no specific ordering with respect to semi-major axis or mass.

Since all planets in Table 5 have been discovered with radial-velocity measurements, it is clear that they are drawn from the part of the parameter space accessible to this method (see Sect. 3.1). In addition to the general selection bias favoring massive planets and short-period orbits, this means that the present sample should be biased towards systems in which the masses increasing with the orbital semi-major axes. A good example is v And, in which all three planets give rise to roughly equal radial-velocity amplitudes K, because the loss of sensitivity with a is compensated by the increase in m_p (see Table 5 and Fig. 15). A system with masses decreasing from the inside out would be much harder to detect, because the outermost planet would be hard to detect by virtue of its low mass combined with large a.

Another shortcoming of the radial-velocity method is that it does not provide any information about the inclinations of the planets' orbits. Knowing these would be important for several reasons:

- If the individual planets in a system have different inclinations, we may misinterpret the overall architecture of the system. (The planet with the largest $m_p \sin i$ is not necessarily the one with the largest mass.)
- The relative inclination of the orbits is an important diagnostic for the dynamical evolution of the system, see Sect. 3.3.
- The gravitational interaction between a pair of planets depends on their masses. Knowing these only modulo factors $\sin i$ limits the ability to model perturbations of the orbits due to these interactions.

Table 5. Parameters of multiple planetary systems, derived from Keplerian fits to the radial-velocity curves

star	m_* [$M_\odot$]	P [days]	K [m s^{-1}]	e	$m_p \sin i$ [$M_{\rm jup}$]	a [AU]	comment
υ And b		4.617	70.15	0.01	0.64	0.058	
υ And c	1.30	241.16	53.93	0.27	1.79	0.805	apsidal lock
υ And d		1276.15	60.62	0.25	3.52	2.543	
55 Cnc b		14.653	71.5	0.03	0.83	0.115	
55 Cnc c	1.03	44.3	11.2	0.40	0.18	0.241	3:1 resonance
55 Cnc d		4400	50.2	0.34	3.69	5.2	
GJ 876 b	0.32	61.020	210.0	0.10	1.89	0.207	2:1 mean motion and secular resonance
GJ 876 C		30.120	81.0	0.27	0.56	0.130	
47 UMa b	1.03	1079.2	55.6	0.05	2.86	2.077	7:3 resonance?
47 UMa c		2845.0	15.7	0.00	1.09	3.968	
HD 168443 b	1.01	58.1	470.0	0.53	7.64	0.295	
HD 168443 c		1770.0	289.0	0.20	16.96	2.873	
HD 37124 b	0.91	153.3	35.0	0.10	0.86	0.543	
HD 37124 c		1942.0	19.0	0.60	1.00	2.952	
HD 12661 b	1.07	263.6	74.4	0.35	2.30	0.823	secular resonance
HD 12661 c		1444.5	27.6	0.20	1.57	2.557	
HD 38529 b	1.39	14.309	54.2	0.29	0.78	0.129	
HD 38529 c		2174.3	170.5	0.36	12.7	3.68	
HD 82943 b	1.05	444.6	46.0	0.41	1.63	1.159	2:1 resonance
HD 82943 c		221.6	34.0	0.54	0.88	0.728	
HD 74156 b	1.05	51.6	112.0	0.65	1.61	0.278	
HD 74156 c		> 2650	125.0	0.35	> 8.21	> 3.82	

From Marcy et al. (2003b)

It will thus be very important to develop techniques that can measure these inclinations, such as astrometry (see Sect. 9). For the moment, keeping the limitations of the Doppler method in mind, we can still learn a lot from the available data.

Individual Systems

In the following we will take a closer look at each of the systems from Table 5 in turn, highlighting some of the important features and peculiarities. This will provide the observational backdrop for the discussion of gravitational interactions and dynamical resonances in Sect. 3.6.

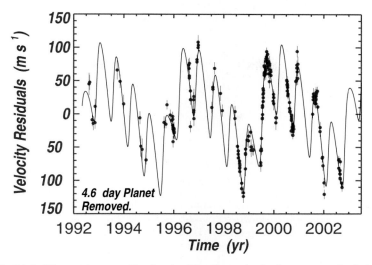

Fig. 15. Lick Observatory residual velocities for υ And after removal of the Keplerian wobble caused by the inner companion, using best-fit orbital parameters of $P = 4.6171$ days and $K = 75\,\mathrm{m\,s^{-1}}$. Two time scales are apparent in the residuals at 3 and 0.7 yr. The solid line shows the theoretical velocity curve caused by the outer two companions. Updated from Butler et al. (1999)

υ Andromedae.

Among the early results from the radial-velocity survey at Lick Observatory was the discovery of a planet orbiting the F8V star υ And with a 4.6-day period (Butler et al. 1997). The real claim to fame for this star came somewhat later with the identification of two additional companions with orbital periods of 241 and $\sim 1,280$ days (Butler et al. 1999, see Figs. 15 and 16). Since this original detection, the observed radial velocities have followed the predictions from a triple Keplerian fit without any indication for gravitational interaction between the planets, or for a fourth planet in the system (Marcy et al. 2003b). However, the apsidal lines of the outer two orbits are nearly aligned with each other (entries for ω in Table 6), hinting at a secular resonance involving these planets (see Sect. 3.6).

As demonstrated in Fig. 16, there are three massive planets in the υ And system in a volume that in the Solar System is populated only by the much smaller terrestrial planets. It is thus not surprising that gravitational interaction between the planets plays a much more significant role systems such as υ And than in the Solar System. Since these interactions depend directly on the masses of the planets, it would be interesting to get limits on their orbital inclinations. The astrometric signature expected from the outermost planet, υ And D, is just at the threshold of being detectable at the precision of the

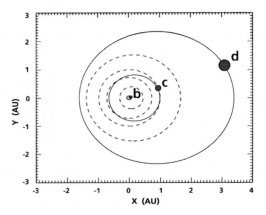

Fig. 16. Orbits of υ And b, c, and d, compared to the inner Solar System. The orbits of the three planets are shown co-planar and face-on; the actual relative orbital inclinations are not known, however. Because of the existence of multiple massive planets close to each other, dynamical interactions are much more important in the υ And system than in the Solar System

Table 6. Parameters of the planets of υ Andromedae, derived from Keplerian fits to the radial-velocity curve

parameter	planet B	planet C	planet D
P [days]	4.6170	241.5	1284
T [JD - 2,450,000]	2.093	160.5	64
e	0.012	0.28	0.27
ω [deg]	73	250	260
K_* [m s^{-1}]	70.2	53.9	61.1
a_p [AU]	0.059	0.83	2.53
$m_p \sin i$ [$M_{\rm jup}$]	0.69	1.89	3.75

Updated from Butler et al. (1999)

Hipparcos data (see also Sect. 9.2). The χ^2 of the Hipparcos measurements is indeed minimized by $m_p = 10.1\,M_{\rm jup}$, but a mass as low as $4.1\,M_{\rm jup}$ (which corresponds to $i = 90°$) or as high as $19.6\,M_{\rm jup}$ would also be allowed at the $2\,\sigma$ level (Mazeh et al. 1999).

55 Cancri.

The same paper that announced the discovery of the first planet around υ And also reported a planet orbiting the star 55 Cnc with a period of 14.65 days (Butler et al. 1997). The velocity residuals after subtracting the best-fit

Keplerian orbit showed a clear long-term trend, suggesting a possible second companion. Continued monitoring of 55 Cnc did indeed reveal two additional periodicities, with $P = 44.3\,\mathrm{d}$ and $P = 12\,\mathrm{yr}$, respectively (Marcy et al. 2002). While the latter can clearly be attributed to another planet in the system, the former is close to the rotation period of the star, and might thus be caused by surface inhomogeneities. There are a number of arguments against the rotational modulation hypothesis (including the requirement that the surface structure would have to be stable over at least 14 years), and thus it seems likely that there are indeed three planets around 55 Cnc, with a 3:1 orbital resonance between the inner two.

Significant excess emission at 60 μm was observed towards 55 Cnc with the Infrared Space Observatory (ISO), suggesting that this star may be a Vega-excess object (Dominik et al. 1998). This interpretation gained support by the putative detection of a scattered-light disk in ground-based near-infrared observations (Trilling and Brown 1998). Observations with the NICMOS instrument on the Hubble Space Telescope failed to detect this disk, however, indicating that the ground-based result is probably spurious (Schneider et al. 2001). From a non-detection at $\lambda = 850\,\mathrm{\mu m}$, Jayawardhana et al. (2002) obtain an upper limit of less than $10^{-3}\,M_\oplus$ in small dust grains associated with 55 Cnc; they suggest that the 60 μm excess results from a nearby sub-millimeter source within the ISO beam.

Gliese 876.

A planet with $m_p \sin i = 2\,M_\mathrm{jup}$ in a 60-day orbit around the M4 dwarf star GJ 876 was independently discovered by the Swiss and Californian planet search teams (Delfosse et al. 1998; Marcy et al. 1998). This planet is remarkable because of the low mass of its host star, which offers interesting opportunities for follow-up observations. The astrometric wobble of GJ 876 caused by the gravitational pull of this planet has indeed been detected with the Fine Guidance Sensors on the Hubble Space Telescope; this marks the first secure astrometric detection of an extrasolar planet (Benedict et al. 2002, see also Sect. 9.2). The inferred inclination $i = 84°$ implies that the mass of the planet m_p is close to its minimum mass $m_p \sin i$.

Continued observations of GJ 876 soon revealed that the radial-velocity data could not be modeled with a single Keplerian orbit; a second planet with a period of 30 days is needed to obtain a satisfactory fit (Marcy et al. 2001a, see Fig. 17). The two orbits have a 2:1 period ratio, and their axes appear to be nearly aligned (see Table 7); this is strong evidence that the two planets are locked in an orbital resonance. Taking the planet–planet interaction into account actually improves the χ^2 of a fit to the radial-velocity data considerably compared to a fit with two Keplerians (Marcy et al. 2003b).

47 UMa.

The two planets orbiting 47 UMa (Butler et al. 1996; Fischer et al. 2002a) have nearly circular orbits, with periods that are not close to any small-integer

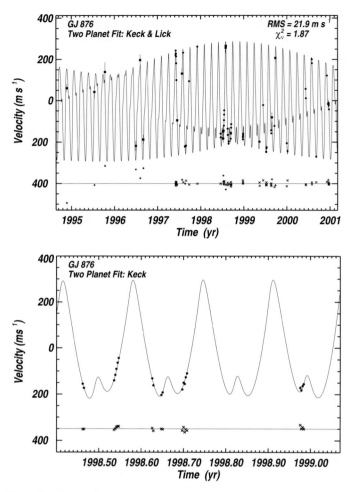

Fig. 17. *Top*: Combined Lick and Keck Observatory velocities for GJ 876, fitted with a model containing two planets in Keplerian orbits. The filled circles represent Keck velocities, and filled squares represent those from Lick. Residuals are shown below the radial-velocity curve. *Bottom*: Zoom on the time interval between 1998.4 and 1999.1, with velocities from Keck. The inflections in the velocities are due to the "beating" between the two planets. From Marcy et al. (2001a)

ratio. There is some resemblance to Jupiter and Saturn, which have similar period and mass ratios as the 47 UMa planets. Modeling the formation of the two planets in the 47 UMa system within the core accretion – gas capture model leads to the conclusion that they can both have formed through this mechanism within $\sim 3\,\mathrm{Myr}$ at their present distances from the star (Kornet et al. 2002).

Table 7. Parameters of the two planets orbiting GJ 876, derived from Keplerian fits to the radial-velocity curve

parameter	inner planet	outer planet
P [days]	30.12 ± 0.02	61.02 ± 0.03
T [JD - 2,450,000]	31.4 ± 1.2	106.2 ± 1.9
e	0.27 ± 0.04	0.10 ± 0.02
ω [deg]	330 ± 12	333 ± 12
K_* [m s^{-1}]	81 ± 5	210 ± 5
$a_* \sin i$ [AU]	0.00022	0.00117
a_p [AU]	0.130	0.208
$m_p \sin i$ [$M_{\rm jup}$]	0.56 ± 0.09	1.89 ± 0.3

From Marcy et al. (2001a)

HD 37124.

Systems like HD 37124 (Vogt et al. 2000; Marcy et al. 2003b) have be called "hierarchical", because the two planets have widely spaced orbits, which makes them structurally and dynamically reminiscent of hierarchical stellar systems. A slightly puzzling aspect of the HD 37124 system is the large eccentricity of the outer planet, whose origin is unknown.

HD 12661.

The planets orbiting HD 12661 have periods of 260 and 1440 d (Fischer et al. 2001, 2003b). Although this means that this system is also hierarchical ($P_1/P_2 = 5.48$), it has been argued that it resides in a secular resonance (Lee and Peale 2003).

HD 168443.

The HD 168443 system is interesting because it contains two very massive companions, with $m_p \sin i = 7.7\,M_{\rm jup}$ and $m_p \sin i = 17\,M{\rm jup}$, respectively (Marcy et al. 2001b; Udry et al. 2002). Beyond the somewhat irrelevant question whether these objects should legitimately be called "planets" or something else, one may ask whether they likely formed in the same way as their lower-mass analogs. If one is willing to speculate that some planets form through core accretion and gas capture, while others form through gravitational instabilities in a gas disk, the planets around HD 168443 would be prime candidates for the second process.

HD 38529.

With periods of 14.3 d and 6.0 yr, the two planets of HD 38529 form an extremely hierarchical system (Fischer et al. 2001, 2003b). The host star has spectral type G4IV and $m_* = 1.39\,M_\odot$, making it the most massive planet-bearing star known. The main sequence progenitor of HD 38529 was probably of spectral type F5V; it would have been difficult to obtain the radial-velocity precision during that stage that can now be obtained for the subgiant.

HD 82943.

The two planets orbiting HD 82943 have periods of 220 and 440 d, respectively, indicating a likely 2:1 mean motion resonance (Santos et al. 2003b; Mayor et al. 2004).

HD 74156.

The parameters of the outer planet in the system HD 74156 are not known precisely yet; the period is at least 6 yr and $m_p \sin i \gtrsim 8\,M_{\rm jup}$ (Marcy et al. 2003b). Since the period of the inner planet is only 51.6 d, this is also a very hierarchical system (Naef et al. 2004).

3.6 Interactions Between the Planets in Multiple Systems

General Formalism

As a first approximation, it is normally assumed that planets orbit their parent stars in Keplerian orbits; gravitational interactions between the planets and tidal effects can be treated as small perturbations. It has been pointed out already that in some extrasolar systems these perturbations are much stronger than in the Solar System, because they contain massive planets in close proximity to the parent star and to each other. This situation is somewhat similar to the ring and moon systems of the giant planets in the Solar System, which also exhibit a wealth of phenomena due to tidal and mutual gravitational interaction (e.g., De Pater and Lissauer 2001; Murray and Dermott 1999).

It is well known that there is no analytic solution to the general three-body problem. Various simplifications have therefore been studied, including the *restricted three-body problem*, in which one of the bodies is assumed to have negligible mass, and *Hill's problem* in which the mass of one body is much larger than the other two. One usually stars by writing the potential as the sum of a part that describes the Keplerian motion of the bodies about the central star, plus a part called *disturbing function*, which contains the *direct terms* accounting for the pairwise interactions among the planets and the *indirect terms* associated with the back-reaction of the planets on the central star. The gradient of the disturbing function describes the additional forces on the planets. One can then proceed by expanding the disturbing function in terms of small parameters; depending on the system under consideration one can consider expansions in the eccentricities, inclinations, or ratios of the planets' masses to the mass of the star. This expansion can now be inserted into *Lagrange's planetary equations*, which form a set of differential equations that express the time derivatives of the orbital elements by partial derivatives of the disturbing function (e.g., Murray and Dermott 1999).

Using this formalism, one thus arrives at a system of coupled differential equations for the orbital elements. To study the long-term evolution of the orbits, one can ignore all short-period terms of the disturbing function; these will average out to zero over sufficiently long time intervals. Solving the simplified set of differential equations, which contains only the *secular* (i.e., long-period) terms, one typically obtains solutions that are periodic in a and e, and contain linear terms for (orbital precession) and ω (precession of the periastron). Additional complications occur, however, if the e orbital periods of the two planets are nearly commensurate, i.e., if their ratio is close to a ratio of two small integers. The terms in the disturbing function corresponding to such a *mean motion resonance* do not average to zero, because the disturbance occurs always at approximately the same orbital phase. The situation is analogous to a simple harmonic oscillator

$$m\ddot{x} + m\omega_0^2 x = F \cos \omega_d t \qquad (10)$$

driven at a forcing frequency ω_d near the natural frequency ω_0. For $\omega_d \neq \omega_0$ the solution to (10) is

$$x = \frac{F}{m(\omega_0^2 - \omega_d^2)} \cos \omega_d t + C_1 \cos \omega_0 t + C_2 \sin \omega_0 t \,. \qquad (11)$$

We see that for $\omega_d \approx \omega_0$ the response can be very large even for a small driving force F. A famous example in the Solar System is the 3:2 mean motion resonance between the orbits of Pluto and Neptune. The angle $\phi \equiv 3\lambda_P - 2\lambda_N - \omega_P$, where λ_P and λ_N are the mean longitudes of Pluto and Neptune, and ω_P Pluto's longitude of perihelion, librates about 180° with a period of 19,670 years (Cohen and Hubbard 1965). This mechanism prevents close encounters of Pluto with Neptune and thus stabilizes the orbit of Pluto. A second type of resonance, called *secular resonance* arises if one of the precession rates ($\dot{\omega}$ or ˙) equals an eigenfrequency of the system.

A slightly different general formalism for calculations of the tidal, rotational, and dynamical evolution of planetary systems has been developed by Mardling and Lin (2002). It involves calculating the evolution of the orbital angular momentum vector and of the Runge–Lenz vector[15] of the inner orbit. Since these vectors are constant for unperturbed orbits, their components vary slowly compared to the orbital period. The secular evolution of the orbital elements can therefore be obtained by time-averaging the rates of change of

---

[15] The Runge–Lenz vector points in the direction of periastron and has a magnitude equal to the eccentricity.

the orbital angular momentum and Runge–Lenz vectors over the inner orbit. The resulting equations are quite complicated, but they can be used to implement fairly efficient and flexible computer programs for dynamical simulations of extrasolar planet systems.

Long-Term Stability

An important question about a multiple planetary system is its long-term stability. It is very difficult to prove that a system is stable in the sense that all planets remain bound for all time. One therefore frequently restricts the analysis to the weaker *Hill stability*, which means that the planets cannot undergo close approaches, which otherwise might disrupt the system. It can be shown that two planets in initially circular co-planar orbits cannot enter each other's Hill spheres for definition) if (Gladman 1993)

$$\Delta \equiv (a_2 - a_1)/a_1 > 2\sqrt[6]{3}(\mu_1 + \mu_2)^{1/3} \approx 2.4(\mu_1 + \mu_2)^{1/3} \,, \qquad (12)$$

where μ_1 and μ_2 are the ratios of the planetary masses to the mass of the central star, and a_1 and a_2 the orbital radii. This criterion gives a useful first indication that systems such as HD 168443 and 47 UMa are stable, unless their inclinations are extremely small, which would make μ_1 and μ_2 much larger than their minimum values (Marcy et al. 2001b; Fischer et al. 2002a). One caveat is that for a large *relative* inclination between the two planets a more stringent criterion than (12) would have to be applied (Ida and Makino 1993).

Orbital Resonances

More detailed studies of the stability of the known multiple systems make use of the analytic formalism sketched above, and of numerical integrations of the orbits. Orbital resonances of different types play an important role in at least five systems (υ And, 55 Cnc, GJ 876, HD 12661, and HD 82943), as indicated in Table 5. Orbital dynamics of extrasolar planets has thus become a rich field, in which many new results can be expected as more multiple systems are discovered, and as improved data become available for the systems that are already known.

Soon after the discovery of υ And c and υ And d, the first stability analyses of this system were carried out (Laughlin and Adams 1999; Lissauer and Rivera 2001). More recent studies have made use of updated planetary parameters from continued monitoring observations. They indicate that the outer two planets occupy nearly edge-on orbits with low relative inclination (Lissauer and Rivera 2001; Chiang et al. 2001; Chiang and Murray 2002). The two planets seem to inhabit a secular resonance, in which $\Delta\omega \equiv \omega_D - \omega_C$ librates about 0°. It is worth pointing out that a detailed analysis is needed to find this possible resonance mechanism in υ And.

The 2:1 mean motion resonance in the GJ 876 system is more obvious, as it is directly reflected in the orbital periods of the two planets. The resonance certainly helps to stabilize the system; stable configurations can be found with both high and low values of the relative inclinations of the two orbits (Rivera and Lissauer 2001; Ji et al. 2002). This is quite remarkable in view of the relatively large orbital eccentricities, but the configuration of GJ 876 remains in fact stable even at much larger eccentricities (Lee and Peale 2003). It has already been pointed out that the interaction between the two planets in the GJ 876 system is sufficiently strong to produce a detectable deviation from a model with two Keplerian orbits (see Sect. 4.1). It should thus be possible to determine the true planet masses (without the $\sin i$ factor) from the strength of this interaction, but this procedure does not yet lead to reliable results, even under the assumption of coplanarity (Laughlin and Chambers 2001; Nauenberg 2002a).

In some cases, the current uncertainties of the orbital parameters prevent clear statements on the dynamical state of a given system, because regular and chaotic orbits can be located close to each other in parameter space (e.g. HD 12661, Kiseleva-Eggleton et al. 2002; Lee and Peale 2003; Goździewski and Maciejewski 2003). In the HD 82943 system, which contains two planets in a 2:1 mean motion resonance, the nominal best-fit Keplerian orbits represent an unstable system, which should self-destruct quickly (Goździewski and Maciejewski 2001). An ad-hoc adjustment of the argument of periastron ω of the inner by about 30° leads to a stable system, however, with a model radial-velocity curve that is nearly indistinguishable from the original one. Similarly, the stability of the 47 UMa system depends critically on the eccentricity of the outer planet, which is poorly constrained by the current data, and on the relative inclination of the two planets, which cannot be determined with radial-velocity measurements (Laughlin et al. 2002; Goździewski 2002).

It has been speculated that planets could also be found in a 1:1 resonance, similar to Jupiter's Trojan asteroids (Laughlin and Chambers 2002; Nauenberg 2002b). For an exact 1:1 resonance the radial-velocity signature would be indistinguishable from that of a single planetary companion, but if there are slight deviations from the exact resonance, the pair of planets can execute horseshoe-type orbits around the Lagrangian points. A good example for this type of motion in the Solar System is given by Saturn's moons Janus and Epimetheus (see e.g. Murray and Dermott 1999). In that case, deviations from the single-planet radial-velocity curve can become quite significant over a few orbital periods. If planets in a 1:1 resonance exist, they could therefore be detected in the near future; it is even possible that one or the other of the known "single" planets will over time turn out to be a Trojan pair.

Commensurabilities between orbital periods can be set up during the early evolution of a planetary system, through orbital migration (see Sect. 3.2). If both planets open a gap in the disk, the outer planet migrates more quickly and approaches the inner planet, until it becomes locked in a 2:1 resonance (Snellgrove et al. 2001). This resonance can be maintained through the subsequent evolution of the system. It is also possible that the planets get locked up in other resonances (e.g., 3:1, 4:1, 5:1, or 5:2), depending on their masses, initial eccentricities, and the time scale for eccentricity damping in the disk (Nelson and Papaloizou 2002). Gathering sufficient statistical information about the incidence of these different resonances might thus provide a useful diagnostic tool of the migration process and of the interactions between the protoplanets and the disk during the formation phase of planetary systems.

4 Radial-Velocity Surveys

The radial-velocity technique has without doubt been the most successful planet detection method so far. In fact, *all* known extrasolar planets around main-sequence stars were discovered in this way[16]. The basis of the radial-velocity method is relatively simple: one obtains a time series of high-resolution spectra of the target star, and searches for periodic variations of the absorption line Doppler shift due to the motion of the star around the center of mass of the star–planet system. It is the exquisite precision of radial-velocity measurements achieved during the past decade that has made the plethora of recent planet detections possible. In this chapter we will discuss the foundations of the method, the design principles of the spectrographs used, and astrophysical limitations of the radial-velocity technique. The properties of the currently known extrasolar planets will be the subject of Sect. 3.

4.1 The Radial-Velocity Technique

Planetary Orbits from Radial Velocities

The orbit of a binary system is defined by seven parameters, the so-called *orbital elements* (see e.g. Batten 1973):

1. P, the orbital period;
2. i, the inclination of the orbital plane with respect to the tangent plane of the sky;
3. Ω, the position angle (measured from North through East) of the line of nodes, which is the intersection of the orbital and tangent planes;
4. ω, the angle between the direction of the ascending node (at which the star crosses the tangent plane while receding from the observer) and the periastron;
5. a, the semi-major axis of the orbit;
6. e, the eccentricity of the orbit;
7. T, the time of passage through periastron.

---

[16] Editor note added in proof: It is not true anymore, recently, planets have been detected by transit observations (see 6.4)

The radial velocity curve V of the primary star in a spectroscopic binary can be expressed as
$$V = \gamma + K_1 \left[\cos(\nu + \omega) + e \cos \omega\right] , \qquad (13)$$
where γ is the radial velocity of the center of mass of the system, K_1 the velocity amplitude, and ν the *true anomaly*, i.e., the position angle measured from periastron. The time dependence of $\nu(t)$ is given implicitly by the relations (see e.g. Heintz 1971; Murray and Dermott 1999)
$$\frac{2\pi}{P}(t - T) = E - e \sin E \qquad (14)$$
and
$$\tan \frac{\nu}{2} = \sqrt{\frac{1+e}{1-e}} \tan \frac{E}{2} ; \qquad (15)$$
the quantity E in these equations is called the *eccentric anomaly*. It is thus clear that the parameters P, T, e, and ω can be determined directly from the shape of the velocity time series. Ω and i, on the other hand, cannot be determined from spectroscopic observations alone. The semi-major axis of the primary around the center of mass is related to K_1 by
$$a_1 \sin i = \frac{P}{2\pi}\sqrt{1-e^2}\, K_1 . \qquad (16)$$

According to Kepler's Third Law,
$$a^3 = \left(\frac{P}{2\pi}\right)^2 G(m_1 + m_2) , \qquad (17)$$
where $a \equiv a_1 + a_2$ is the semi-major axis of the relative orbit of the two components. Using $m_1 a_1 = m_2 a_2$ and (16) and (17), we can derive the relation
$$\frac{(m_2 \sin i)^3}{(m_1 + m_2)^2} = \frac{P}{2\pi G} K_1^3 (1 - e^2)^{3/2} . \qquad (18)$$

The left-hand side of this equation is called the *mass function* of the system. If the secondary is a planet, we can use $m_2 \ll m_1$ to simplify (18). This gives
$$m_p \sin i \approx \left(\frac{P}{2\pi G}\right)^{1/3} K_* m_*^{2/3} \sqrt{1-e^2} , \qquad (19)$$
i.e., we can derive $m_p \sin i$ from the radial-velocity data provided that the mass of the central star m_* is known. In more convenient units one can write
$$m_p \sin i\, [M_{\rm jup}] \approx 3.5 \cdot 10^{-2} K_*[{\rm m\,s^{-1}}]\, P^{1/3}[{\rm yr}] ; \qquad (20)$$
this means that Jupiter causes a $12.5\,{\rm m\,s^{-1}}$ wobble in the radial velocity of our Sun.

The quantity $m_p \sin i$ is frequently referred to as the "minimum mass" of the planet. In each individual case, the actual mass of the planet may be considerably larger than this lower limit inferred from the radial-velocity technique. In a statistical sense, however, this uncertainty is not as severe as one might think. In a set of randomly oriented orbits $\cos i$ is uniformly distributed between 0 and 1. (It is more likely to observe an object nearly equator-on than nearly pole-on.) This means that in 87% of all cases $\sin i \geq 0.5$, and that in only 0.5% of all cases $\sin i \leq 0.1$. Therefore distributions of $m_p \sin i$ from radial velocity surveys are fairly representative of the true distribution of planetary masses (see also Sect. 3.2).

Multiple Systems

To first approximation, systems with several planets can be represented by a linear superposition of the individual Keplerian orbits. This approximation is a good one if the planetary masses are small, so that their mutual interaction can be neglected. In the case of massive planets, however, this approximation can break down on time scales that are not very long compared to the orbital periods, so that a treatment of the full many-body problem including dynamical resonances is required. In practice, this is best done in several steps (Laughlin and Chambers 2001; Rivera and Lissauer 2001). The starting point is a set of Keplerian fits for each planet; the corresponding orbital elements are called the *osculating elements* at the starting epoch. One can then perform a self-consistent integration of the many-body problem, compute a synthetic radial-velocity curve of the central star from the solution, compare this synthetic curve to the observations, and calculate the corresponding χ^2 value. This procedure can be repeated for different sets of osculating elements; the Levenberg–Marquardt method (e.g. Press et al. 1992) can be used to find the osculating elements that minimize the χ^2.

In this context, it is important to realize that the interaction between the planets depends on their masses and the relative inclination of their orbital planes. If sufficiently precise radial-velocity data are available, it is therefore possible to derive these parameters from dynamical analyses of multiple systems. The uncertainties are fairly large, however, because the parameter space to be searched has many dimensions, especially if the planets are not assumed a priori to be in coplanar orbits. Direct measurements of the relative orientations of the orbits with astrometric methods (see Sect. 9) can provide much better constraints on the dynamical evolution of multiple systems.

4.2 Limitations of the Radial-Velocity Precision

The Principle of Precise Doppler Spectroscopy

To detect the reflex motion of stars orbited by extrasolar planets, it is necessary to determine their radial velocity variations with stunning precision:

a measurement error of $3\,\mathrm{m\,s^{-1}}$ means that the wavelength shift of the stellar absorption spectrum has to be determined to one part in 10^8. The resolving power of modern high-resolution spectrographs is typically of order $R \equiv \lambda/\Delta\lambda \lesssim 100,000$; a precision of $1/1,000$ resolution element is therefore required. This is possible only by taking spectra with high signal-to-noise, and averaging over many spectral lines. Several conditions must be met to reach the desired precision of a few $\mathrm{m\,s^{-1}}$:

- The target star must have a sufficient number of absorption lines. This excludes main-sequence stars of spectral earlier than roughly F5 V, which have fewer lines than the cooler stars.
- The stellar absorption lines must be narrow. This again excludes stars with early spectral types and young stars, because they show too much rotational broadening.[17]
- The stellar photosphere must be sufficiently stable. This excludes active (e.g., flaring) stars and pre-main-sequence objects.
- The spectrograph used must be extremely stable, or a suitable calibration technique has to be applied.

For the first three reasons, radial-velocity surveys have concentrated mostly on F, G, and K main-sequence stars. M dwarfs and even brown dwarfs are now also attracting much interest for searches of low-mass planets, because the detection limit for m_p scales with $m_*^{2/3}$ (19). Many K giants are also suitable for precise radial-velocity monitoring, and giant planets have been found orbiting some of them (Frink et al. 2002; Sato et al. 2003).

Photon Noise, the Fundamental Limit

To understand the fundamental limit of the attainable radial-velocity precision, consider first one pixel on the detector of the spectrograph. The intensity change ΔN (measured in detected photo-electrons) in this pixel due to a small variation of the radial velocity ΔV can be written as (Connes 1985; Bouchy et al. 2001)

$$\Delta N \equiv N - N_0 = \frac{\partial N_0}{\partial \lambda} \Delta\lambda = \frac{\lambda}{c} \frac{\partial N_0}{\partial \lambda} \Delta V. \qquad (21)$$

Solving for ΔV, we obtain

$$\Delta V = \frac{c}{\lambda} \frac{N - N_0}{\partial N_0 / \partial \lambda}. \qquad (22)$$

---

[17] The rotation rate of main-sequence stars is linked to their structure. Stars with $m \lesssim 1.4\,M_\odot$ have outer convection zones; the interplay of convection with rotation leads to differential rotation and drives a dynamo. Magnetic breaking reduces the stellar rotation rate. This leads to a drastic difference in the typical rotation rates between stars earlier and later than F5 V.

In the photon noise-limited case the measurement error on N is proportional to $\sqrt{N}$; the Doppler precision is therefore inversely proportional to $|\partial N_0/\partial\lambda|\,\lambda/\sqrt{N_0}$, which can be taken as a "figure of merit" of the pixel in the stellar spectrum under consideration. When we combine the data from all pixels i in the spectrum, they should get weights $w(i)$ that are proportional to the square of this figure of merit:

$$w(i) \equiv \frac{\lambda^2(i)[\partial N_0(i)/\partial\lambda(i)]^2}{N_0(i)+\sigma_D^2} , \qquad (23)$$

where we have also included a potential contribution to the noise from the detector σ_D. The radial-velocity change computed from the full spectrum is then

$$\Delta V = \frac{\sum \Delta V(i) w(i)}{\sum w(i)} = c\,\frac{\sum [N(i)-N_0(i)]\sqrt{\frac{w(i)}{N_0(i)+\sigma_D^2}}}{\sum w(i)} . \qquad (24)$$

One can easily verify that the associated measurement uncertainty $\sigma_{\Delta V}$ can be expressed as

$$\sigma_{\Delta V} = \frac{c}{\sqrt{\sum w(i)}} . \qquad (25)$$

It is now convenient to introduce a "quality factor" Q defined by

$$Q \equiv \frac{\sqrt{\sum w(i)}}{\sqrt{\sum N_0(i)}} = \frac{\sqrt{\sum w(i)}}{\sqrt{N_{\text{tot}}}} , \qquad (26)$$

where N_{tot} is the total number of detected photons. In the high-flux limit, where detector noise is negligible, Q in independent of the stellar flux; it represents the sharpness and richness in spectral lines of the spectrum. With this definition we can finally write

$$\sigma_{\Delta V} = \frac{c}{Q\sqrt{N_{\text{tot}}}} . \qquad (27)$$

This formulation is well suited for modeling the influence of stellar spectral type, rotational line broadening, and spectrograph resolution on the attainable velocity precision (see Fig. 18). For $v\sin i \lesssim 6\,\text{km}\,\text{s}^{-1}$ the line profiles are broadened by the rotation, which leads to a linear decrease of the average $\partial N_0/\partial\lambda$ and therefore of Q (see (23) and (26)). For larger values of $v\sin i$, neighboring spectral lines start to become blended, which leads to $Q \propto (v\sin i)^{-1}$. At low spectral resolution ($R \lesssim 50{,}000$) all lines are blended and $Q \propto R$. When the resolution is increased to match the intrinsic (broadened) line width, Q reaches a constant value. Better spectral resolution is therefore beneficial, but only up to $R \approx 100{,}000$.

Limitations due to Stellar Variability

For radial-velocity measurements with a precision of a small fraction (of order $1/1{,}000$) of the line width, physical processes in the stellar photosphere or

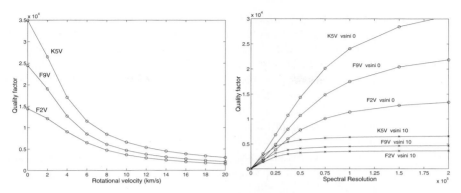

Fig. 18. Quality factor for radial-velocity measurement in the spectral range 3800 Å to 6800 Å. *Left panel*: Dependence on rotational broadening $v \sin i$ for K5 V, F9 V, and F2 V stars, for infinite spectral resolution. *Right panel*: Dependence on spectrograph resolution for K5 V, F9 V, and F2 V stars with $v \sin i = 0$ and $v \sin i = 10\,\mathrm{km\,s^{-1}}$. From Bouchy et al. (2001)

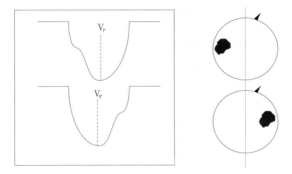

Fig. 19. Illustration of the effect of star spots on line profiles of a rotating star. From Queloz (1999)

chromosphere that affect the line profiles have to be considered carefully. An obvious example are starspots (Fig. 19). When a starspot (or group of spots) rotates into view, it hides part of the approaching side of the star. This causes a bump in the blue wings of absorption lines, which corresponds to a redward shift of the line centroid. When the spot rotates across the meridian, the bump moves from the blue wing to the red wing of the line, now causing a blueshift of the line. If the spot or spot group is long-lived, it will rotate periodically into and out of view, thus mimicking the periodic signal of a planet. To avoid this kind of misinterpretation, one should not rely on the line centroid (or on cross-correlating the observed spectra with a template) alone, but also check for variations of the line depths and shapes. Only if all lines vary synchronously and without changing their profiles is the interpretation of radial-velocity variations as the signature of a planet tenable.

Another indicator for the presence of star spots is of course photometric variability. A case in point is the G0 V star HD 166435 (Queloz et al. 2001). Observations with the ELODIE spectrograph at the Observatoire de Haute Provence revealed low-amplitude radial velocity variations with a period of 3.7987 days, suggestive of a possible planetary companion. Photometric observations uncovered variations with the same period and a one-quarter cycle phase shift, however, as expected for dark photospheric spots. This interpretation is also supported by a detailed analysis of the spectroscopic data, which revealed variations of the line profiles and a loss of coherence of the radial-velocity signal on time scales longer than ~ 30 days (which gives an indication of the time over which star spots are stable). Photometric variability has also been observed in HD 192263 (Henry et al. 2002), which had been thought to host a 0.75 $M_{\rm jup}$ planet in a 24-day orbit[18]. With a careful analysis it is thus frequently possible to separate true planetary companions from photospheric effects.

There remain difficult cases, however, in which the planetary hypothesis is plausible, but very hard to establish beyond reasonable doubt. A good example for this category is ε Eridani, which shows radial-velocity variations with amplitude $K = 19\,{\rm ms}^{-1}$ and period $P = 6.9\,{\rm yr}$ (Hatzes et al. 2000). These variations can be fit reasonably well with a Keplerian orbit, but the star also displays variations of the Ca II H and K lines indicative of magnetic activity. Further observations will be required to attribute the seven-year variations to either a companion or stellar activity.

In any case, even low-level stellar variability produces background noise, which limits the ultimate precision that can be attained with Doppler observations. The activity of cool stars is directly related to their rotation rate and thus to their age. A high rotation rate usually implies a stronger dynamo and thus stronger magnetic activity (spots, X-ray emission, chromospheric lines, ...). Magnetic breaking reduces the rotation rate and thus the activity. The time scale for this process depends on the mass of the star; low-mass stars (spectral type M) take the most time to slow down and thus show pronounced activity even at fairly old ages. The typical radial-velocity noise due to spots in G dwarfs decreases from $\approx 30\ldots 50\,{\rm m\,s}^{-1}$ at the age of the Hyades ($\sim 625\,{\rm Myr}$) to $\lesssim 5\,{\rm m\,s}^{-1}$ at the age of the Sun; convective perturbations of the radial velocity can have a similar magnitude (Saar and Donahue 1997).

Good indicators for activity in cool stars are the profiles of the Ca II H and K resonance lines at 3968.5 Å and 3933.7 Å, which consist of a narrow chromospheric emission component (in active stars) superposed on a very broad photospheric absorption line (see Fig. 20, left panels). For a quantitative analysis of these line profiles, one usually uses an "activity index" $R'_{\rm HK}$, which is defined as the ratio of the chromospheric emission in the cores of the Ca II H and K lines to the total bolometric emission of the star (Noyes et al. 1984).

---

[18] Note of the Editor added in proof: Santos et al. 2003a,b claims that the photometric and line profile are not synchronized and that the planet interpretation still hold.

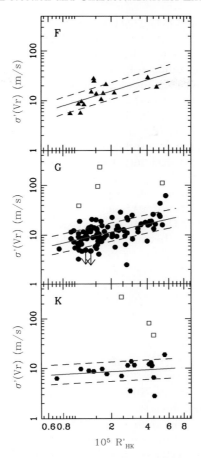

Fig. 20. Plots of the reduced radial-velocity rms, $\sigma'(\text{Vr})$, as a function of the activity index R'_{HK} for F, G, and K dwarfs (for details see text). The solid lines represent the best linear least-squares fit to the data; the rms around the fit is indicated by the dashed lines. Open squares represent stars with planetary systems with $\sigma'(\text{Vr})$ computed without subtracting the orbital solution. From Santos et al. (2000)

Plots of the radial-velocity jitter in F, G, and K dwarfs versus R'_{HK} show that these quantities are indeed correlated (Saar et al. 1998; Santos et al. 2000, see also the right panels in Fig. 20). The same plots also demonstrate that those stars for which planetary companions have been announced clearly stand out from the general distribution; this is another argument that the observed radial-velocity variations cannot be explained by stellar activity.

In addition to this clearly established general correlation of chromospheric activity with radial-velocity variability, the activity indices themselves show temporal variations. Saar and Fischer (2000) find that in 30% of the stars observed in the Lick survey the activity measured in the Ca II $\lambda 8662$ line (which is one of the lines of the Ca II infrared triplet) is correlated with simultaneously measured radial velocities, i.e., the level of activity itself produces a shift of the radial velocity. They argue that the main cause of these effects is modification of the mean line bisector shape brought on by long-term, magnetic activity-induced changes in the surface brightness and convective patterns. Taking out this trend before analyzing the radial velocities thus reduces the residual scatter. On the other hand, Paulson et al. (2002) find a similar correlation between $R'_{\rm HK}$ and the radial velocity in only five of 82 stars in the Hyades. It thus remains to be seen to which extent activity indices can be used not only to pre-select intrinsically "quiet" stars, but also to remove activity-induced systematic effects from radial-velocity data.

Stellar pulsations can also produce radial-velocity variations that could be mistaken for the signature of a planet. Again, photometry is a good way of double-checking whether this may be the case. For radial pulsations, there is a straightforward relation between the amplitude ΔR of the radius changes and the radial-velocity curve,

$$\Delta R = \int_0^{T/4} V\,dt\,. \tag{28}$$

Using this relation and

$$\frac{\Delta L}{L} = 2\frac{\Delta R}{R} + 4\frac{\Delta T}{T}\,, \tag{29}$$

which follows directly from the Stefan–Boltzmann Law $L = 4\pi R^2 T_{\rm eff}^4$, it is possible to predict the photometric variability from the observed radial-velocity variations. For example, in the case of 51 Peg, the amplitude of the radial-velocity curve (59 m s^{-1}, Mayor and Queloz 1995) implies $\Delta R/R = 0.5\%$ and thus $\Delta L/L = 1\%$ (for $\Delta T/T = 0$). The observed photometric stability of better than 0.1% therefore rules out radial pulsations; the assumption that ΔR would be compensated by ΔT such that $\Delta L/L \leq 0.1\%$ is too contrived. No such direct case can be made against non-radial pulsations, which do not necessarily imply detectable photometric variations. However, several strong arguments are also available against this interpretation:

- Only modes with very low amplitudes ($\ll 1\,{\rm m\,s^{-1}}$) and periods $\lesssim 1\,{\rm h}$ are detected in the Sun, and expected for Sun-like stars.
- Mechanisms that could excite a high-order non-radial pulsation mode should also excite many other similar modes. There is no plausible mechanism that could selectively excite one single mode, and thus mimic a planetary signal.
- Detailed studies of stars for which planets have been published have not revealed any changes of the line shapes (e.g. Gray 1998).

Taken together, these arguments rule out pulsations as a plausible explanation for the observed radial-velocity variations.

The situation is somewhat more complicated in the case of giant stars. All evolved stars exhibit intrinsic variability to some degree. A well-known example is the K2 giant α Bootis, which has a complicated variability pattern with an amplitude of $\sim 200\,\mathrm{m\,s^{-1}}$ on time scales down to a few days (Hatzes and Cochran 1994). Giant stars have therefore not been targeted by the planet surveys. A survey of nearby K giants aimed at assessing their suitability as astrometric reference stars has shown, however, that many of these stars have fairly stable radial velocities (Frink et al. 2001). In fact, about 2/3 of the stars observed are drawn from a distribution with a mean radial velocity scatter of $\sim 20\,\mathrm{m\,s^{-1}}$. There is also a correlation with color, which implies that a suitable color selection criterion can reduce the contamination by photospherically unstable targets. The companion of the K2 III giant ι Draconis with a minimum mass of 8.9 AU discovered serendipitously in the survey mentioned demonstrates that detecting planets around giants is indeed possible (Frink et al. 2002).

4.3 Spectrograph Design

Cross-Dispersed Echelle Spectrographs

Precise radial-velocity measurements require a large spectral range covered with high spectral resolution ((27), Fig. 18). These requirements are met best with cross-dispersed echelle spectrographs (Schroeder 2000; Vogt 1987; Baranne 1999). This type of instrument takes its name from the arrangement of two separate dispersing elements. The first is a grating used in high orders, which is responsible for the spectral resolution. To avoid that the overlapping orders of the main grating fall on the same pixel on the detector, a second low-dispersion element (prism, grism or combination of the two) is used in the orthogonal direction. This leads to a format which uses most of the real estate on a CCD chip for a long high-resolution (typically $R \equiv \lambda/\Delta\lambda = 50,000\ldots 100,000$) spectrum.

Extraordinary measures have to be taken, of course, to obtain a long-term reproducibility of order 1/1,000 pixel for measurements with these instruments. The first requirement is to build the spectrograph as stable as possible. Changes in the spectrograph point spread function (i.e., the observed profile of an infinitely narrow line) due to flexure or thermal expansion can significantly alter the measured position of line centroids, and thus introduce noise in the radial-velocity measurements. In an air-filled spectrograph one also has to take into account changes of the observed "air" wavelength with pressure ($90\,\mathrm{m\,s^{-1}\,mbar^{-1}}$) and temperature ($200\ldots 300\,\mathrm{m\,s^{-1}\,K^{-1}}$, depending on the observatory elevation). The key to success is referencing all observations to a stable standard, and to eliminate all systematic errors that can enter this process.

When a stellar radial-velocity measurement has been obtained, it must be transformed from the observatory reference system into an inertial reference frame. Getting a sufficiently precise ephemeris for the motion and rotation of the Earth is no problem, but timing the observation requires some care. As the radial velocity of the Earth can change significantly (up to $2.4\,\mathrm{m\,s^{-1}\,min^{-1}}$) while the shutter is open, we need to know the photon-weighted midpoint of the exposure. Taking this simply as the midpoint between opening and closing the shutter can produce a very significant error if the weather is partly cloudy; it is therefore advisable to monitor the photon flux during the exposure with a separate high-speed photometer.

The Simultaneous Thorium Technique

The classical way of providing good wavelength calibration of astronomical spectra is the simultaneous observation of the star and an emission spectrum from an arc lamp. Spectrographs based on this principle have been used for many years by the Swiss planet search team (ELODIE and CORALIE, Baranne et al. 1996). The new HARPS spectrograph to be installed at ESO's 3.6 m telescope on La Silla has been designed to reach a Doppler precision of $1\,\mathrm{m\,s^{-1}}$ (Pepe et al. 2000)[19]. These instruments use two optical fibers to couple both the star light and the light from the Thorium lamp to the spectrograph input; each "science exposure" thus contains truly simultaneous calibration information. The observable quantities are thus the wavelength differences $\lambda_s(f_1, t_1) - \lambda_T(f_2, t_1)$ between stellar absorption features and Thorium lines. (The argument indicate which fiber was used and the time of the exposure.) In addition, a "calibration exposure" is taken in which Thorium light is coupled into both fibers. The observed Doppler shift between the two Thorium spectra $\lambda_T(f_1, t_2) - \lambda_T(f_2, t_2)$ reflects systematic effects induced by the two different paths through the spectrograph. The double difference $[\lambda_s(f_1, t_1) - \lambda_T(f_2, t_1)] - [\lambda_T(f_1, t_2) - \lambda_T(f_2, t_2)]$ therefore provides a reference of the stellar spectrum to the Thorium lines, which is free of both temporal drifts and systematic differences between the two fibers. Advantages of the Thorium technique are a large usable spectral range and relatively high transmission ($\sim 80\%$ for a well-adjusted fiber).

In addition to providing a convenient way of coupling the telescope to the spectrograph, the optical fibers fulfill the important role of stabilizing the stellar light on the spectrograph slit. In a classical spectrograph, which is attached directly to the telescope, slight displacements of the star with respect to the spectrograph slit can lead to serious shifts of the observed wavelength (see Fig. 21). Keeping the star centered on the slit with the precision required for planet surveys is beyond the capabilities of telescope guide systems. In a fiber-fed spectrograph this problem is reduced substantially by the "scrambling" effect of the fiber: an off-axis illumination of the fiber input still leads to

---

[19] Editor note added in proof: HARPS is installed and available to the community since October 2003

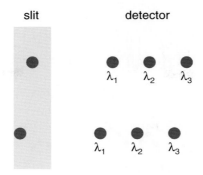

Fig. 21. Stars well-centered (*top*) and poorly centered (*bottom*) on the spectrograph slit. It is apparent that motion of the light centroid perpendicular to the slit (along the dispersion direction) leads to an apparent shift of the observed wavelength

a circularly symmetric output from the fiber (see e.g., Fig. 7 in Queloz 1999). If this azimuthal averaging of the fiber is insufficient, one can further improve the uniformity of the slit illumination by employing a *double scrambler*, which also performs some radial redistribution of the light (Queloz et al. 1999).

The Iodine Absorption Method

An alternative approach to coping with long-term drifts of the spectrograph and unstable illumination of the slit is passing the starlight through an absorbing medium before entry into the spectrograph, thereby superimposing reference absorption lines that experience the same instrumental shifts as the stellar spectrum. It was first suggested to use telluric absorption lines (i.e., absorption lines originating in the Earth's atmosphere) as wavelength references (Griffin and Griffin 1973); modern versions of this technique use an absorption cell in front of the spectrograph slit (Campbell and Walker 1979). A long-term precision of $\sim 15\,\mathrm{m\,s^{-1}}$ has been achieved with an absorption cell filled with hydrogen fluoride gas (Campbell et al. 1988). The main drawbacks of the HF molecule are the fairly small wavelength range covered by the absorption band (~ 100 Å), its corrosive nature, and its lethal effect on humans.

After an extensive search for a better absorbing medium, Marcy and Butler (1992) concluded that gaseous iodine is the molecule of choice. It combines the advantages of strong line absorption coefficients (requiring only a short absorption length and low pressure), large wavelength coverage (from 5,000 to 6,300 Å), chemical stability, and low risk to human health. A 10 cm long cell filled with gaseous I_2 at a pressure of 1/100 atm was built for the Hamilton Echelle Spectrograph (Marcy and Butler 1992); similar cells have also been installed at other telescopes. Observations with an iodine cell produce a spectrum in which stellar and iodine lines are heavily blended with each other. The data analysis therefore requires sophisticated modeling of the observed

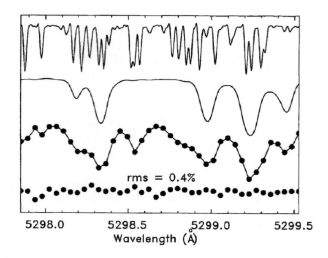

Fig. 22. Modeling of a spectrum observed through an iodine cell. *Top*: The template iodine spectrum. *Second*: The template stellar spectrum (in this case τ Ceti, G8 V). Note how rich this spectrum is; the figure shows only ∼1.6 Å of the ∼850 Å range used for the Doppler analysis. *Third*: The points are in observation of τ Ceti made through the iodine absorption. The solid line is a model of the observation, composed of the template iodine and stellar spectra. The free parameters consist of the spectrograph point spread function and the Doppler shift of the template star relative to the template iodine. *Bottom*: 10 times the difference between the model and the observation. The model and the observation differ by 0.4% rms. From Butler et al. (1996)

spectrum (see Fig. 22). The data reduction software needs three inputs: the observed spectrum I_{obs}, a high-resolution iodine template spectrum T_{I_2}, and a high-resolution spectrum I_S of the target star (obtained by deconvolving a spectrum with very high signal-to-noise, taken without the iodine cell). The spectra taken through the iodine cell are then modeled as (Butler et al. 1996)

$$I_{\mathrm{obs}}(\lambda) = k\left[T_{I_2}(\lambda) I_S(\lambda + \Delta\lambda)\right] * \mathrm{PSF} , \qquad (30)$$

where k is a normalization constant and PSF the spectrograph point spread function. The operator $*$ denotes convolution as defined in (238). The Doppler shift $\Delta\lambda$ is obtained by a χ^2 minimization procedure that adjusts this parameter together with twelve others describing the wavelength scale and the shape of the spectrograph point spread function. This technique has consistently produced Doppler measurements with a long-term stability of $3\,\mathrm{m\,s}^{-1}$ (Butler et al. 1996).

Absolute Astronomical Accelerometry

The simultaneous Thorium and Iodine absorption methods rely on the "random" positions of absorption or emission lines in the spectrum of a molecule or atom. In comparison, it should in principle be advantageous to use the regular transmission comb of an interferometer as a wavelength reference (e.g., Ge et al. 2002). A variation of this concept is the "Absolute Astronomical Accelerometer" proposed by Connes (1985). This instrument is based on two control loops. First, the variable path difference of a tunable Fabry–Perot interferometer is adjusted such that its transmission maxima track the variable Doppler shift of a star. This tracking ability is the main advantage of this concept over the methods described above; it eliminates systematic errors due to the relative shifts of stellar and calibration lines caused by the annual and diurnal variation of the observatory velocity in an inertial reference frame. The second loop involves a tunable laser, which tracks one of the Fabry–Perot transmission peaks. The net result is that the wavelength of the laser line tracks the radial velocity of the star. The beam from the tunable laser is then mixed with a stabilized laser; the change in the beat frequency is the signal that contains the desired information about changes in the stellar Doppler shift. A prototype instrument has been built and tested in the laboratory. It remains to be seen whether wavelength references based on interferometric approaches can reach or surpass the long-term radial-velocity stability that has been demonstrated with gas cells and lamps.

4.4 Radial-Velocity Surveys

The First Planet Detections

In the August 1, 1995 issue of *Icarus* appeared a paper summarizing the results from a 12-year search for Jupiter-mass companions to 21 nearby stars. No planets were found, with limits in the range $m \sin i \leq 1 \ldots 3 \, M_{\text{jup}}$ for any possible planets with orbital periods up to 15 years (Walker et al. 1995). In retrospect this team was remarkably unlucky; as we now know, their precision and sample size gave them a $> 50\%$ chance to actually discover the first planet around a Solar-type star. But because the sample picked by Walker et al. happened to contain no massive short-period planet, the honor of the first discovery went to Mayor and Queloz (1995), who announced the planet orbiting 51 Peg only three months later. The radial-velocity variations of this star were almost immediately confirmed by Marcy and Butler (1995), who had also started a long-term planet search. This survey soon uncovered two additional planets (Marcy and Butler 1996; Butler and Marcy 1996), providing a first glimpse of the unanticipated diversity of giant planets.

Recent Surveys

Fueled by the unexpected discoveries of giant planets with short orbital periods, the ongoing surveys intensified their efforts, and several new radial-velocity projects were started. More than 2,000 stars are now being monitored with Doppler precisions in the range $3\ldots 15\,\mathrm{m\,s^{-1}}$. Among the projects that have contributed to the list of known planets are: ELODIE at the Observatoire de Haute Provence (Udry et al. 2001); its improved sister CORALIE at the Swiss Euler Telescope on La Silla (Udry et al. 2000); the Hamilton Echelle Spectrograph at Lick Observatory (Marcy and Butler 1998; Cumming et al. 1999); HIRES at the Keck Observatory (Vogt et al. 2000); the Advanced Fiber-Optic Echelle at Whipple Observatory (Nisenson et al. 1999); the Anglo-Australian Telescope (Butler et al. 2001); the Coudé Echelle Spectrometer at ESO's 3.6 m Telescope on La Silla (Endl et al. 2002); the McDonald Observatory (Cochran et al. 2000); and the Lick bright K giant survey (Frink et al. 2002).

Several attempts have been made to analyze the combined published results from all surveys, in order to obtain statistical information on the distribution of planet masses and periods, and to assess the fraction of stars that have planetary companions (e.g., Nelson and Angel 1998; Tabachnik and Tremaine 2002; Lineweaver and Grether 2002). While such compilations can provide much useful information, of course, one has to keep in mind that it is extremely difficult to estimate the completeness of the underlying data. Many of the long-term surveys have improved their observing techniques over the years, which leads to complicated sensitivity limits as a function of orbital period. Furthermore, the temporal sampling may vary widely from star to star, because "interesting" targets were followed much more frequently than others. A slightly enhanced level of stellar activity may also have an adverse influence both on the detection threshold for planets and on the enthusiasm of the observers to obtain many data points. One finally has to keep in mind that a star without a published planet is not necessarily a star without a detected planet – the observers may just have chosen to wait with the publication until they can get a satisfactory orbital fit. In spite of all these caveats, the amount of information on extrasolar planets that has been gathered with the radial-velocity method is now large enough to enable interesting statistical conclusions. We will come back to this point in the following chapter.

5 Gravitational Microlensing

The detection and monitoring of gravitational microlensing events towards the Galactic bulge and the Large Magellanic Cloud has been used successfully as a tool to study the composition and mass distribution of the Galaxy (Paczyński 1986, 1996; Alcock 2000). The light curves of lensing events involving the linear motion of a point-like lens in front of a point-like source have a characteristic shape; any deviations from this shape can be used to infer parameters not described by this simple geometry: parallax (affecting the relative path of source and lens), resolution of the stellar disk, or the presence of companions to the source or lens. In the present context, we are mostly interested in the last of these effects: the detection of "binary lenses" with low-mass secondaries.

The monitoring of gravitational microlensing events is arguably the only method that is capable of detecting Earth-like planets from the ground; this is the most important driver behind the further development of this technique. So far, however, no secure planet detection has been made in this way (Sackett 2000). This chapter introduces the theory of gravitational microlensing, first for a single lens, then for the more complicated case of binary lenses. We will then discuss the available results from current microlensing monitoring experiments.

5.1 Theory of Gravitational Microlensing

Gravitational Lensing by a Single Pointlike Lens

According to the General Theory of Relativity, light from a background source passing by a foreground star of mass M with an impact parameter (minimum distance) $r \gg R_S$ is deflected by an angle

$$\alpha = \frac{4GM}{c^2 r} = \frac{2R_S}{r}, \tag{31}$$

where we have introduced the *Schwarzschild radius* $R_S \equiv 2GM/c^2$. Derivations of (31) can be found in any textbook on General Relativity.[20] It was realized soon that light bending could lead to multiple images of the same source (Eddington 1920; Chwolson 1924; Einstein 1936); this effect is called "gravitational lensing".

To describe the geometry of a gravitational lens, we use the following notation: θ is the observed position of the source, θ_S the direction to the source in the absence of lensing, D_S and D_L denote the distances of the source and lens from the observer, and $D_{LS} \equiv D_S - D_L$ the distance from the lens to the source. Simple geometry (see Fig. 23) then leads to the relation:

$$\theta_S D_S = r \frac{D_S}{D_L} - D_{LS} \alpha(r). \tag{32}$$

---

[20] The famous observational verification of (31) for the bending of light in the gravitational field of the Sun during the total eclipse of 1919 (Dyson et al. 1920) played an important role for the popularization and acceptance of General Relativity. Modern precision measurements of light deflection provide tests of extensions and alternatives of General Relativity.

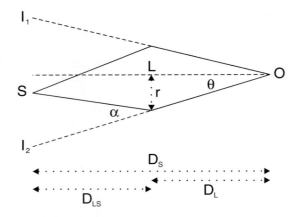

Fig. 23. Geometry of gravitational lensing. Rays from the source S are bent in the lens plane by an angle α, so that the observer O sees two images I_1, I_2

Using $r = D_L \theta$, this can also be written as

$$\theta_S = \theta - \frac{D_{LS}}{D_S}\alpha(r). \tag{33}$$

It is now convenient to introduce the characteristic angle

$$\theta_E \equiv \sqrt{\frac{4GM}{c^2}\frac{D_{LS}}{D_L D_S}}\;; \tag{34}$$

the corresponding characteristic length in the lens plane $r_E \equiv \theta_E \cdot D_L$ is called the *Einstein radius*. Inserting this definition and (31) in (33) gives the *lens equation*

$$\theta^2 - \theta_S \theta - \theta_E^2 = 0. \tag{35}$$

This quadratic equation has the two solutions

$$\theta_{1,2} = \frac{1}{2}\left(\theta_S \pm \sqrt{4\theta_E^2 + \theta_S^2}\right), \tag{36}$$

which give the positions of the two images seen by the observer. If the source, lens, and observer lie on a straight line, $\theta_S = 0$, and (36) indicates the presence of two images at positions $\pm\theta_E$. In this case, however, the line containing source, lens, and observer is a symmetry axis. We can use any plane containing this line to draw Fig. 23, and get two images at $\pm\theta_E$ in that plane. The image is thus a circle with radius θ_E around the direction from the observer towards the lens, the so-called "Einstein ring". If $\theta_S \neq 0$, the two solutions of (36) satisfy the inequalities $\theta_1 \geq \theta_E$ and $-\theta_E < \theta_2 < 0$. This means that the two images lie on opposite sides of the observer-lens axis, one of them inside and one of them outside the Einstein ring.

We will see below (40) that if $\theta_S \gg \theta_E$ one of the images is very faint, and the brightness of the other image is hardly affected by the lens. We therefore expect that the lens has noticeable effects only when $\theta_S \lesssim \theta_E$. Then it follows immediately from (36) that the separation of the two images is

$$\theta_1 - \theta_2 = \sqrt{4\theta_E^2 + \theta_S^2} \approx 2\theta_E. \tag{37}$$

If we insert typical numbers for observations of stars in the Galactic bulge into (34), we obtain

$$\theta_E = 1.1 \text{ mas} \cdot \left(\frac{M}{M_\odot}\right)^{1/2} \left(\frac{7 \text{ kpc}}{D_S}\right)^{1/2} \left(\frac{D_{\text{LS}}}{D_L}\right)^{1/2}. \tag{38}$$

The separation of the two images is thus of order a few milliarcseconds. This is usually too small to be resolved, and all we can observe is the combined brightness of the two images (but see Sect. 5.4). It is this situation that is usually called "microlensing".

To compute the observed flux from the images we first note that the surface brightness is not changed by the lensing process.[21] What is affected, though, is the solid angle of the image $\Delta\Omega$ subtended on the sky. The flux of an infinitesimally small source is simply given by the product of surface brightness and solid angle. The ratio of the observed flux to the flux in the absence of lensing (the "amplification" A, which should more aptly be called "magnification") is thus simply given by $\Delta\Omega/\Delta\Omega_S$. Equation (33) defines a mapping from θ_S to θ; the area distortion of this mapping is given by the determinant of the Jacobi matrix J. We therefore get:

$$A_{1,2} = \left(\frac{\Delta\Omega_S}{\Delta\Omega_{1,2}}\right)^{-1} = \frac{1}{|\det J|}\bigg|_{\theta_1,\theta_2} = \left|\frac{\partial\theta_S}{\partial\theta}\right|^{-1}_{\theta_1,\theta_2}. \tag{39}$$

For the calculation of this expression, we introduce the quantity $u \equiv \theta_S/\theta_E$, the source-lens separation in units of the Einstein radius. We have to keep the two-dimensional nature of the lens mapping in mind. Because of the symmetry of the lens, nothing is changed in the direction perpendicular to the plane of

---

[21] If the curvature radius of space-time is large compared to the wavelength, it can be shown that the photon phase space density along each photon's world line, or equivalently the quantity $I(\nu)/\nu^3$, is conserved. Among the well-known direct consequences are that the bolometric surface brightness of galaxies $I_{\text{bol}} \propto (1+z)^{-4}$, and that the spectrum emitted by a blackbody (such as the cosmic microwave background) remains a blackbody spectrum with observed temperature $T_{\text{obs}} = T_{\text{em}}/(1+z_{\text{em}})$. For gravitational lensing in our Galaxy we are interested in the special case $z = 0$, i.e., ν is the same for the emitter and observer. For a detailed discussion see Chapter 22 of Misner et al. (1973).

Fig. 23, and the Jacobian can easily be evaluated using polar coordinates; it is given by

$$A_{1,2} = \left| \frac{\theta_{1,2}}{\theta_S} \cdot \frac{d\theta_{1,2}}{d\theta_S} \right| = \frac{u^2 + 2}{2u\sqrt{u^2 + 4}} \pm \frac{1}{2}. \tag{40}$$

That the first equality is true is also obvious from Fig. 24. The total combined brightness of the two unresolved images is thus

$$A = A_1 + A_2 = \frac{u^2 + 2}{u\sqrt{u^2 + 4}}. \tag{41}$$

So far we have dealt with a static configuration of a background source being lensed by a foreground object at a normalized projected separation u. Differential Galactic rotation and peculiar motions lead to a relative motion of source and lens, however, with a typical magnitude

$$\dot\theta = \frac{v}{D_L} = 12 \, \text{mas yr}^{-1} \left(\frac{v}{200 \, \text{km s}^{-1}} \right) \left(\frac{3.5 \, \text{kpc}}{D_L} \right), \tag{42}$$

where v is the relative perpendicular velocity of the lens with respect to the source. The typical time scale t_E of a microlensing event is given by the time needed to move by one Einstein radius

$$t_E \equiv \frac{\theta_E}{\dot\theta} = 0.13 \, \text{yr} \left(\frac{M}{M_\odot} \right)^{1/2} \left(\frac{D_L}{3.5 \, \text{kpc}} \right)^{1/2} \left(\frac{D_{LS}}{D_S} \right)^{1/2} \left(\frac{v}{200 \, \text{km s}^{-1}} \right)^{-1}. \tag{43}$$

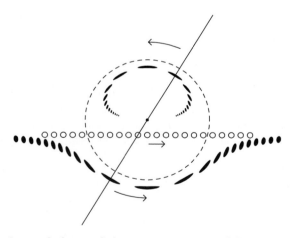

Fig. 24. Location and shape of the two images in a Schwarzschild lens. In this drawing, the lensing mass is indicated with a dot at the center of the Einstein ring, which is marked with a dashed line; the source positions are shown with a series of small open circles; and the locations and the shapes of the two images are shown with a series of dark ellipses. At any instant the two images, the source, and the lens are all on a single line. From Paczyński (1996)

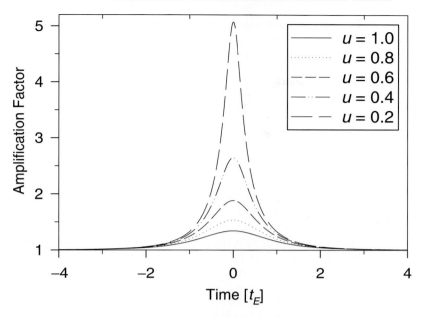

Fig. 25. Microlensing light curves for five different values of the minimum impact parameter. From (41) it follows directly that at the time of crossing the Einstein ring ($u = 1$) the amplification $A = 3/\sqrt{5} = 1.34$, and that the maximum amplification $A(u_{\min}) \approx 1/u_{\min}$

To first order the motion is linear and can thus be parameterized by:

$$u(t) = \sqrt{(t - t_{\min})^2/t_E^2 + u_{\min}^2}\,. \tag{44}$$

Here $t_{\min}$ and $u_{\min}$ are the time of closest approach, and the corresponding impact parameter. Substituting (44) in (41) gives an analytic description of the amplification as a function of time, i.e., of the light curve of the lensing event. Such light curves are shown in Fig. 25 for five different values of $u_{\min}$. It is important to note that the light curves of all microlensing events involving a point source and a pointlike lens that moves at a constant rate are completely determined by four parameters: $t_{\min}$, t_E, $u_{\min}$ and the brightness of the source at $u \to \infty$. Of these parameters, only t_E is related to the properties of the lens. It is therefore only possible to derive a combination of lens mass, distance and transverse velocity in this simple situation.

Lensing Anomalies and Binary Lenses

Real astrophysical sources and lenses are not point sources, of course, and the relative motion is not necessarily rectilinear. We may thus observe lensing "anomalies", i.e., deviations from the simple model described in the previous section, and attempt to obtain additional information from them. For planet searches, we are mostly interested in binary lenses with very small mass ratio $\mu \equiv m_2/m_1$. The binary lens equation is a straightforward generalization of (33):

$$\vec{\theta}_S = \vec{\theta} - \frac{D_{LS}}{D_S} \left(\vec{\alpha}_1(\vec{r}_1) + \vec{\alpha}_2(\vec{r}_2) \right) , \qquad (45)$$

where the two indices 1, 2 refer to the two binary components. We have now written all two-dimensional quantities explicitly as vectors, because there is no plane of symmetry anymore. The analysis of binary lenses is considerably more complicated than that of single lenses. We should expect, however, that source positions for which the determinant of the Jacobi matrix $|\det J| = 0$ have special significance: according to (39) the amplification is infinite for these positions. The locus of the positions for which this condition holds is called a "caustic". The caustic of a point source lens consists only of the point $u = 0$ (see (40)), corresponding to the appearance of an Einstein ring when observer, lens, and source are perfectly aligned. In contrast, the caustics of binary lenses are extended and complicated in shape (Schneider and Weiß 1986; Erdl and Schneider 1993). In the lens plane, the condition $|\det J| = 0$ defines the "critical curves". When the source crosses a caustic at location θ_S, two new highly amplified images appear with positions θ on the corresponding critical curve (or two existing images brighten, merge and disappear if the caustic is crossed in the opposite direction). An example for the case of equal masses of the two binary components is shown in Fig. 26. The left panel shows the structure of the caustic and critical curve, and five possible relative paths of a source with respect to the lens. The source has not been assumed to be pointlike, but rather a uniform disk of diameter $r_s = 0.05\,r_E$. The brightness at any time therefore has to be computed by integrating the amplification as given by (39) over the area subtended by the disk. For each source position along the path, the brightness has been calculated in this way, and plotted versus time in the right panel.

It is apparent from Fig. 26 that a wide variety of light curves are possible for binary lenses (see also Alcock et al. 2000a). The mass ratio μ, the projected binary separation b (in units of r_E), and the angle of the source trajectory with the binary axis are additional free parameters that have to be fit to the observational data. The example of the first binary lens detected, OGLE #7, shows that this can be done fairly well if observations with good sampling and signal-to-noise exist, especially when data points close to caustic crossings are available. OGLE #7 was observed independently by two collaborations (Udalski et al. 1994; Bennett et al. 1995), and the fits to the two disjoint data sets agree well with each other (Alcock et al. 2000a). Additional "anomalous" effects can complicate the interpretation, however. Parallax (due to the annual motion of the Earth around the Sun) leads to a non-linear relative motion of the source and lens, and the orbital motion of the binary can change the caustic structure itself on a time scale comparable to t_E. Confusion, i.e., blending

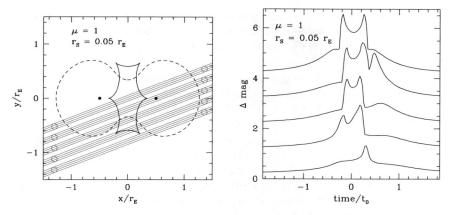

Fig. 26. Microlensing by a binary consisting of two identical point masses, $m_1 = m_2 = 0.5m$, separated by one Einstein ring radius, r_E. The closed figure drawn with a thick solid line in the left panel is the caustic located in the source plane. The closed figure drawn with a thick dashed line is the critical curve. A source placed on a caustic creates an image located on the critical curve. Five identical sources are moving along the straight trajectories, as marked. All sources have radii equal to $r_s = 0.05 r_E$, as shown with small open circles. The corresponding light curves are shown in the right panel. The top light curve corresponds to the top trajectory. The sharp spikes are due to caustic crossings by the source. (The light curves are shifted by one magnitude for clarity of the display.) From Paczyński (1996)

with unrelated stars, may adversely affect the photometric measurements, or the source may be a binary star. It is thus necessary to explore the available parameter space fully to avoid misinterpretations of complicated light curves.

5.2 Planetary Systems as Gravitational Lenses

Typical Sizes and Time Scales

A star that has a planetary companion can act as a binary lens with an extreme mass ratio $10^{-6} \lesssim \mu \lesssim 10^{-3}$. Mao and Paczyński (1991) suggested that for $\mu = 10^{-3}$ and a projected separation $\approx r_E$ the detection efficiency should be a few percent, opening the possibility of detecting planets in microlensing surveys. The question about detection thresholds, optimized observing strategies, and the number of expected planet detections has since attracted much interest. Most of the simulations have been done for source stars in the Galactic bulge ($D_S \approx 7\ldots 8\,\mathrm{kpc}$) lensed either by bulge ($D_L \approx 6\,\mathrm{kpc}$) or disk ($D_L \approx 3\ldots 4\,\mathrm{kpc}$) stars. It is useful to consider first a few typical numbers for these parameters. The Einstein radius

$$r_E \approx 4\,\mathrm{AU} \left(\frac{M}{M_\odot}\right)^{1/2} \tag{46}$$

is of the order of the orbital radius of Jupiter; this is favorable for the detection of Solar System analogs. The Einstein radius of the planet is related to θ_E by $\theta_p = \sqrt{\mu}\,\theta_E$. One should expect that the influence of the planet is significant over an area with this radius; this is frequently true (Gould and Loeb 1992), but there are important exceptions to this rule (Griest and Safizadeh 1998, see "high-magnification events" below). Assuming for the moment the scaling with $\sqrt{\mu}$, we can derive the planet anomaly duration directly from (43):

$$t_p \approx 2\,\text{days} \left(\frac{m_p}{M_{\text{jup}}}\right)^{1/2} \left(\frac{v}{200\,\text{km s}^{-1}}\right)^{-1}. \tag{47}$$

This implies that monitoring with good temporal sampling is required, especially for the detection of Earth-like planets, for which the typical time scale is only a few hours.

A similar scaling argument can be used to estimate the probability that a given planet will actually be detected. At any given time, the probability for amplification by the planet is $\propto (\theta_p/\theta_E)^2 = \mu$, but the total area swept by the planetary Einstein ring while the source sweeps across the Einstein ring of the lensing star is $\propto \theta_p/\theta_E = \sqrt{\mu}$. We thus expect detection probabilities of a few per cent for Jupiter-mass planets ($\sqrt{\mu} \approx 0.03$).

Another important number is the radius of the planetary Einstein ring projected back to the source plane,

$$\tilde{r}_p \equiv \sqrt{\mu}\, r_E \frac{D_S}{D_L}. \tag{48}$$

Numerical values are $\tilde{r}_p \approx 50\,R_\odot$ for $M = M_\odot$ and $m_p = M_{\text{jup}}$, and $\tilde{r}_p \approx 3\,R_\odot$ for $M = M_\odot$ and $m_p = M_\oplus$. These numbers can be compared with the radii of clump giants ($\sim 13\,R_\odot$) and stars near the main-sequence turn-off in the bulge ($\sim 3\,R_\odot$). We see that the effect of the non-zero source size can be safely neglected for Jupiter-like planets, because the star is always much smaller than the planet's Einstein ring radius. For Earth-like planets, however, the radius of the background star is comparable or larger than $\tilde{r}_p$. This means that turn-off stars are much better suited for low-mass planet searches than giants, because in the latter case the planetary anomalies will be strongly smeared out by the large size of the source (see Fig. 30).

Light Curves and Detection Limits

The complicated caustic structure of binary lenses gives rise to a large variety of possible light curves, as discussed above (see Fig. 26); the same is true in the planet case ($\mu \ll 1$). The binary signature is most obvious during caustic crossings, as illustrated in Fig. 27. This figure shows the light curve of a system composed of eight planets with $\mu = 10^{-5}$ located along a straight line, with a source moving with zero impact parameter along this line. Each peak corresponds to the crossing of a planetary caustic; the figure thus demonstrates

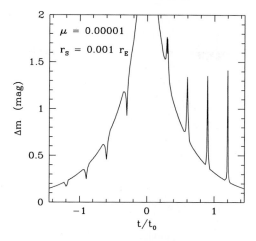

Fig. 27. Simulated light curve of a (very artificial) planetary system, which is made of a star and eight planets, each with mass fraction $\mu = 10^{-5}$, and all located along a straight line. The source with a radius $r_S = 10^{-3} r_E$ is moving along the line defined by the planets, with zero impact parameter. The planets are located at the distances from the star: $b = 0.57, 0.65, 0.74, 0.86, 1.16, 1.34, 1.55, 1.76$ in the lens plane, which corresponds to the disturbances in light variations at the times $t/t_0 = -1.2, -0.9, -0.6, -0.3, 0.3, 0.6, 0.9, 1.2$, as shown in the figure. Note that planetary disturbances create local light minima for $b < 1 (t/t_0 < 0)$ and local maxima for $b > 1$ $(t/t_0 > 0)$. From Paczyński (1996)

the effect of Earth-like planets at different separations from the parent star. For more massive planets, significant anomalies can occur even if the source does not cross a caustic (Bolatto and Falco 1994).

To explore the range of possible light curves expected in more realistic cases, one can compute the magnification for every point in the lens plane, and consider "random" paths of the source across this magnification pattern. It is convenient to consider the anomaly $\delta \equiv (A - A_0)/A_0$, where A_0 is the magnification in the absence of planets. This is frequently better than working with A itself, because δ is frequently quite small, especially for small μ. The amplification and corresponding anomaly generated by a Jupiter-like planet are shown in Fig. 28.

The calculation of the magnification pattern and light curves can be repeated for many combinations of the parameters μ (mass ratio), b (projected separation in units of the Einstein radius) and for different source trajectories (see Sackett 1999; Wambsganss 1997). Massive planets are easier to detect, because their anomaly contours cover a larger area on the sky, which makes it more likely that they are intersected by the source trajectory. For a given mass ratio μ, the anomalous regions are largest when $b \approx 1$, i.e., when the star–planet separation is of the order of the Einstein radius (see Fig. 2 in Gaudi and Sackett 2000).

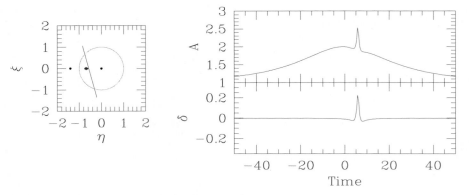

Fig. 28. A background point source travels along a trajectory that just misses the caustic caused by a Jupiter-like planet with mass ratio $\mu = 0.001$ located at $1.3\,r_E$ from its parent star. The three panels show the trajectory and the corresponding amplification A and anomaly δ. The time is given in days. From Sackett (1999)

In the case of planets ($\mu \ll 1$), and if $b \neq 1$, the structure and location of the caustics is quite simple (Griest and Safizadeh 1998). The pointlike single-source caustic becomes a small wedge-shaped caustic still located near the center of the Einstein ring, and one or two new "planetary" caustics appear at locations that depend on the position of the planet. For $b > 1$ there is one planetary caustic located between the lens and the planet, for $b < 1$ there are two planetary caustics on the opposite side of the lens (see Fig. 29).[22] The location x_c of the planetary caustics is approximately given by

$$x_c \approx \left(b^2 - 1\right)/b\,. \tag{49}$$

The caustics are located inside the Einstein ring if $|x| < 1$, i.e., if $1/2\left(\sqrt{5}-1\right) < b < 1/2\left(\sqrt{5}+1\right)$ or $0.618 < b < 1.618$. This region for b is called the "lensing zone"; it has considerable importance for planet searches that are follow-up observations of microlensing surveys. A microlensing event is recognized when the amplification A exceeds a certain value; frequently a source position on the Einstein ring ($u = 1$) corresponding to $A = 1.34$ (41) is used as a detection threshold. If the planetary caustics are located inside the Einstein ring, there is a chance that they will be crossed by the source during the course of the event; if they are located far outside the Einstein ring, however, there will only be a small anomaly during the lensing event. From (46) we therefore

---

[22] We can now also understand Fig. 27 better. For the planets at $b < 1$, the source passes between the two planetary caustics, located at the opposite side from the star. The amplification in this region is negative, leading to the dips. For the planets at $b > 1$, the source passes through the planetary caustic, which causes the peaks.

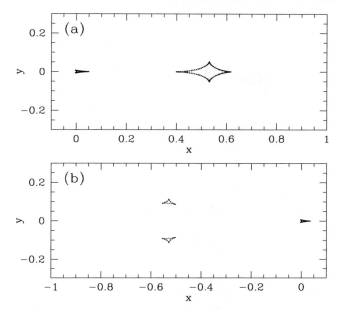

Fig. 29. Caustics for $\mu = 0.003$, showing the central caustic near the origin and the larger planetary caustics. The top panel is for a planet at $b = 1.3$, the bottom panel for $b = 1/1.3 = 0.769$. From Griest and Safizadeh (1998)

see that microlensing is most sensitive to planets with projected separations in the range $2.5 \ldots 6.5$ AU around Solar-type stars, or at $1.4 \ldots 3.5$ AU from $0.3\,M_\odot$ dwarfs. Planets with larger orbital radii may spend some fraction of their orbital time in the lensing zone, depending on the orbital inclination.

The detection of planets in the lensing zone can be hampered by the smearing caused by the non-zero source radius R_S, as mentioned above. If the source covers the entire planetary caustic, $\delta \propto (\tilde{r}_p/R_S)^2$. On the other hand, the typical time scale is longer than given by (47), by a factor $R_S/\tilde{r}_p$. The effect of the finite source size is shown in Fig. 30 for a few typical cases. It is evident from this figure that the ability to detect planets with $\mu \lesssim 10^{-4}$ depends critically on the source radius; the peak deviation from the single-lens light curve is strongly reduced in particular for giant stars.

For each of the light curves in Fig. 30 (and similar light curves for equally probable orientations and impact parameters) we can now ask the question whether the planet would be detected by a photometric monitoring program. The answer will generally be "yes" if $|\delta|$ exceeds a certain threshold (set by the photometric precision) for a minimum time (determined by the time sampling of the photometry). Representative detection probabilities for a model planet system with one planet per factor of 2 in distance from the lens star are listed in Table 8, to illustrate the effect of the source radius. In the real world

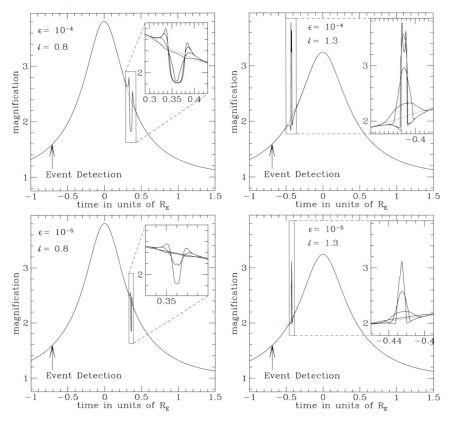

Fig. 30. Theoretical microlensing light curves that show planetary deviations are plotted for mass ratios $\mu = 10^{-4}$ and 10^{-5} and separations of $b = 0.8$ and 1.3. The main plots are for a normalized stellar radius $r_S \equiv (R_S/r_E) \cdot (D_L/D_S) = 0.003$ while the insets show light curves for radii of 0.006, 0013, and 0.03 as well. The amplitude of the maximum deviation from the dotted single-source light curve decreases with increasing r_S. For each of these light curves, the source trajectory is at an angle of $\arcsin 0.6 = 36°\!.9$ with respect to the star–planet axis. The impact parameter $u_{\min} = 0.27$ for the $b = 0.8$ plots and $u_{\min} = 0.32$ for the $b = 1.3$ plots. For these parameters the source trajectory crosses the x-axis near x_c. From Bennett and Rhie (1996)

additional complications are caused by varying seeing conditions, which lead to night-to-night variations of the photometric precision, and by gaps in the data during daytime or due to clouds.

Estimation of Planet Parameters from Microlensing

The next question that needs to be addressed is whether it is possible to use microlensing light curves not only to detect planets, but also to determine

Table 8. Planetary detection probability during microlensing events

r_S	μ	$p(2)\%$	$p(4)\%$	$p(10\%)$	$p(20\%)$
0.003	10^{-4}	0.188	0.144	0.094	0.052
0.006	10^{-4}	0.238	0.159	0.085	0.043
0.013	10^{-4}	0.201	0.118	0.052	0.014
0.03	10^{-4}	0.120	0.035	0.012	0.000
0.003	10^{-5}	0.060	0.034	0.014	0.004
0.006	10^{-5}	0.052	0.026	0.005	0.002
0.013	10^{-5}	0.019	0.008	0.001	0.000
0.03	10^{-5}	0.002	0.000	0.000	0.000

Probabilities p are shown as a function of the threshold for $|\delta|$ and for different values of the normalized source star radius $r_S \equiv (R_S/r_E) \cdot (D_L/D_S)$ and the mass ratio μ. Idealized "factor of 2" planetary systems with one planet per factor of 2 in distance from the lens star are assumed. A planet is considered to be detected if $|\delta|$ is larger than the threshold for a period of time longer than $t_E/400$. The r_S values of 0.003 and 0.006 correspond to a turn-off source star with disk and bulge lenses, respectively, while the r_S values of 0.013 and 0.03 correspond to a giant source with disk and bulge lenses. From Bennett and Rhie (1996)

some of their parameters. First of all, the orbital velocity of planets in the lensing zone is small compared to the typical transverse velocity of the lens with respect to the source ($v \approx 200\,\mathrm{km\,s^{-1}}$). This means that their mass ratio $\mu = (\theta_p/\theta_E)^2$ can be estimated roughly from the duration of the planetary anomaly t_p and the duration of the main event t_E, namely

$$\mu \approx (t_p/t_E)^2 . \tag{50}$$

The time difference between the peak of the anomaly and the main peak gives an indication of the location of the planetary caustic within the Einstein ring, and thus an estimate of the projected separation b. It is clear from (49), however, that planets with separations b and $1/b$ give rise to caustics at nearly identical positions. The degeneracy between these two cases can be broken by high-quality light curves, because the structure of the magnification pattern near the caustic is different for $b < 1$ and $b > 1$ (compare e.g., the left ($b = 0.8$) and right ($b = 1.3$) panels in Fig. 30). A second degeneracy exists because the source can pass either between the star and the caustic or further from it. This degeneracy is more difficult to break with observational data, but its influence on the estimates for b and µ is relatively small (Gaudi and Gould 1997).

Finite-source effects create additional complications, because they make the duration of the anomaly longer. If we use (50), we will therefore overestimate µ, potentially by a large factor. The differences between the light curves of a large source crossing the caustic of a low-mass planet and of a point source crossing a higher-mass caustic can be very subtle; in some cases photometry with precision much better than 1% (and sufficient time resolution) would be needed to distinguish between these cases. If this cannot be done, µ may be uncertain by a factor $\sim 1/\delta_{\max}$, or as large as a factor of 20 for anomalies with a maximum deviation $\delta_{\max} = 5\%$ (Gaudi and Gould 1997). If additional information is available, it can be used, however, to alleviate this unsatisfactory situation. For example, the typical time t_S it takes the source to cross a caustic is given by its angular diameter, divided by the relative proper motion of the lens and source. The angular diameter can be estimated from dereddened colors and magnitudes, so it is possible to determine t_S if the relative proper motion can be measured. If $t_S < t_p$, (50) can be used safely to estimate µ, otherwise not. Another possibility is using multi-color (e.g., visible and near-IR) light curves to get a handle on finite-source effects. This idea is based on the fact that stellar limb darkening is generally much stronger at shorter wavelengths. If the source is large compared to the caustic structure, one therefore expects noticeable changes in the color as the center and the limb of the star are amplified by different factors. This color change may be much larger and easier to measure than the details in the shape of the light curve (Gaudi and Gould 1997).

The discussion in the previous paragraphs has tacitly assumed that an observed "blip" in a microlensing light curve is due to a planetary companion of the lens. In practice, it may not at all be easy to establish that this is indeed the case. For example, if the *source* is a binary with a magnitude difference $\Delta V = 5 \ldots 10$ (such as a clump giant primary with a G … M main-sequence secondary), a single lensing star may produce a light curve mimicking a planetary anomaly. An analysis of the likelihood of such events shows that they may be a significant contaminant in samples of putative planetary lenses, unless precautions are taken to distinguish between the two possibilities; binary sources can be recognized in multi-color light curves, or by spectroscopic follow-up observations (Gaudi 1998).

If all goes well, it is thus possible to identify planetary microlensing anomalies, and to extract µ and b from the observations. What we would really like to know are the mass of the planet m_p and its orbital radius a. With reasonable statistical assumptions about the distribution and velocities of lenses and sources and the measured value of t_E one can estimate the Einstein radius r_E and the lens mass m_L to a factor of ~ 5. It is thus possible to determine $m_p = \mu \cdot m_L$ and a lower limit to $a \geq b \cdot r_E$ with the same uncertainty.

Searching for Planets in High-Magnification Events

We have seen above (Fig. 29) that a binary lens with $\mu \ll 1$ gives rise to a small central caustic as well as one or two larger "planetary" caustics. The size of the central caustic along the x-axis is given to a good approximation by (Griest and Safizadeh 1998)

$$u_c \approx \frac{\mu b}{(b-1)^2} , \qquad (51)$$

provided that b is not close to unity. Since u_c is of order μ (and not $\sqrt{\mu}$ as the size of the planetary caustics) one should think that the central caustic is rather unimportant for planet searches. The arguments and calculations discussed in the previous sections were indeed done for the planetary caustics. Griest and Safizadeh (1998) pointed out, however, that observations that concentrate on high-magnification events offer a good chance to detect planets due to the proximity of the source path to the central caustic. The argument is based on (41): if $A \gg 1$, then $u \approx 1/A$. If there is a shift or distortion of the caustic of size du due to a planet, we will observe an anomaly $\delta \equiv dA/A \approx -Adu$. For planets in the lensing zone $0.618 \leq b \leq 1.618$ we thus expect deviations of order

$$\delta \approx u_c A \approx \mu A , \qquad (52)$$

where we have somewhat pessimistically set $b/(b-1)^2 \approx 1$. For example, because all values of $u_{\min}$ are equally probable, 5% of all microlensing events will have $u_{\min} \leq 0.05$ and $A_{\max} \geq 20$. A $\mu = 10^{-3}$ planet will thus produce a 2% anomaly near the peak of such events, which should not be too hard to detect. More detailed simulations show indeed that the detection probability for such planets is close to 1 (Griest and Safizadeh 1998).

The very high detection rate for planets during high-magnification events has two important consequences. First, if no anomaly is observed in a well-sampled light curve, the presence of Jupiter-like planets in the lensing zone can be safely excluded. Monitoring of a modest number of such events can thus establish the abundance of such planets. Second, if the lensing star has multiple planets in the lensing zone, each one of them will cause a detectable distortion of the central caustic. The resulting light curve will likely be complicated and difficult to interpret, but there is a good chance that systems with multiple planets can be found in this way (Gaudi et al. 1998). This is not the case in the "traditional" approach, because it would require a fortuitous alignment of two planets for the source to cross the planetary caustics of both of them.

Gravitational Lensing of Planets

We have discussed above that binary sources may be a significant problem for searches for planetary companions to the lens. We can of course turn the argument around and ask whether it is possible to take advantage of the situation if it is not the lens, but rather the source that is accompanied by a planet. In this case the star and the planet are both amplified by the same lens, but if there are any caustic crossings, they will occur at different times for the star and the planet. The peak amplification of the planet is larger because of the smaller radius of the planet. For Jupiter-size planets in close orbits (0.05 AU) with near-unity albedo, the maximum fractional deviation of the light curve above that expected when the source star does not have a planetary companion can get close to 1% (Graff and Gaudi 2000). The typical time it takes the planet to cross the caustic is $\lesssim 1$ h. It should be possible to search for these "blips" near the caustic crossing time of the parent star, but a large amount of observing time on big telescopes would be needed. Planets with much larger orbital radius a cannot be detected in this way, because the amount of light reflected by the planet scales with a^{-2}. (Note that now the anomaly is caused by the light, not the mass of the planet.)

Observations of lensed planets with future giant telescopes (which are needed to get good SNR and time resolution for fairly faint sources) could reveal a number of interesting effects. The shape of the illuminated fraction of the planet has a strong influence on the light curve; crescents can produce higher peak amplification than half or full disks, partly offsetting the smaller fraction of reflected light (Ashton and Lewis 2001). Reflection by condensed particles in the planetary atmosphere leads to partial polarization of the light from the planet (Seager et al. 2000); the amplification of the planet with respect to the star during a caustic crossing may in favorable cases enhance the total polarization to a detectable level (a fraction of a percent, Lewis and Ibata 2000). Lensing of crescent-shaped planets should also lead to characteristic polarization signatures. Studies of these phenomena could provide information on the particles in the planets' atmospheres and complement observations of "reflected light" from planets in the Solar neighborhood (Sect. 6.6).

5.3 Microlensing Planet Searches

Search Strategies

The probability that a given star in the Galactic bulge is being lensed at a given time is very low (about 10^{-6}). Surveys that monitor a very large number of stars are therefore necessary to detect the occasional brightening of a star due to microlensing. Several such projects were launched in the 1990s, and have now detected more than 1,000 microlensing events: EROS (Aubourg et al. 1993; Derue et al. 2001), OGLE (Udalski et al. 1993, 2000), MACHO (Alcock et al. 1993, 2000b), and MOA (Bond et al. 2001). The temporal sampling – typically one observation per night – of these surveys is inadequate to find planetary anomalies directly. They issue alerts of ongoing events, however, allowing more frequent follow-up observations by teams that have formed specifically for this purpose, for example PLANET (Albrow et al. 1998), GMAN (Alcock et al. 1997), and MPS (Rhie et al. 1999). The PLANET collaboration, for example, uses 1 m class telescopes located in Chile, Tasmania, Australia, and South Africa to achieve round-the-clock coverage of selected ongoing events. It is clear that southern sites are preferred for the monitoring of fields in the Galactic bulge, but because of the large number of small telescopes in the northern hemisphere, and their favorable longitude distribution, searches for Jupiter-like planets from the north are also feasible (Tsapras et al. 2001).

As we have seen in Sect. 5.2, even Earth-mass planets produce large anomalies under favorable conditions (caustic crossings, small background star). Monitoring projects that achieve $\sim 1\%$ photometric precision with $\sim$hourly sampling therefore have sufficient *sensitivity* to detect planets over a large range of masses, but the *efficiency* depends sensitively on the planet mass, and on the detailed photometric performance – most detectable planetary anomalies result from non-caustic crossing events (Gaudi and Sackett 2000). The predicted number of planets that should be detected by monitoring projects depends strongly on the assumptions made, including how planetary masses and separations vary with lens mass (Peale 1997). Since the detection efficiency for any given event depends strongly on $u_{\min}$, it is important to know the distribution of this parameter in the observed sample of microlensing events (Gaudi and Sackett 2000). The chances to find planets with follow-up monitoring can be substantially increased if the original survey produces a large number of high-amplification events (Bond et al. 2002a).

The two-step strategy – search for microlensing using wide-field cameras, and follow-up with targeted observations of "alerted" events – offers currently the best chances to detect planets through microlensing anomalies. This approach has the disadvantage that the high-quality follow-up data are obtained only after the alert; on the rising wing of the event only the monitoring observations are available. The lack of densely sampled data for the first part of the light curve hampers the ability to discriminate between planets and other types of anomalies. In the future it may be possible to use one and the same experiment to detect and monitor microlensing events by conducting frequent observations of a large sample of stars. In such a survey with uniform time coverage it is of course possible to reconstruct the past behavior of any "interesting" star. Next generation of dedicated survey telescopes equipped with wide-field cameras, such as the VLT Survey Telescope, could conduct an efficient survey for low-mass planets, whose anomalies last only a few hours. The largest difficulty of such a project are the gaps in the light curves during daytime and periods of bad weather.

This problem could be overcome by an orbiting telescope, for example the proposed Galactic Exoplanet Survey Telescope (GEST), Bennett and Rhie 2000). A diffraction-limited 1.5 m telescope with a 1° field-of-view and a gigapixel CCD array could monitor $\sim 2 \cdot 10^8$ stars in the Galactic bulge, and observe $\sim 12,000$ microlensing events during a 2.5 yr mission lifetime. A mission like GEST could detect 10 to 20 Earth-mass planets at 1 AU separation if all stars have such companions. The detection efficiency is even better at somewhat larger separations, and thousands of gas-giant planets could be found. A microlensing survey from space would thus be a powerful way to determine the abundance of terrestrial and giant planets in the Galaxy.

Putative Planet Detections

A few claims of microlensing planet detections have appeared in the literature, but none of them has stood up to further scrutiny. It is nevertheless instructive to take a look at a few examples, because we can see some of the difficulties that anyone attempting to establish the planetary nature of a microlensing anomaly will face.

MACHO 97-BLG-41 was a very unusual event with a complicated light curve (see Fig. 31), which clearly indicates a multiple lens, but cannot be modeled with static binary models. Bennett et al. (1999) interpreted the caustic structure as coming from a triple system consisting of a stellar binary with $\sim 1.8\,\mathrm{AU}$ separation, orbited by a Jovian planet ($m = 3.5 \pm 1.8 M_{\mathrm{jup}}$) at $\sim 7\,\mathrm{AU}$. The PLANET collaboration showed, however, that their own data on this event could be modeled by a normal binary, whose orbital motion changes the orientation and separation of the two stars between the times of the two caustic crossings (Albrow et al. 2000b). Furthermore, this model also provides a stunningly good fit to the MACHO/GMAN data (Fig. 31, right panel), on which Bennett et al. (1999) had based their claim of a planet detection. This alone does not disprove the existence of a Jovian planet in this system, but the PLANET data are inconsistent with the particular model of Bennett et al. (1999), and the existence of a simple, plausible binary model that explains all data on MACHO 97-BLG-41 strongly suggests that this is the correct interpretation of this event.

The microlensing event MACHO 98-BLG-35, which reached a peak amplification factor of almost 80, was monitored intensely by the MOA, MPS, and PLANET teams. Based on the MPS and MOA data, Rhie et al. (2000) reported evidence for a planet with mass fraction $4 \cdot 10^{-5} \leq q \leq 2 \cdot 10^{-4}$.

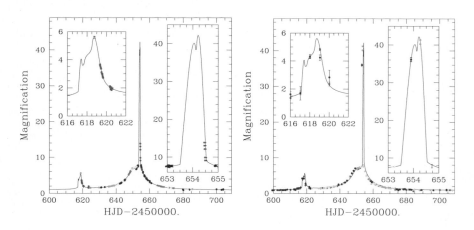

Fig. 31. Rotating binary model for MACHO 97-BLG-41, with data points from PLANET (*left panel*) and from MACHO/GMAN (*right panel*). The fit was obtained using only the PLANET data; the MACHO/GMAN data did not enter the fit and were simply superposed to the model in the right-hand panel. Nevertheless, the model reproduces the MACHO/GMAN data extremely well even in regions where no PLANET data are available. From Albrow et al. (2000b)

A reanalysis of the same observations with an improved photometric algorithm, and inclusion of additional PLANET data, showed, however, that a planet with these parameters can be ruled out (Bond et al. 2002b). This reanalysis revealed apparent lower-level anomalies, which can be fitted by models with one, two or three planets, all with masses $q < 3 \cdot 10^{-5}$. The fact that several different models give fits of similar quality raises the suspicion that they actually fit noise in the data. The problem is that the inclusion of planets lowers the χ^2 by a formally significant amount, but systematic deviations mimicking small planetary anomalies may well be present in the light curves (Gaudi et al. 2002). The evidence for one or more planets associated with MACHO 98-BLG-35 therefore remains tentative at best.

A convincing fast anomaly was observed in MACHO 99-BLG-47 (Albrow et al. 2002). According to (50) the short duration of the anomaly should indicate small secondary mass, but in this case the light curve can be modeled better with a binary star, in which both components have nearly equal masses and either a very small or very large separation (compared to θ_E). Albrow et al. (2002) also show that this interpretation is much more likely than a solution with a planet, which would require rather extreme parameters for the peak amplification, event duration, and blending.

We are thus led to conclude that the ambiguities and degeneracies mentioned in Sect. 5.2 can easily conspire with observational uncertainties to mimic planetary anomalies. Establishing a secure planet detection will require an extremely careful analysis. For Jupiter-mass planets, the main challenge is exhausting the full parameter space in modeling complex light curves (e.g., due to rotating binaries). Terrestrial planets tend to produce anomalies at the threshold of statistical significance, and small random wiggles in the light curves can easily give rise to spurious detections.

Limits on the Abundance of Planets

Whereas microlensing observations have not been successful yet at making a convincing planet detection, they can nevertheless be used to establish useful statistical limits on the abundance of massive planets in the Galaxy. The starting point is the exclusion of planetary anomalies at a certain significance level from a well-sampled microlensing light curve. High-magnification events are well-suited for this purpose: since *every* planet in the lensing zone gives rise to an anomaly close to the peak, the absence of any such anomaly proves the absence of planets in the lensing zone (to a certain mass threshold, which depends on the photometric errors). This argument was used by Rhie et al. (2000) to conclude that there could not be any Jupiter-mass planets in the lensing zone of the event MACHO 98-BLG-35, which has already been discussed above. Similar, somewhat weaker constraints could be placed on the existence of companions in OGLE 1998-BUL-14 (Albrow et al. 2000a).

This argument can of course be extended from individual cases to a combined analysis of a well-understood sample of microlensing events (Gaudi et al. 2002). The first step in this analysis is the selection of a clean sample of events, based on criteria that reject events with sparse light curves, poorly determined parameters, or non-planetary anomalies. For each "good" event, one then searches for deviations of the light curves from the best-fitting point-source/point-lens (PSPL) model. This is done through an exhaustive search of the parameter space of possible binary models and source trajectories, followed by a χ^2 analysis. For each set of parameters $\mathcal{P}$ a synthetic light curve is computed and compared with the data, giving $\chi^2_\mathcal{P}$. If $\chi^2_\mathcal{P}$ was significantly smaller than χ^2_PSPL, i.e., $\chi^2_\mathcal{P} - \chi^2_\mathrm{PSPL} < -\Delta\chi^2_\mathrm{thresh}$, we would conclude that we have found a planet with parameters $\mathcal{P}$. On the other hand, if $\chi^2_\mathcal{P}$ is significantly larger than χ^2_PSPL, we can rule out the existence of such a planet. By integrating over the possible source trajectories we can then determine the probability with which we would have detected a planet with given projected separation b and mass ratio μ. After repeating this procedure for each event in the sample, we can determine the maximum fraction of stars f that can have planets with parameters b and μ, which is still consistent with the non-detections at a certain confidence level (see Fig. 32). The reliability of the result of this procedure clearly depends on the correct modeling of subtleties like finite-source effects, and on the adoption of a realistic threshold $\Delta\chi^2_\mathrm{thresh}$ at which differences in χ^2 are regarded "significant".

Five years of photometric data collected by the PLANET collaboration have been analyzed in this way (Albrow et al. 2001; Gaudi et al. 2002). Of all observed events, 43 fulfill the selection criteria used by the authors and form the basis of the statistical arguments. At 95% confidence, less than 25% of the lenses have companions with mass ratio $\mu = 10^{-2}$ in the lensing zone (see Fig. 32). With the help of a model for the mass, velocity and space distribution of bulge lenses, this result can be converted to a statement about Jupiter-mass companions of M dwarfs in the Galactic bulge: less than 33% of the $\sim 0.3\,M_\odot$ stars have companions with $m_p \geq M_\mathrm{jup}$ and $1.5\,\mathrm{AU} < a < 4\,\mathrm{AU}$.

5.4 Astrometric and Interferometric Observations of Microlensing Events

The Photocenter of a Single Lens

Our discussion of gravitational microlensing has so far focused on the observable change in the combined brightness from all images. A look at Fig. 24, however, suggests that the change of position with time may also be detectable. If the resolution is insufficient to separate the two images, an astrometric observation will measure the position of the "center of light". To compute the deviation $\Delta\theta$ from a straight-line motion, we add the positions of the two images, weighted by their respective brightness, and subtract the position of the source in the absence of lensing, $\theta_E u$. From (36) and (40) we thus get

$$\Delta\theta = \frac{1}{A}\left(\theta_1 A_1 + \theta_2 A_2\right) - \theta_E u = \frac{u}{u^2 + 2}\theta_E\,. \tag{53}$$

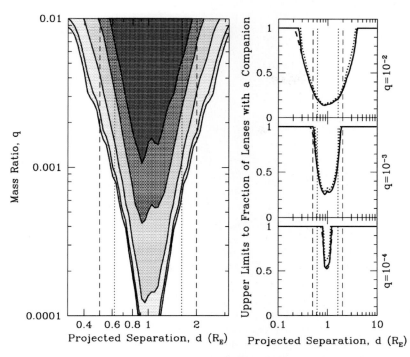

Fig. 32. *Left panel*: Exclusion contours (95% confidence level) for the fractions of primary lenses with a companion derived from the PLANET sample of 43 events, as a function of the mass ratio and projected separation of the companion. Solid black lines show exclusion contours for $f = 75\%$, 66%, 50%, 33%, and 25% (outer to inner). The dotted and dashed vertical lines indicate the boundaries of the lensing zone and extended lensing zone, respectively. *Right panel*: Cross sections through the left panel, showing for three different mass ratios the upper limit to the fraction of lenses with a companion as a function of projected separation. The solid line is derived from the point-source efficiencies with a threshold of $\Delta\chi^2_{\mathrm{thresh}} = 60$. The dotted line is derived from the point-source efficiencies with a threshold of $\Delta\chi^2_{\mathrm{thresh}} = 100$. The dashed line is finite-source efficiencies with a threshold of $\Delta\chi^2_{\mathrm{thresh}} = 60$. The dotted vertical lines indicate the boundaries of the lensing zone $0.6 \leq d \leq 1.6$. The dashed vertical lines indicate the extended lensing zone, $0.5 \leq d \leq 2$. From Gaudi et al. (2002)

The function $u/(u^2 + 2)$ has a maximum for $u = \sqrt{2}$; the corresponding astrometric deviation is

$$\Delta\theta_{\max} = \frac{1}{2\sqrt{2}} \theta_E \approx 0.4 \,\mathrm{mas} \,, \qquad (54)$$

where we have used the numerical estimate from (38). All microlensing events with $u_{\min} \leq \sqrt{2}$ therefore produce astrometric signatures with a peak amplitude that depends only on θ_E and has a value ($\sim 0.4\,\mathrm{mas}$) that is well within

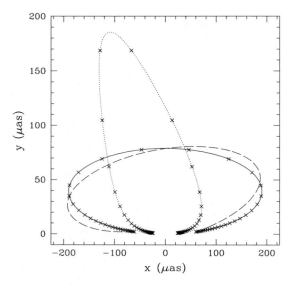

Fig. 33. Astrometric microlensing of a single star. The solid line shows a simple single-lens curve with $u_{\min} = 0.3$, and $t_E = 40$ days. The curve is plotted over one year, with x's marking each week, so only the 5 or so weeks at largest y have magnification greater than 1.34. The dashed line shows the same with an example parallax effect included, while the dotted line shows the effect of blending (blend fraction $f_b = 60\%$). From Safizadeh et al. (1999)

reach of precise astrometric methods (see Sect. 9). It is not difficult to show (e.g. Boden et al. 1998) that the two-dimensional motion $\Delta\vec{\theta}$ is an ellipse with eccentricity

$$e = \sqrt{\frac{2}{u_{\min}^2 + 2}} \; ; \tag{55}$$

for very small $u_{\min}$ the motion becomes nearly one-dimensional ($e \to 1$). The solid line in Fig. 33 shows this ellipse for the case $u_{\min} = 0.3$; note that the two axes in this figure have different scales, and that the motion of the photocenter is fastest at the time of closest approach. Parallax and blending (i.e., contributions from the lens or from unrelated nearby stars to the total light) lead to distortions of this simple shape, as illustrated in Fig. 33. The combined analysis of photometric an astrometric information can help to resolve some of the degeneracies between the possible source, lens, and planet parameters pointed out in Sect. 5.1 and 5.2 (Han 2002).

Planetary Signatures

It is to be expected, of course, that planetary companions of the lens lead to modifications of the astrometric signature, similar to those of the light curve.

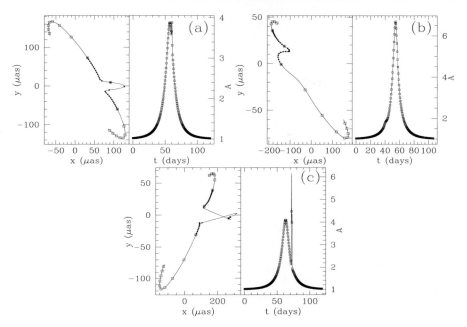

Fig. 34. Some examples of planetary astrometric and photometric curves. All examples assume $\mu = 10^{-3}$, with a primary lens angular Einstein radius of $550\,\mu$as, which corresponds to a Saturn-mass planet orbiting a $0.3\,M_\odot$ star. Panel (**a**) has $b = 1.3$, panel (**b**) has $b = 0.7$, and panel (**c**) shows a caustic-crossing event with $b = 1.3$. In the astrometric panels one square is plotted per week, so the durations of the deviations are of order a few days. Dots are plotted every 12 h during the deviation. From Safizadeh et al. (1999)

This is indeed the case, as can be seen in Fig. 34, which displays astrometric and photometric curves for a Saturn-mass planet. The excursions due to the planets are much shorter in duration than the overall microlensing events; this should simplify distinguishing them from the global distortions of the astrometric motion due to parallax and blending effects. The signature of the planet is particularly strong if a caustic-crossing occurs.

As for microlensing light curves, finite-source effects tend to smear out the planetary signal; this is important especially for low-mass planets (Safizadeh et al. 1999, see Fig. 35). If the source star is not too large, caustic-crossing events reach peak deviations of a few hundred µas, but only for a very short time. The peak amplitude of events for which no caustic crossings occur is much smaller. Still, the detection probability for Saturn-mass planets (for which finite-source effects are not important) in the lensing zone is quite high, provided that an astrometric accuracy of a few µas can be achieved. It should thus be possible in principle to search for planetary events with the Space Interferometry Mission (Sect. 9.6). The best observing strategy will probably

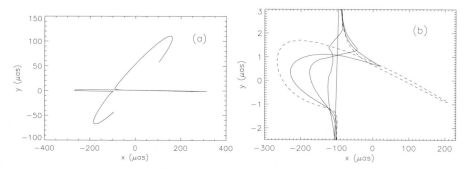

Fig. 35. Astrometric motion for Earth-mass caustic crossing. Panel (**a**) shows the center-of-light motion for a point source, crossing a caustic associated with an Earth-mass planet at $b = 0.825$. The primary lens is $0.3\,M_\odot$ at $D_L = 4\,\text{kpc}$, and source at $D_S = 8\,\text{kpc}$. Panel (**b**) shows a close-up view of the planetary deviation, with finite-size source. The dotted line plots the center-of-light motion for a $1\,R_\odot$ source. The solid lines depict the center-of-light motion for more realistic sizes typical of Galactic bulge stars, namely 3, 5, 9, and $30\,R_\odot$. Note the extreme anisotropy of the axes on the graph. For $t_E = 40\,\text{days}$ the duration of the deviation is about $20\,\text{h}$, with the center of the source spending roughly $90\,\text{min}$ inside the caustic. From Safizadeh et al. (1999)

be obtaining dense temporal sampling close to the peak of high-magnification events, because this gives a relatively high probability of caustic crossings at a time that can be predicted several days in advance.

Resolution of the Individual Images

The separation of the two individual images in a microlensing event (37) and (38) is comparable to the resolution achievable with a long-baseline interferometer. For example, an interferometer with a baseline length of $200\,\text{m}$ operating in the H band ($1.6\,\mu\text{m}$) has a fringe spacing $\lambda/B = 1.6\,\text{mas}$. It thus seems possible to fit "binary" models to interferometric data and to determine the separation and flux ratio of the two images (Delplancke et al. 2001). The principal challenge of such observations is the relative faintness of the lensed stars (even while they are being magnified), which necessitates using a brighter star within the isoplanatic radius to co-phase the interferometer (see Sect. 9.5). For small impact parameter u, the images become noticeably distorted (see Fig. 24), approaching the Einstein ring for $u \to 0$. The VLT interferometer could provide sufficient sensitivity and uv plane coverage to produce true images showing these effects in favorable cases.

Interferometric imaging could also reveal the appearance and disappearance of image pairs during the crossings of planetary caustics. Measuring the individual motions and fluxes of the images in planetary microlensing events could certainly remove most of the ambiguities and uncertainties that arise in the interpretation of light curves. Detailed simulations will be needed to determine the resolution, response time, imaging speed, and sensitivity required for such observations. The capabilities of the present interferometer arrays are likely not sufficient, but obtaining "movies" of planetary lensing events could be an interesting addition to the science case for a future large interferometric facility (e.g., Ridgway and Roddier 2000).

6 Planetary Transits and Searches for Light Reflected by Planets

If a planetary system happens to be oriented in space such that the orbital plane is close to the line-of-sight to the observer, the planets will periodically transit in front of the stellar disk. Photometric or spectroscopic observations of these eclipses can be used to infer orbital and physical parameters of the planets. The first part of this chapter deals with the basic parameters of transits that can be derived from simple geometric considerations. Summaries of ongoing observing programs, and of photometric space missions that are currently under development, follow in the next sections. The last part of the chapter discusses the prospects of detecting the light reflected by extrasolar planets without spatially resolving them from their parent stars.

6.1 The Geometry of Transits

The Probability of Transits

The first question that we would like to answer is about the probability that eclipses occur in a set of planetary systems with randomly oriented equatorial planes. For simplicity, the following discussion is restricted to circular orbits, although the case of strongly eccentric orbits would certainly be relevant, too (see Sect. 3.3). It is obvious that eclipses occur if and only if

$$a \cos i \leq R_* + R_p ,\qquad(56)$$

where a and i are the orbital radius and inclination, and R_* and R_p the radii of the star and planet. In a set of randomly oriented orbits $\cos i$ is distributed between 0 and 1, as already mentioned in Sect. 4.1. The probability p_{trans} that transits occur therefore follows directly from (56),

$$p_{\text{trans}} = \frac{R_* + R_p}{a} \approx \frac{R_*}{a} .\qquad(57)$$

Numerical values of the transit probability for the planets in the Solar System are listed in Table 9. Typical values range from $\approx 5 \cdot 10^{-3}$ for the terrestrial planets to a few times 10^{-4} for the gas giants. Together with the time between

Table 9. Transit probabilities, maximum durations, and depths for the planets in the Solar System as seen by a distant observer

planet	probability	duration [h]	depth
Mercury	$1.2 \cdot 10^{-2}$	8	$1.2 \cdot 10^{-5}$
Venus	$6.4 \cdot 10^{-3}$	11	$7.6 \cdot 10^{-5}$
Earth	$4.7 \cdot 10^{-3}$	13	$8.4 \cdot 10^{-5}$
Mars	$3.1 \cdot 10^{-3}$	16	$2.4 \cdot 10^{-5}$
Jupiter	$8.9 \cdot 10^{-4}$	30	$1.1 \cdot 10^{-2}$
Saturn	$4.9 \cdot 10^{-4}$	40	$7.5 \cdot 10^{-3}$
Uranus	$2.4 \cdot 10^{-4}$	57	$1.3 \cdot 10^{-3}$
Neptune	$1.5 \cdot 10^{-4}$	71	$1.3 \cdot 10^{-3}$
Pluto	$1.2 \cdot 10^{-4}$	82	$2.7 \cdot 10^{-6}$

successive transits of each planet – which is obviously equal to the orbital period P – these numbers elucidate the main difficulty of searches for planetary occultations: thousands of stars have to be monitored for many years in order to find a few eclipses, if no prior knowledge about the orbital inclination for individual systems is available. The two main exceptions to this rule will be discussed in Sect. 6.4. These are searches for "hot Jupiters", which have small a and P and thus high transit probability and short intervals between occultations, and searches in binary systems with known inclination of the binary orbit, which assume that potential planets may likely be coplanar with the stellar pair.

Transit Duration

The next question to be addressed is how long transit events for a given planet last. The transit duration t_{trans} is given by the expression

$$t_{\text{trans}} = \frac{P}{\pi} \arcsin\left(\frac{\sqrt{(R_* + R_p)^2 - a^2 \cos^2 i}}{a}\right), \tag{58}$$

where P is the orbital period of the planet. (The expression within the square root follows from Pythagoras' Theorem, the projection of the relevant segment of the orbit beeing approximately straight.) For $a \gg R_* \gg R_p$ (58) can be simplified to

$$t_{\text{trans}} = \frac{P}{\pi}\sqrt{\left(\frac{R_*}{a}\right)^2 - \cos^2 i}. \tag{59}$$

The maximum values for the Solar System (corresponding to $i = 90°$) are again tabulated in Table 9; they range from a few hours to a few days, which is quite favorable for monitoring campaigns.

A mass measurement with the radial-velocity technique is extremely desirable in any case, because the size does not change very much over the mass range $1\,M_{\rm jup} \lesssim m \lesssim 100\,M_{\rm jup}$.

It has been pointed out that Kepler's Law and (58), (60), and (67) are four equations that relate the four observables P, $t_{\rm trans}$, $t_{\rm flat}$, and $\Delta\mathcal{F}$ to the four quantities R_p/R_*, a/R_*, $a\cos i/R_*$, and the density of the star ρ_*, which can therefore be determined directly from high-SNR photometric data (Seager and Mallén-Ornelas 2002). If in addition a stellar mass-radius relation is assumed, one can solve for M_*, R_*, a, i, and R_p. This analysis neglects limb darkening and thus does not provide the best possible estimates of these parameters, but it can be useful to pre-select candidate planetary transits from photometric monitoring campaigns for telescope time-consuming radial-velocity follow-up.

Planetary Radii and Transmission Spectra

Planetary transits offer a unique opportunity to obtain information on the planet's atmosphere through transmission spectroscopy, i.e., by measuring the radius of the planet as a function of wavelength. The observable quantity is $\mathcal{R}(\lambda)$, the ratio of the flux during the transit to that outside of transits:

$$\mathcal{R}(\lambda) \equiv \frac{\mathcal{F}_{\rm trans}(\lambda)}{\mathcal{F}_0(\lambda)}. \tag{71}$$

The integrated light of a star–planet system consists of three separate contributions: (1) the light from the star that reaches the observer directly, (2) starlight that is reflected by the illuminated part of the planetary disk, and (3) thermal emission from the planet. Separating these three components, (71) can be written as

$$\mathcal{R} = \frac{\mathcal{F}_0 + \delta\mathcal{F}}{\mathcal{F}_0} = 1 + \frac{\delta\mathcal{F}_{\rm direct} + \mathcal{F}_{\rm therm} + \mathcal{F}_{\rm refl}}{\mathcal{F}_0}, \tag{72}$$

where it is implicitly understood that all quantities depend on λ. Quantitative estimates of the relative importance of these three contributions show that for observations of transits of "hot Jupiters" in the visible wavelength range the second and third term can be neglected; thermal emission from the planet has to be taken into account only at $\lambda \gtrsim 2.5\,\mu{\rm m}$, and the variation of the reflected light is small because only a small illuminated crescent is visible around the time of transit (Brown 2001). Several separate effects contribute to $\delta\mathcal{F}_{\rm direct}$, however, which have to be treated correctly. First of all, the shape of stellar lines changes during the transit, as the planet blocks light from different parts of the stellar disk. This effect will be discussed below (Sect. 6.3). Intrinsic variations of the flux on the time scale of a few hours are only of order 10^{-5} for the Sun (see Sect. 6.2), but larger effects occur in younger and magnetically more active stars. It is possible to discriminate against them, however, because they don't repeat consistently from one transit to the next, and because they show a wavelength dependence only in the vicinity of strong stellar lines.

Fig. 38. Two rays separated by δz passing tangentially through a planetary atmosphere with scale height H. The opacity along the higher ray is approximately $\exp(-\delta z/H)$ smaller than along the lower ray

In regions of the spectrum away from prominent stellar lines, $\mathcal{R}(\lambda)$ is therefore affected mostly by $(\delta\mathcal{F}/\mathcal{F})_{\text{atmos}}$, the part of the obscuration due to rays that pass through the atmosphere of the planet. The characteristic fractional coverage of the atmosphere projected against the stellar disk is given by the area of an annulus one atmospheric scale height H thick around the planet, divided by the area of the stellar disk. We therefore have

$$\left(\frac{\delta A}{A}\right)_{\text{atmos}} = \frac{2\pi R_p H}{\pi R_*^2} = \frac{2R_p(kT/g\mu)}{R_*^2} \;, \qquad (73)$$

where T and g are the temperature and surface gravity of the planet, and μ the mean molecular weight of the atmospheric constituents. For an atmosphere of H_2 with $g = 10^3\,\text{cm}\,\text{s}^{-2}$, $T = 1,400\,\text{K}$, $R_p = 1.4\,R_{\text{jup}}$, and $R_* = R_\odot$ the numerical value is $\delta A/A = 2.4 \cdot 10^{-4}$.

If we assume that the most important opacity sources are well-mixed in the atmosphere, we can now estimate the variation of $\delta\mathcal{F}/\mathcal{F}$ with wavelength. If σ_1 and σ_2 are the opacities per gram of material at λ_1 and λ_2, the optical depth at λ_1 along a ray 1 will be approximately equal to the optical depth at λ_2 along a ray 2 if these rays are separated by $\delta z = H \ln(\sigma_1/\sigma_2)$ (see Fig. 38). The difference between the occulted flux at the two wavelengths can therefore be written as (Brown 2001)

$$\left(\frac{\delta\mathcal{F}}{\mathcal{F}}\right)_1 - \left(\frac{\delta\mathcal{F}}{\mathcal{F}}\right)_2 \approx \ln\left(\frac{\sigma_1}{\sigma_2}\right) \times \left(\frac{\delta A}{A}\right)_{\text{atmos}} . \qquad (74)$$

The opacity in strong molecular or atomic absorption lines may be $\sim 10^4$ times more than in the nearby continuum, which means that the two rays in Fig. 38 have to be separated by almost ten scale heights to make the line optical depth along the upper ray equal to the continuum optical depth along the lower ray. According to (73) and (74) this results in an observed line depth of $\approx 2 \cdot 10^{-3}$ with respect to the stellar flux.

Oblateness, Rings, Moons, and Starspots

With very precise photometry, it should be possible to search for deviations from the expected shape of the transit light curve given by (70). The giant planets in the Solar System are significantly non-spherical because of their fast rotation rates, and their rotation axes are strongly inclined with respect to the normal of their orbital planes. The ingress and egress in light curves of transits by such a planet are asymmetric, unless $i = 90°$ exactly (Hui and Seager 2002; Seager and Hui 2002). The expected deviations from the light curve of an eclipse by a spherical planet are a few times 10^{-5}, which may be detectable with photometric space missions (see Sect. 6.5).

The transits of giant extrasolar planets also offer a chance to look for rings and moons around them. An opaque ring would potentially have a large cross-section, but its projected area is strongly reduced if it lies in the orbital plane of the planet. Giant moons would also produce characteristic dips or discontinuities in transit light curves, depending on their orbital parameters (Sartoretti and Schneider 1999). The space missions designed to detect transits by Earth-like planets will by definition also be sensitive to moons with radius $R_s \approx 1 R_\oplus$. An alternative way of looking for moons of transiting planets is timing of the eclipses. A satellite of mass m_s in an orbit with radius a_s around a planet with mass m_p and orbital radius and period a_p and T_p will give rise to shifts in the occultation times of order

$$\tau \approx \frac{a_s m_s}{m_p} \frac{T_p}{2\pi a_p}. \tag{75}$$

With sub-second timing of the transit times of a Jupiter-like planet, the COROT satellite should be capable of detecting satellites similar to the Galilean moons (Sartoretti and Schneider 1999).

An interesting question thus concerns the dynamical stability of moons around close-in planets. It is well-known from the Earth–Moon system that tidal interaction causes a satellite to move inward or outward, depending on whether its orbital period is shorter or longer than the rotation period of the planet. For hot Jupiters, the planetary rotation rate is regulated by tidal interaction with the star (not with the moon), which keeps the torque on the moon strong. This leads to a loss of moons with masses $\gtrsim 10^{-6} M_\oplus$ around planets with orbital radii $a \leq 0.1$ AU (Barnes and O'Brien 2002). These considerations therefore predict that moons should not be found around those planets that are most easily detected in transit surveys.

Finally, the transit light curves might show bumps if the planet happens to move across a star spot. This effect will be difficult to detect, however, because the amplitude cannot exceed the fractional area of the stellar disk covered by the spot, and because the variations are erratic and do not repeat from one transit to the next.

6.2 Photometric Error Sources

The precision of photometric observations is ultimately limited by photon noise and (for fainter stars) by the sky background. These errors can be reduced by increasing the exposure time T, because they scale with $T^{-1/2}$.

There are a number of systematic effects, however, which frequently prevent one from reaching the theoretical limit. The most important of these error sources will be discussed in the following sections.

Stellar Noise

Looking at our own Sun, we can identify many distinct mechanisms that cause variations of the emitted flux: oscillations, sunspots, flares, prominences, and variability of the granulation. Stellar activity varies strongly with spectral type and age; for example, many M dwarfs display flares with amplitudes up to one magnitude or even more (Allard et al. 1997 and references therein). Many types of binaries also show periodic brightness variations, either due to eclipses or to distortions of the stellar shape. Transit searches that monitor many thousands of stars are therefore very good at detecting variable stars (e.g., Street et al. 2002), but one has to be able to distinguish these from the planetary transits one is looking for. Fortunately, transits have a short duration and a very characteristic shape, so that it is possible to search for these events in light curves that are otherwise flat within the noise. The Sun shows variations up to $\sim 0.15\%$ on time scales close to the rotation period due to spots, but there is very little power on time scales shorter than a day, which are typical for planetary transits (Borucki and Summers 1984 and references therein). Therefore, at least around Solar-type stars, it should be possible to clearly discriminate transits events from stellar variability, even for planets as small as $\sim 1\,R_\oplus$.

Atmospheric Noise

The Earth's atmosphere limits the precision that can be reached in ground-based photometric observations. The most important effects to consider are scintillation, changes of the extinction with time and zenith angle, and seeing variations. Scintillation is strongest for small apertures and short integration times (see Sect. 7.8). Quantitative estimates (e.g., Dravins et al. 1998) scaled to the parameters relevant for planet transit searches indicate that scintillation noise is normally not a limiting factor. Variations of the extinction due to changes in the air mass during the observations, or to non-photometric conditions, are more difficult to deal with. CCD photometry generally offers substantial advantages over classical one-channel or two-channel photometers. If the CCD field is sufficiently large, many stars can be measured simultaneously. It is then possible to perform differential photometry by dividing the observed brightness of each star by that of a set of calibration stars or by the median of all stars in the field. This eliminates extinction fluctuations and variations due to changes in the air mass to first order. If there are enough stars in the field, it it possible to perform an even better photometric calibration by taking color terms into account. Because of the limited dynamic range of the CCD, the useful range between the brightest and faintest stars spans not more than ~ 4 magnitudes. Somewhat ironically, CCD photometry is therefore more difficult for very bright stars, since for these the contrast to the potential comparison stars within the field-of-view is usually much too large.

Changes in the width and shape of the point spread function due to short-term or night-to-night variations of the seeing (see Sect. 7.4) can be a serious error source, in particular in crowded regions of the sky. Images of stars than can be cleanly separated in a good night may be blended together when the seeing is bad, which can introduce severe photometric errors. Three main photometric methods are currently in use: aperture photometry, which measures the flux within a circle of specified radius at the location of each star; psf-fitting photometry, which fits a model of the point spread function with variable intensity at the position of each star (e.g. DAOPHOT, Stetson 1987); and image subtraction techniques, in which a reference image of the field is subtracted from each frame before the photometry is carried out (Alard and Lupton 1998). In each technique it is possible to take psf variations into account. Image subtraction algorithms generally appear to give better results than psf-fitting (Mochejska et al. 2002). If the stars are well separated from each other, aperture photometry gives excellent results; a precision of ~ 0.2 mmag for light curves of bright stars binned and averaged over 4.5 h has been demonstrated with this technique (Everett and Howell 2001).

It was suggested early on that it might be easier to look for the characteristic color changes (see Fig. 37) than for the absolute brightness changes; it was assumed that the most important errors would cancel in a differential measurement between two filters (Rosenblatt 1971). This is not true, however, because scintillation and variable extinction are intrinsically chromatic, which makes it quite difficult to detect the color changes, which are approximately ten times smaller than the photometric variations.

Instrumental Noise

While modern CCD cameras do not contribute significantly to the noise of ground-based photometric measurements, instrumental effects have to be taken into account at the $\sim 10^{-5}$ precision level that can be achieved in space. The most troublesome difficulty is the effect of intra-pixel variations of the quantum efficiency; if a point source of constant intensity is scanned across a pixel, variations in the electron count by several percent may be observed. The best strategy to mitigate this problem is avoiding steep gradients in the focal plane illumination by defocusing the telescope, so that the light from each star is sampled by $\sim 5 \times 5$ pixels. Laboratory tests have shown that a photometric precision of 10^{-5} can indeed be achieved with a careful calibration of systematic effects (Robinson et al. 1995). A difficulty of the defocusing approach is the increased sensitivity to contamination by faint background stars. There is a requirement on the stability of the telescope pointing to avoid background stars moving onto and off the pixels used for each target star. In addition, faint eclipsing binaries near the target star may mimic planetary transits; these false detections have to be eliminated by follow-up observations (see Sect. 6.4).

6.3 HD 209458

Light Curve and System Parameters

Photometric follow-up observations of stars with known radial-velocity variations lead to the discovery of the first transiting planet, HD 209458 B (Henry et al. 2000; Charbonneau et al. 2000). This detection of dips in the light curve with the "right" shape and at the "right" times provides an important confirmation of the existence of extrasolar planets, dispelling the last doubts about the interpretation of the radial-velocity variations. In addition, the transit light curve provides for the first time access to physical parameters of a planet other than $m \sin i$, as explained in Sect. 6.1. Observations of HD 209458 with the STIS spectrograph on the Hubble Space Telescope have produced an exquisite transit light curve, with typical photometric errors of $\sim 1.1 \cdot 10^{-4}$ for each 60 s data point (Brown et al. 2001, see Fig. 39). This corresponds to a signal-to-noise of ~ 150 for the transit depth of 1.64%.

The mass of HD 209458 has been estimated to be $m_* = 1.1 \pm 0.1\, M_\odot$ from its metallicity and location in the HR diagram, using stellar evolutionary models (Mazeh et al. 2000). With this mass as input value, the transit light curve can be used to determine the radii R_* and R_p, and the orbital inclination

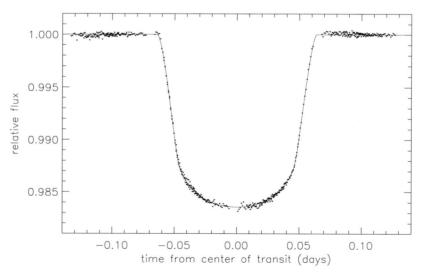

Fig. 39. Phased light curve of the transits of HD 209458 B from observations with the STIS spectrograph on the Hubble Space Telescope. From Brown et al. (2001)

Table 10. Parameters of HD 209458 and its planet, compiled from Henry et al. (2000); Mazeh et al. (2000); Queloz et al. (2000a); Robichon and Arenou (2000) and Brown et al. (2001)

parameter	value
m_*	$1.1 \pm 0.1\, M_\odot$
R_*	$1.146 \pm 0.050\, R_\odot$
$v \sin i_*$	3.7 ± 0.5
P	3.524739 ± 0.000014
a	0.046 ± 0.001 AU
$i_{\rm orb}$	$86°\!.68 \pm 0°\!.14$
m_p	$0.69 \pm 0.05\, M_{\rm jup}$
R_p	$1.347 \pm 0.060\, R_{\rm jup}$
ψ	$0 \pm 30°$

The table lists mass, radius, and projected rotation velocity of the star, period, radius, and inclination of the orbit, mass and radius of the planet, and the angle between the equatorial plane of the star and the planetary orbit

$i_{\rm orb}$. The orbital period P, orbital radius a, and planetary mass m_p are known from the radial-velocity observations. Values for these parameters are listed in Table 10, together with the stellar rotation velocity $v \sin i_*$, and the angle ψ between the equatorial plane of the star and the planet's orbit (from an analysis of spectroscopic data during transit, see below).

The Radius of HD 209458 b

Knowledge of both the mass and radius of HD 209458 B is a key for comparisons with physical models of planets. A first important conclusion is that it is made predominantly of hydrogen; a rocky or icy planet with a mass of $m_p = 0.69\, M_{\rm jup}$ would have a radius smaller than HD 209458 B by a factor of 3 to 4 (Burrows et al. 2000). A more detailed analysis shows that the radius of HD 209458 B is also larger than that expected for an isolated $0.69\, M_{\rm jup}$ planet; this is a consequence of the retardation of contraction by the stellar irradiation. An interesting conclusion is that HD 209458 B must have migrated to its present position early on (or even been born there); if it had dwelled more than $\sim 10^7$ years at a distance $\gtrsim 0.5$ AU, it would already have contracted during that time to a radius smaller than the presently observed value (Burrows et al. 2000). The exact radius–age relation for a given mass and external irradiation depends sensitively on the Bond albedo (for definition see Sect. 6.6);

uncertainties in this quantity therefore limit our current ability to compare the observational data to detailed model calculations. The most recent models that take into account the irradiation of HD 209458 B by its parent star predict a radius ∼20% lower than the observed value; an additional source of energy might therefore be required (Baraffe et al. 2003).

The effective planetary radius is a function of wavelength due to variations of the atmospheric opacity (see Sect. 6.1). For the parameters of HD 209458 B, dramatic variations in the occulted area are expected due to alkali metal lines in the visible (Seager and Sasselov 2000), and to H_2O absorption in the near-IR (Hubbard et al. 2001). The HST STIS data mentioned above show indeed that the transit of HD 209458 is deeper at the wavelength of the 589 nm Na resonance doublet than in the adjacent continuum; the difference is $(2.32 \pm 0.57) \cdot 10^{-4}$ with respect to the stellar flux (Charbonneau et al. 2002). These measurements constitute the first detection of an extrasolar planet atmosphere, and confirm the important role of alkali metal absorption for the spectra of "hot Jupiters". A detection of the first overtone band of CO near $2.3\,\mu m$ may also be possible with the same technique (Brown et al. 2002). A quantitative interpretation of the observed wavelength dependence of the planetary diameter is only possible with the inclusion of non-LTE effects in the modeling of the upper atmosphere (Barman et al. 2002).

It has been suggested that even planets that do not show photometric eclipses could exhibit "transits" of absorption features. Because of the strong stellar irradiation and interaction with the stellar wind, the mass loss from planets like 51 Peg b may be appreciable and lead to and "exosphere" similar to the tails of comets. Because of its larger size, this exosphere would periodically eclipse the star for a larger range of inclinations than the planet itself. A search toward 51 Peg with the Short Wavelength Spectrometer of the ISO satellite did not result in the detection of any absorption lines from atoms, molecules or their ionization or dissociation products from a transiting cloud (Rauer et al. 2000). Similar observations toward 51 Peg and HD 209458 in the visible have not lead to any detections, either (Bundy and Marcy 2000; Moutou et al. 2001). The existence of extended exospheres around strongly irradiated planets thus remains speculative.

Rossiter–McLaughlin Effect

Planetary eclipses affect not only the observed brightness of the star, but also the shape of spectral lines. A prograde planet orbiting in the equatorial plane transits in front of the approaching side of the star during the first half of the transit, and in front of the receding side during the second half. This causes successive dips in the blue-shifted and red-shifted wings of the spectral lines analogous to those due to star spots. If the apparent "radial velocity" of the star is determined from the centroids of the lines, the selective occultation causes an anomaly, whose shape and amplitude depend on the stellar rotation velocity and the geometry of the transit. This effect has been observed in

HD 209458, and used to place an upper limit of 30° on the angle between the orbital plane and the stellar equatorial plane (Queloz et al. 2000a). In addition to providing information on the transit geometry, observations of the expected spectroscopic anomalies could also help to confirm future detections of transiting planets from photometric surveys.

6.4 Photometric Planet Searches

Requirements of Wide-Field Searches

Whereas the detection of transits in the HD 209458 system resulted from follow-up observations of a known planet discovered by radial-velocity surveys, a number of projects are also underway to conduct searches for transiting planets through photometric monitoring of large numbers of stars. The immediate goal of these surveys is the detection of transiting "hot Jupiters". The probability that a planet in a 0.05 AU orbit will show transits is $\sim 10\%$ (57); if a few percent of all Sun-like stars have such planets, one should therefore expect a few detections per 1,000 stars monitored. The expected transit depths ($\sim 1\%$, (60)), durations (~ 3 h, (59)), and repeat rates (once every ~ 4 days) are all favorable for ground-based surveys.

A wide-field search for transiting planets has to accomplish two separate tasks (1) identification of transit candidates, i.e., stars with regularly spaced brightness dips that are consistent with planetary transits and (2) confirmation that the features in the light curve are indeed due to a transiting planet. To identify transit candidates, it is first necessary to obtain a large number of exposures of the target field, and to perform a careful photometric analysis as described in Sect. 6.2. Then one has to search for small dips in the many resulting light curves. The most straightforward approach is performing an automated search for individual dips that have a depth and duration consistent with the expectations for a planetary transit. If there are least three dips in the light curve of one star, a candidate has been found, provided also that the differences between the times at which the dips occur are small multiples of one common interval (the orbital period). For light curves with only two dips, one can predict the epochs at which further transits should occur, and thus establish more candidates with follow-up observations at these times.

If the signal-to-noise ratio in the light curves is too small for a straightforward detection of individual transits, it is still possible to search for the periodic signal of multiple transits. One possibility is folding each light curve with a set of trial periods, and cross-correlating the folded time series with a template transit light curve. If the cross-correlation coefficient for a star–period pair exceeds a certain pre-defined threshold, one has found a transit candidate. This method has the disadvantage that the sensitivity of the search is reduced for transits with a duration that is significantly different from the one assumed in the template. This difficulty can be avoided by using a "box-fitting" algorithm (Kovács et al. 2002). This algorithm searches for signals that alternate periodically between two discrete levels, with much less time spent at the lower level. This is a fair description of planetary transits for the purposes of a search algorithm.

Once a sample of transit candidates has been established, careful follow-up observations have to be performed to discriminate true planets from false alarms. The most important source of contamination are stellar eclipsing binaries, which can mimic planetary transits in a number of ways:

- Eclipses at grazing incidence. If the impact parameter $d_{\min}$ is in the range $R_1 - R_2 < d_{\min} < R_1 + R_2$, the secondary (radius R_2) will never be fully in front of the primary (radius R_1). This produces a shallow eclipse, which could be mistaken for the transit of a smaller object.
- Eclipses of an evolved primary by a main-sequence secondary. The diameter ratio between an evolved star and a low-mass main-sequence star can be similar to that between a main-sequence star and a giant planet; this leads to eclipses of identical depth and similar shape for these two types of systems.
- Eclipses in triple systems. The light curve of a hierarchical triple system, in which the primary is a Solar-type star, and the secondary an eclipsing pair of late-type dwarfs can be very similar to that of a Solar-type star orbited by a giant planet.

In principle, these can all be recognized as false alarms from the light curves alone, since the details of the transit (total duration, duration of ingress and egress, limb darkening profile and color variation) differ from those expected for planets. While this may be a possibility for very accurate photometry from space missions, observations from the ground will normally not reach the required precision. It is therefore necessary to follow-up the candidates with other techniques, which in many cases turns out to be much more difficult than finding the candidates in the first place. Medium-resolution spectroscopy can be used to determine the spectral type and luminosity class of the primary for each candidate event, and thus to weed out eclipses of evolved stars by main-sequence secondaries. Radial-velocity monitoring with relatively low precision (a few times $100\,\mathrm{m\,s^{-1}}$) can rule out eclipses by stellar secondaries at grazing incidence; a null result at this level means that there cannot be a stellar-mass secondary in an edge-on orbit of a few AU or less. For a clear confirmation of the planetary nature of the secondary, a positive detection of the radial velocity variation of the primary is needed. However, this normally requires radial velocity measurements with a precision of a few times $10\,\mathrm{m\,s^{-1}}$. This is quite difficult in the case of faint primaries, but necessary to reject cases like eclipses in triple systems.

Planet candidates from photometric transit data without mass determinations from radial-velocity (or astrometric) measurements are also of very limited use because the radius carries very little information on the physical nature of the transiting object (see Fig. 40). For example, a radius of $0.15\,R_\odot$ can correspond to a low-mass star of $\sim 0.1\,M_\odot$, to a brown dwarf, or to a

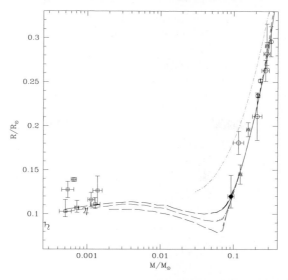

Fig. 40. Mass vs. radius for observed low-mass stars and giant planets and theoretical isochrones. Eclipsing binaries are shown as *circles* (OGLE-TR-122b in black), interferometric data as *open triangles*. ? isochrones for masses from 0.06 to 1.4 $M_\odot$ are plotted for 5 Gyr (*solid*) and 0.1 Gyr (*dash-dotted*). *Dashed lines* represent the Baraffe et al. (2003) CON models for masses for 0.5, 1 and 5 Gyr, from top to bottom (Figure added during proof by Editor)

planet with a mass as low as $2\,M_{\mathrm{jup}}$, depending on age. If one assumes that the companion is several billion years old, the range of possible masses for a given radius is substantially smaller. Because of the shallowness of the mass–radius relation, however, observational errors and uncertainties in the radius of the primary still lead to a very large allowed range for the companion mass. Follow-up observations are therefore required for a proper interpretation of planet candidates discovered by photometric surveys.

Searches in the Field and Toward the Galactic Bulge

The best method to look for transiting planets around nearby stars is monitoring with a small wide-angle telescope, which can produce simultaneous light curves of thousands of stars. The Vulcan project at the Lick Observatory on Mt. Hamilton uses a lens with 12 cm aperture to observe 6,000 stars brighter than 13th magnitude in a $7° \times 7°$ field with a 4096×4096 pixel CCD (Borucki et al. 2001). A similar setup is used by the STARE group (Brown and Charbonneau 2000). The detection efficiency of these surveys is somewhat limited because of the unavoidable gaps in the light curves during daytime and periods of bad weather. It would therefore be highly desirable to establish a network of small telescopes with good longitude distribution, which could perform continuous monitoring of many square degrees over several months. The planets detected by this network would be relatively nearby ($\lesssim 300\,\mathrm{pc}$) and could therefore be studied in detail with follow-up spectroscopy.

An alternative approach to simultaneous monitoring of many stars is using wide-field cameras on large telescopes (e.g., Mallen-Ornélas et al. 2003). Modern all-purpose cameras with CCD mosaics typically cover of order $30' \times 30'$, and exposure times of a few minutes at 4 m class telescopes are sufficient for $\lesssim 1\%$ photometry on stars between 16th and 18th magnitude. In the Galactic plane, several 10,000 stars can be monitored in this way. The Galactic bulge possesses an even higher stellar density, but it is necessary to look at even fainter stars ($19 \leq m_I \leq 21$), because the brighter bulge stars have already evolved off the main sequence and therefore have diameters that are too large for planetary transit detection. This means that a $5' \times 5'$ field-of-view is sufficient, but a $\sim 10\,\mathrm{m}$ aperture is required for an efficient search for planets transiting bulge stars (Gaudi 2000). However, the microlensing experiments surveying large fields toward the bulge also observe many thousands of disk main sequence stars located in the same sky area. Transits of disk stars with depths of a few percent, corresponding to estimated companion radii down to $\sim 1.5\,R_{\mathrm{jup}}$ have indeed been identified in the OGLE microlensing data set (Udalski et al. 2002).

A substantial problem of deep surveys is the difficulty of confirming planet candidates and measuring their masses, due to the faintness of the primaries and the potentially very high false alarm rate due to blending and grazing eclipses of stellar binaries. Spectroscopic observations of a number of the OGLE candidates from Udalski et al. (2002) have been used to *reject* most of them as low-mass stellar companions (Dreizler et al. 2002), but two candidates still remained in this study. It has recently been claimed that one of them (OGLE-TR-3) and another OGLE candidate (OGLE-TR-56) are indeed transiting extrasolar planets (Dreizler et al. 2003; Konacki et al. 2003), but these "planet detections" appear to be highly dubious[23]. In neither case are the radial-velocity data good enough for a secure detection of the claimed variations; this leaves many alternative interpretations of the eclipses open. More importantly, both papers claim a detection in a part of parameter space where planets are known to be exceedingly rare: the OGLE candidates have periods of $P = 1.1899\,\mathrm{d}$ and $P = 1.2119\,\mathrm{d}$, respectively, whereas the shortest-period planet known from radial-velocity surveys has $P = 2.5486\,\mathrm{d}$ (Udry et al. 2003a). Since the radial-velocity technique is very sensitive to short-period planets, the complete absence of planet-mass objects with periods close to those of OGLE-TR-3 and OGLE-TR-56 from the radial-velocity samples means that they must be very rare. The probability that the first *two* objects

---

[23] Editor note added in proof: Several planets on short orbits have been since detected in OGLE fields after submission of the manuscript. OGLE-TR-3 has been shown to be a grazing binary system

found with transit searches should reside in a sparsely populated region of parameter space is exceedingly small.

Searches in Open Clusters

Old open clusters are interesting targets for transit searches, since they contain numerous main-sequence stars with uniform distance, age and metallicity (Janes 1996; Quirrenbach et al. 2000). The clusters NGC 2682, NGC 6819, and NGC 7789 were selected for a pilot study with the 1 m Nickel Telescope at Lick observatory and the 1 m Jacobus Kapteyn Telescope at La Palma. The relative small fields of these telescopes (about $6' \times 6'$) allowed monitoring of only a few hundred stars, but the possibility of performing relative photometry at the $\lesssim 0.3\%$ level in the relatively crowded cluster fields was successfully demonstrated (Quirrenbach et al. 2000). This project has therefore been continued at the 2.4 m Isaac Newton Telescope, also at La Palma (Street et al. 2000). This survey, and a similar search in NGC 6791, have detected large numbers of variable stars in the cluster fields (Street et al. 2002; Mochejska et al. 2002). The data reduction has been refined to the point where an accuracy at the theoretical limit is reached for the vast majority of all stars. For more than 10,000 stars in the field of NGC 6819, the precision is sufficient to detect transits down to $1\,R_{\rm jup}$.

Under the assumption that $\sim 1\%$ of all stars have "hot Jupiter" companions, and that $\sim 10\%$ of these produce transits, several transits should have been detected by the observations of NGC 6819. One apparent transiting object with a radius similar to HD 209458 B has indeed been identified (Street et al. 2003). Improvements of the transit detection algorithm will be needed to identify more candidates, to measure the frequency of hot Jupiters in the cluster, and compare it to that in the Solar neighborhood. Measurements in a larger sample of open clusters could help to explore the relation of planet formation to metallicity and other environmental factors.

The Globular Cluster 47 Tucanae

The WFPC2 instrument on the Hubble Space Telescope was used in July 1999 to monitor a field in the globular cluster 47 Tuc continuously for 8.3 days (Gilliland et al. 2000). The core of the cluster was placed on the PC chip of WFPC2; a total of $\sim 34,000$ main-sequence stars with a typical projected distance from the core of $1'$ could thus be monitored. The noise in the light curves of stars in the magnitude range $17.1 \leq m_V \leq 21.5$ ranged from $\sim 0.3\%$ to $\sim 3\%$. Assuming a frequency of "hot Jupiters" identical to that in the Solar neighborhood, some $15\ldots 20$ transits should have been detected, but none were actually found. It has thus been established with extremely high significance that hot Jupiters are much less prevalent in 47 Tuc than in the Solar neighborhood.

Two possible reasons for this difference come to mind immediately. The lack of massive close-in planets in 47 Tuc could either be a metallicity effect, or related to the high density of stars in the cluster. 47 Tuc is a massive cluster with a stellar density of $n \sim 10^5 \, \mathrm{pc}^{-3}$ in the core, and $n \sim 10^4 \, \mathrm{pc}^{-3}$ (corresponding to $\sim 10^3 \, M_\odot \, \mathrm{pc}^{-3}$) at $1'$ from the core. The metallicity of 47 Tuc is $[\mathrm{Fe/H}] = -0.7 \, \mathrm{dex}$, but the abundance of α elements with respect to Fe is $[\alpha/\mathrm{Fe}] = 0.4 \, \mathrm{dex}$ (Salaris and Weiss 1998). Extrapolating the metallicity dependence of the incidence of planets (Sect. 3.4) to even lower values, it certainly appears plausible that "hot Jupiters" are exceedingly rare around stars as metal-poor as those in 47 Tuc. On the other hand, disruption of planet formation by close stellar encounters should also play an important role. It is likely that close-in planets with $a \lesssim 0.3 \, \mathrm{AU}$ survive for a Hubble time at a density $n \sim 10^4 \, \mathrm{pc}^{-3}$, which corresponds to the "typical" environment of the stars monitored in the WFPC2 observations (Davies and Sigurdsson 2001). One has to consider the history, of the cluster and the planetary system, however. A cluster that starts its life with a certain density will expand when gas that was not used up by star formation is expelled by the winds of OB stars or supernova explosions (Goodwin 1997; Kroupa et al. 2001). This leads to a decrease of the cluster density by a factor of $\sim$10 or more. It is thus quite possible that all protoplanetary disks or young planetary systems were destroyed in an early high-density phase of 47 Tuc (Bonnell et al. 2001). It is also possible that the ionizing flux of young massive stars destroys all circumstellar disks in the cluster before they can form planets (Armitage 2000); this process seems indeed to be at work near θ^1 Ori C in the Orion Trapezium (Johnstone et al. 1998). More observations in environments that cover different combinations of density and metallicity are needed to settle the question which of these factors is responsible for the lack of close-in planets in 47 Tuc.

Eclipsing Binaries

An interesting way to increase the odds of finding transiting planets are observations of eclipsing binary systems (Schneider and Chevreton 1990). This idea is based on the two assumptions that (a) planets form around close binaries with similar properties to those around single stars and (b) the orbital plane of the planet(s) will be roughly coplanar with the binary orbit. Our current understanding of binary star formation is certainly not good enough to provide compelling arguments either in favor of or against these hypotheses, and observational data do not exist.[24] These uncertainties notwithstanding,

---

[24] Note that planets around members of wide multiple systems have been found by the radial-velocity method, but close binaries are not suited as targets for this technique because of the large radial-velocity amplitude of the binary orbit.

an extensive observing campaign has been carried out to search for planetary transits in the CM Draconis system (Deeg et al. 1998; Doyle et al. 2000). This is arguably the best-suited binary for such a project, because both components are M4.5 dwarfs, which gives them a combined disk area of ~ 12% of the Solar disk (see Lacy 1977 for details). This means that a transit of a planet with $3.2\,R_\oplus$ would cause a 1% photometric dip. Furthermore, the inclination is nearly edge-on, $i = 89.82°$, which implies that planets in coplanar orbits with radii up to 0.35 AU would actually cause eclipses (see (56)).

Several 1 m class telescopes have been used to obtain a long time series of photometric measurements of CM Dra, which comprises more than 25,000 data points covering over 1,000 h with an rms precision of 0.2% to 0.7% (Deeg et al. 1998; Doyle et al. 2000). A single transit from a planet significantly larger than $3\,R_\oplus$ would produce a signal that could easily be detected above the noise. Considering the time coverage of the light curve, the detection probability for such planets with orbital periods between 7 and 60 days is > 90%, but no such events were found. Smaller planets could still be detected if they transit several times. The search for multiple transits from the same planet is much more complicated in this case than for a single star, because the orbital motion of the binary causes distortions of the individual dips, and a non-periodic signal from the repeated transits. Instead of doing a periodogram analysis, one therefore has to generate simulated light curves and cross-correlate them with the data, which is time-consuming and CPU intensive. Doyle et al. (2000) conclude that the detection probability for $2.5\,R_\oplus$ planets with periods up to 10 days in the CM Dra light curve from 1994 to 1998 was 80%. They found a few planet "candidates", i.e., assumed sets of planetary radius, orbital period and epoch that would produce several transits matching observed dips near the noise level. Follow-up observations around the predicted transit times of seven of the best candidates in 1999 showed that six of them were not real, however, leaving only one good candidate. Further unpublished observations of this candidate gave somewhat contradictory results: two rather convincing transit dips at the right time followed by three non-transit events that should have been there (L. Doyle, priv. comm.). It is presently unclear whether this is an inconsistency due to instrumental problems, or a detection of real transits of a non-coplanar object whose orbital nodes subsequently precessed from the line-of-sight. One can thus summarize that a fair part of the parameter space for planets $\gtrsim 2.5\,R_\oplus$ with periods between 7 and 60 days in orbits coplanar with the binary has been searched in CM Dra with negative results; no strong statement is currently possible for similar planets in non-coplanar orbits.

A second way to look for the signature of planets in the CM Dra light curve is by timing the eclipse minima (Deeg et al. 2000). This analysis is complementary to the search for transits; it is less sensitive to small planets, but it can detect planets with longer periods and in non-coplanar orbits. The amplitude of the timing residuals τ for a planet with mass m_p orbiting a binary with component masses $m_{1,2}$ at an orbital radius a is given by[25]

$$\tau = \frac{m_p a \sin i}{(m_1 + m_2)c}.\qquad(76)$$

The low mass of the components of CM Dra ($m_1 + m_2 = 0.44\,M_\odot$) is favorable, and the light curve obtained for the transit search covers more than 80 primary and secondary eclipses of the binary pair. A periodogram analysis of the 41 eclipses with the best-determined minimum times yields no peak above ~ 3 s, from which the presence of large planets (e.g., $m_p \sin i \geq 1\,M_{\rm jup}$ for $a = 2\,{\rm AU}$) can be ruled out according to (76). There is weak evidence for excess power at periods around 1,000 days, which would correspond to a circumbinary planet with $m_p \sin i \approx 1.5 \ldots 3\,M_{\rm jup}$ at $a \approx 1.1 \ldots 1.45\,{\rm AU}$ (Deeg et al. 2000). The significance of this feature in the power spectrum is at present uncertain, but it demonstrates the possibility of using transit timing in eclipsing binaries for searches of Jupiter-like planets.

6.5 Photometric Space Missions

As we have seen in Sect. 6.2, the Earth's atmosphere is the most severe source of photometric noise. Observations from orbiting observatories can thus reach much better precision, as already demonstrated by the HST light curve of HD 209458 (Fig. 39), obtained with an instrument that was not specifically built as a precise photometer. A number of photometric space missions will be launched in the near future; they will provide new opportunities for observations of extrasolar planets.

COROT, MONS, and MOST

Three small photometric satellites are expected to be launched over the next few years: the Canadian MOST spacecraft (Matthews et al. 2000), the Danish MONS (Christensen-Dalsgaard 2002), and the French/ESA project COROT (Baglin et al. 2002). These missions were initially conceived with the primary objective of studying the internal structure of stars through asteroseismology; they have therefore been designed to perform exquisite photometry of very bright stars (better than 10^{-6} for periodic signals like stellar oscillation modes). With the discovery of hot Jupiters it has become obvious that these satellites will also be able to perform precise measurements of the light curves of planetary transits, and to detect their reflected light. Observations of extrasolar planets have therefore been added to their scientific goals. Two different

---

[25] Note the difference between (75) and (76). In the first case, the invisible body orbits one component of the eclipsing system, and the second factor in the timing equation is the inverse of the orbital velocity of the eclipsing body. In the latter case, the invisible body orbits the eclipsing system at a large distance, and the velocity in the denominator is the speed of light.

modes are possible: a survey mode, which uses a wide-field camera to search for new transiting planets, and a targeted mode to get detailed light curves for stars that are already known to host planets. COROT, for example, will monitor about 6,000 to 12,000 stars in the range $m_V = 11\ldots16.5$. It should be able to detect planets with radii as small as $R_p = 1.6\,R_\oplus$, provided that their orbital periods are not longer than 50 days (Rouan et al. 2000). During the first two years of its lifetime, the planet program of MOST will consist of pointed observations of ε Eri, τ Boo, 51 Peg, HD 38529, and HD 209458. More targets may be added later.

Kepler and Eddington

ESA's Eddington (Favata 2002) and NASA's Kepler (Borucki et al. 1997; Koch et al. 1996, 1998) missions are more ambitious than the relatively small satellites described in the last section: they will search for transit of Earth-like planets and thus measure the frequency of other potentially habitable worlds. This requires both a large field-of-view and a sizeable aperture. Eddington will cover a field of at least 6 square degrees with a 1.2 m telescope. Kepler will even observe a 12° field with a 0.95 m aperture; this requires a focal plane covered with 42 CCD detectors.

If most stars have Earth-like planets, both Kepler and Eddington are expected to produce many reliable detections. In addition to detecting up to 50 "Earth twins", each of these missions would be able to measure the reflected light from hundreds of hot Jupiters, and observe transits for ∼100 of them. The success of such an ambitious program clearly depends on a reliable data processing pipeline, which has to find the transiting objects and discriminate them from low-amplitude variables. This is particularly difficult for small planets in short-period orbits; for these objects many transits will be observed, but each one will have a very low signal-to-noise. Understanding the properties of the light curves (after phasing with trial periods and co-adding of the individual transits), and the statistical significance of low-level transit detections, is therefore an important task (Deeg et al. 2000; Jenkins et al. 2002). If the Kepler and Eddington missions can be carried out successfully, they will likely be first to give us an answer to the question whether Earth analogs are common or rare around stars that are similar to our Sun.

6.6 Searches for the Light Reflected by the Planet

Definition of Albedo

For a discussion of the properties of starlight reflected by a planet, we first have to clarify the notion of albedo. The best-known concept is that of the *Bond albedo A* defined by

$$A \equiv \frac{P_\text{refl}}{P_\text{incid}}, \tag{77}$$

where P_{incid} is the total power of light incident on a surface, and P_{refl} that reflected by it. The Bond albedo governs the equilibrium temperature T_{eff} of a planet heated by its parent star. Balancing radiation losses with internal energy production and insolation gives

$$4\pi R_p^2 \sigma T_{\mathrm{eff}}^4 = P_{\mathrm{int}} + \pi R_p^2 (1-A) L_* / (4\pi a^2) \; , \tag{78}$$

where R_p is the radius of the planet, a its orbital radius, P_{int} the internal energy production per time interval, and L_* the stellar luminosity. If the internally generated heat is negligible compared to the insolation, the equilibrium temperature is given by

$$T_{\mathrm{eff}} = \left[\frac{(1-A) L_*}{16 \pi \sigma a^2} \right]^{1/4} . \tag{79}$$

While the Bond albedo is needed to estimate the temperature of a planet, the quantity that is most useful to describe the reflected light is the *geometric albedo* p_λ. It is defined as the reflectivity of a planet at wavelength λ, measured at full phase, i.e., $\alpha = 0$, where α is the angle star–planet–observer. (We neglect the possibility of eclipses in this section.) The observed intensity of the reflected light $\mathcal{F}_\lambda(\alpha)$ can then be written as

$$\mathcal{F}_\lambda(\alpha) = p_\lambda \mathcal{F}_{\mathrm{incid}} \left(\frac{R_p}{d} \right)^2 \phi_\lambda(\alpha) \; , \tag{80}$$

where $\mathcal{F}_{\mathrm{incid}}$ is the stellar flux incident on the planet, and $\phi_\lambda(\alpha)$ the *phase function*, which describes the phase dependence of the scattering and is normalized such that $\phi_\lambda(0) = 1$. The flux from a star at distance d observed on the Earth is related to the flux incident on a planet at orbital distance a by the inverse square law

$$\mathcal{F}_* = \mathcal{F}_{\mathrm{incid}} \left(\frac{a}{d} \right)^2 . \tag{81}$$

Dividing (80) by (81), we thus obtain an expression for the intensity ratio $\epsilon_\lambda(\alpha)$ between the planetary and stellar spectra:

$$\epsilon_\lambda(\alpha) \equiv \frac{\mathcal{F}_\lambda(\alpha)}{\mathcal{F}_*} = p_\lambda \left(\frac{R_p}{a} \right)^2 \phi_\lambda(\alpha) . \tag{82}$$

For circular orbits, the phase angle α can be computed from

$$\cos\alpha = -\sin i \sin 2\pi\Phi \; , \tag{83}$$

where $\Phi \in [0,1)$ is the orbital phase, measured from the time of maximum recessional velocity of the star.[26]

---

[26] Note that different definitions of the zero point of the orbital phase are used in the literature. The one adopted here is customarily used for spectroscopic binaries, whereas in eclipsing binaries the time of the primary eclipse is normally chosen as the zero point of the orbital phase. In analogy to the latter definition, some authors count the phase from the time of inferior conjunction. For circular orbits, the difference between the two definitions is 0.25.

The functions p_λ and $\phi_\lambda(\alpha)$ depend on the properties of the planetary atmosphere. For reference purposes, it is useful to consider the case of a *Lambert sphere*, which scatters all incoming photons isotropically. It can be shown that for a Lambert sphere

$$p_\lambda = 2/3 , \tag{84}$$

and

$$\phi_\lambda(\alpha) = \frac{\sin\alpha + (\pi - \alpha)\cos\alpha}{\pi} . \tag{85}$$

Inserting these two relations in (82), we get

$$\epsilon_\lambda(\alpha) = \frac{2}{3}\left(\frac{R_p}{a}\right)^2 \left(\frac{\sin\alpha + (\pi - \alpha)\cos\alpha}{\pi}\right) . \tag{86}$$

This equation can be used together with (83) to calculate simple models of the phase variations. Inserting typical numbers of "hot Jupiters" in (86), we expect variations of order

$$\Delta\epsilon = 6 \cdot 10^{-5} \left(\frac{R_p}{R_{\rm jup}}\right)^2 \left(\frac{a}{0.05\,{\rm AU}}\right)^{-2} \tag{87}$$

for planets with edge-on orbits. (For $i < 90°$ the variations are smaller, because we cannot probe the full range of phase angles.) With the precision of the upcoming photometric space missions (Sect. 6.5), it should thus be possible to detect the starlight reflected by hot Jupiters through their phase variations.

For more realistic predictions of the brightness of the reflected light, one has to compute detailed models of the structure of the planetary atmosphere (see below). The presence of strong atomic and molecular absorption features also opens the possibility to obtain spectroscopic information on the planetary atmospheres. It should finally be pointed out that the atmospheric light scattering processes lead to a high degree of polarization of the reflected light at phase angles near or slightly below $90°$. The expected signature of a planet in the stellar polarization fraction is only a few times 10^{-6}, however, which is below the current detection limit (Seager et al. 2000).

The Signature of Spectral Features in Reflected Light

It is obvious from (80) that the light reflected by a planet carries spectral imprints both from the stellar photosphere (wavelength dependence of $\mathcal{F}_{\rm incid}$) and from the planet's atmosphere (parameterized by p_λ and $\phi_\lambda(\alpha)$). For observations with low spectral resolution, the description in (80) is sufficient, but at a resolving power $R \gtrsim 1{,}000$ we have to take a more careful look at the implications of the Doppler effect. First of all, the radial velocity v_p of the planet is given by

$$v_p(\Phi) = K_p \cos 2\pi\Phi = -K_* \frac{m_*}{m_p} \cos 2\pi\Phi \;, \tag{88}$$

where K_p and K_* are the radial-velocity amplitudes of the planet and the star, respectively. If the geometric albedo of the planet can be regarded as constant over a small spectral range in the vicinity of a stellar absorption line, a "ghost image" of this line appears in the reflected spectrum with (time-dependent) amplitude and Doppler velocity given by (82) and (88). The velocity amplitude K_p is typically many $\mathrm{km\,s^{-1}}$ ($K_p = 30\,\mathrm{km\,s^{-1}}$ for $m_* = 1\,M_\odot$ and $a = 1\,\mathrm{AU}$), which is much larger than the line width of old G dwarfs, and much more than the resolution achievable with an Echelle spectrograph. It thus appears promising to search for the reflected "ghost spectrum" in high-signal-to-noise spectra of stars with known planets. The difference of two such spectra taken at different orbital phases should reveal two ghost spectra (one of them positive, the other negative) at the velocities predicted by (88). One important difficulty of such searches is the unknown orbital inclination of the planet. Combining (19) with (88), we obtain

$$v_p(\Phi) = \left(\frac{2\pi G}{P}\right)^{1/3} m_*^{1/3} \sin i \, \cos 2\pi\Phi. \tag{89}$$

This means that only an upper limit to K_p is known a priori. The need to perform a search for the weak reflected line over the velocity range up to this limit increases the probability of a false detection; therefore a higher signal-to-noise ratio is necessary than for the detection of an equally faint line at a known position.

A second consideration concerns the width of the reflected lines. For a planet orbiting in the equatorial plane, the observed Doppler width v_{refl} of the reflected lines is given by

$$v_{\mathrm{refl}} = 2\pi R_* \sin i \left| \frac{1}{P_{\mathrm{rot}}} - \frac{1}{P_{\mathrm{orb}}} \right| = v_* \sin i \left| 1 - \frac{P_{\mathrm{rot}}}{P_{\mathrm{orb}}} \right| \;, \tag{90}$$

where P_{rot} is the rotational period of the star, P_{orb} the orbital period of the planet, and $v_* \sin i$ the Doppler width of the directly observed stellar lines. In the case of "hot Jupiters", P_{rot} and P_{orb} may be similar to each other; it has also been argued that a convective envelope with mass $m \approx 0.01\,M_\odot$ could become tidally locked in less than the age of the system (Marcy et al. 1997). In this case, there is no relative motion between any point on the stellar surface and the planet. The reflected spectrum therefore does not show any broadening due to stellar rotation, in agreement with (90). The broadening caused by the rotation of the planet (which is certainly tidally locked) is small, because of the small planetary radius. The width of the reflected lines is therefore dominated by convective motions in the stellar photosphere, and may thus be substantially smaller than $v_* \sin i$.

The spatial structure of a random process can be described by *structure functions*. The structure function $D_x(R_1, R_2)$ of a random variable x measured at positions R_1, R_2 is defined by

$$D_x(R_1, R_2) \equiv \left\langle |x(R_1) - x(R_2)|^2 \right\rangle \tag{91}$$

(see also (246)). In words: the structure function measures the expectation value of the difference of the values of x measured at two positions R_1 and R_2. For example, the temperature structure function $D_T(R_1, R_2)$ is the expectation value of the difference in the readings of two temperature probes located at R_1 and R_2. In the following paragraph, a simple argument based on dimensional analysis will be used to derive structure functions for the Kolmogorov model.

The Structure Function for Kolmogorov Turbulence

The only two relevant parameters (in addition to l_0 and L_0) that determine the strength and spectrum of Kolmogorov turbulence are the rate of energy generation per unit mass ε, and the kinematic viscosity ν. The units of ε are $\text{J s}^{-1} \text{kg}^{-1} = \text{m}^2 \text{s}^{-3}$, and those of ν are $\text{m}^2 \text{s}^{-1}$. Under the assumption that the turbulence is homogeneous and isotropic, the structure function of the turbulent velocity field, $D_v(R_1, R_2)$, can only depend only on $|R_1 - R_2|$, and can therefore be written as:

$$D_v(R_1, R_2) \equiv \left\langle |v(R_1) - v(R_2)|^2 \right\rangle$$
$$= \alpha \cdot f\bigl(|R_1 - R_2|/\beta\bigr) , \tag{92}$$

where f is some as yet unspecified dimensionless function of a dimensionless argument. It is immediately clear that the dimensions of α must be velocity squared, and those of β length. Since α and β depend only on ε and ν, it follows from dimensional analysis that

$$\alpha = \nu^{1/2} \varepsilon^{1/2} \quad \text{and} \quad \beta = \nu^{3/4} \varepsilon^{-1/4} . \tag{93}$$

In addition, the structure function must be independent of ν in the inertial range, because dissipation does not play a role here. This is possible only if f has the functional form

$$f = k \cdot \bigl(|R_1 - R_2|/\beta\bigr)^{2/3} \tag{94}$$

with a dimensionless numerical constant k, because only in this case the dependence on ν drops out in the expression of the structure function:

$$D_v(R_1, R_2) = \alpha \cdot k \cdot \bigl(|R_1 - R_2|/\beta\bigr)^{2/3} = C_v^2 \cdot |R_1 - R_2|^{2/3} , \tag{95}$$

where $C_v^2 \equiv \alpha \cdot k/\beta^{2/3} = k \cdot \varepsilon^{2/3}$. We have thus derived the important result mentioned above, namely a universal description of the turbulence spectrum. It has only one parameter C_v^2, which describes the turbulence strength.

Structure Function and Power Spectral Density of the Refractive Index

The turbulence, with a velocity field characterized by (95), mixes different layers of air, and therefore carries around "parcels" of air with different temperature. Since these "parcels" are in pressure equilibrium, they must have different densities ρ, and therefore different indices of refraction n. The "parcels" are carried along by the velocity field of the turbulence. The temperature fluctuations therefore also follow Kolmogorov's Law with a new parameter C_T^2:

$$D_T(R_1, R_2) = C_T^2 \cdot |R_1 - R_2|^{2/3} \; ; \tag{96}$$

note that this is completely analogous to (95). From the Ideal Gas Law, and $N \equiv (n-1) \propto \rho$, it follows that the structure function of the refractive index is

$$D_n(R_1, R_2) = D_N(R_1, R_2) = C_N^2 \cdot |R_1 - R_2|^{2/3} \; , \tag{97}$$

with C_N given by

$$C_N = \left(7.8 \cdot 10^{-5} P[\text{mbar}]/T^2[\text{K}]\right) \cdot C_T \,. \tag{98}$$

It should be noted that (97) contains a complete description of the statistical properties of the refractive index fluctuations, on length scales between l_0 and L_0. It is possible to calculate related quantities such as the power spectral density Φ from the structure function D. Now we write $R \equiv R_1 - R_2$, and use the relation between the structure function and the covariance (247), and the Wiener-Khinchin Theorem (245). In this way we obtain from (97):

$$C_N^2 \cdot R^{2/3} = D_N(R) = 2 \int_{-\infty}^{\infty} d\kappa \left[1 - \exp(2\pi i \kappa R)\right] \Phi(\kappa) \,. \tag{99}$$

Calculating $\Phi(\kappa)$ from this relation is a slightly non-trivial task[27]; the result is:

$$\Phi(\kappa) = \frac{\Gamma(\tfrac{5}{3}) \sin \tfrac{\pi}{3}}{(2\pi)^{5/3}} C_N^2 \kappa^{-5/3} = 0.0365 \, C_N^2 \kappa^{-5/3} \,. \tag{100}$$

We have thus obtained the important result that the power spectrum of Kolmogorov turbulence follows a $\kappa^{-5/3}$ law in the inertial range.[28]

---

[27] See Tatarski (1961). Note that his definition of the power spectral density has an additional factor $\frac{1}{2\pi}$, and that his ω corresponds to $2\pi\kappa$.

[28] Note: We have defined $R = |R_1 - R_2|$ and κ as one-dimensional variables, and consequently used a one-dimensional Fourier transform in (99). Sometimes three-dimensional quantities $\vec{R}$ and $\vec{\kappa}$ are used instead. Then a three-dimensional Fourier transform with volume element $4\pi |\vec{\kappa}|^2 d|\vec{\kappa}|$ has to be used in (99), and the result is a power spectrum $\Phi(|\vec{\kappa}|) \propto |\vec{\kappa}|^{-11/3}$.

7.2 Wave Propagation Through Turbulence

The Effects of Turbulent Layers

We now look at the propagation of an initially flat wavefront through a turbulent layer of thickness δh at height h. The phase shift produced by refractive index fluctuations is

$$\phi(x) = k \int_{h}^{h+\delta h} dz\, n(x,z) \,, \tag{101}$$

where $k \equiv 2\pi/\lambda$ is the wavenumber corresponding to the observing wavelength. For layers that are much thicker than the individual turbulence cells, many independent variables contribute to the phase shift. Therefore the Central Limit Theorem implies that ϕ has Gaussian statistics.

We will now use the statistical properties of the refractive index fluctuations, which were calculated in Sect. 7.1, to derive the statistical behavior of the wavefront $\psi(x) = \exp i\phi(x)$. We first express the coherence function $B_h(r)$ of the wavefront after passing through the layer at height h in terms of the phase structure function (see Sect. 11 for definitions):

$$\begin{aligned} B_h(r) &\equiv \langle \psi(x)\psi^*(x+r) \rangle \\ &= \left\langle \exp i[\phi(x) - \phi(x+r)] \right\rangle \\ &= \exp\left(-\tfrac{1}{2}\left\langle |\phi(x) - \phi(x+r)|^2 \right\rangle\right) \\ &= \exp\left(-\tfrac{1}{2}D_\phi(r)\right) \,. \end{aligned} \tag{102}$$

Here we have used the fact that $[\phi(x) - \phi(x+r)]$ has Gaussian statistics with zero mean, and applied the relation

$$\langle \exp(\alpha\chi) \rangle = \exp\left(\tfrac{1}{2}\alpha^2 \langle \chi^2 \rangle\right) \tag{103}$$

for Gaussian variables χ with zero mean, which can easily be verified by carrying out the integral over the distribution function.

Calculation of the Phase Structure Function

The next step is the computation of $D_\phi(r)$. We start with the covariance $B_\phi(r)$, which is by definition (240):

$$\begin{aligned} B_\phi(r) &\equiv \langle \phi(x)\phi(x+r) \rangle \\ &= k^2 \int_{h}^{h+\delta h} \int_{h}^{h+\delta h} dz'\, dz''\, \langle n(x,z')n(x+r,z'') \rangle \\ &= k^2 \int_{h}^{h+\delta h} dz' \int_{h-z'}^{h+\delta h-z'} dz\, B_N(r,z) \,. \end{aligned} \tag{104}$$

Here we have introduced the new variable $z \equiv z'' - z'$, and the covariance $B_N(r, z)$ of the refractive index variations. For δh much larger than the correlation scale of the fluctuations, the integration can be extended from $-\infty$ to ∞, and we obtain:

$$B_\phi(r) = k^2 \delta h \int_{-\infty}^{\infty} dz \, B_N(r, z) \,. \tag{105}$$

Now we can use (247) again, first for $D_\phi(r)$, then for $D_N(r, z)$ and $D_N(0, z)$, and get:

$$D_\phi(r) = 2\big[B_\phi(0) - B_\phi(r)\big]$$

$$= 2k^2 \delta h \int_{-\infty}^{\infty} dz \, \big[B_N(0, z) - B_N(r, z)\big]$$

$$= 2k^2 \delta h \int_{-\infty}^{\infty} dz \, \Big[\big(B_N(0, 0) - B_N(r, z)\big) - \big(B_N(0, 0) - B_N(0, z)\big)\Big]$$

$$= k^2 \delta h \int_{-\infty}^{\infty} dz \, \big[D_N(r, z) - D_N(0, z)\big] \,. \tag{106}$$

Inserting from (97) gives:

$$D_\phi(r) = k^2 \delta h \, C_N^2 \int_{-\infty}^{\infty} dz \, \Big[\big(r^2 + z^2\big)^{1/3} - |z|^{2/3}\Big]$$

$$= \frac{2\Gamma(\frac{1}{2})\Gamma(\frac{1}{6})}{5\Gamma(\frac{2}{3})} k^2 \delta h \, C_N^2 \, r^{5/3}$$

$$= 2.914 \, k^2 \delta h \, C_N^2 \, r^{5/3} \,. \tag{107}$$

This is the desired expression for the structure function of phase fluctuations due to Kolmogorov turbulence in a layer of thickness δh.

Wavefront Coherence Function and Fried Parameter

We are now in a position to put the results of the previous sections together. Inserting (107) into (102), we get:

$$B_h(r) = \exp\left[-\tfrac{1}{2}\left(2.914 \, k^2 C_N^2 \, \delta h \, r^{5/3}\right)\right] \,. \tag{108}$$

This expression can now be integrated over the whole atmosphere. In the process, we also take into account that we are not necessarily looking in the vertical direction. Introducing the zenith angle z, this leads to:

$$B(r) = \exp\left[-\tfrac{1}{2}\left(2.914 \, k^2 (\sec z) r^{5/3} \int dh \, C_N^2(h)\right)\right] \,. \tag{109}$$

To simplify the notation, it is now convenient to define the *Fried parameter* r_0 by

$$r_0 \equiv \left[0.423\, k^2 (\sec z) \int dh\, C_N^2(h) \right]^{-3/5}, \qquad (110)$$

and we can write

$$B(r) = \exp\left[-3.44 \left(\frac{r}{r_0}\right)^{5/3}\right] , \quad D_\phi(r) = 6.88 \left(\frac{r}{r_0}\right)^{5/3}. \qquad (111)$$

We have thus derived fairly simple expressions for the wavefront coherence function and the phase structure function. They depend only on the Fried parameter r_0, which in turn is a function of turbulence strength, zenith angle, and wavelength. The significance of the Fried parameter will be discussed further in Sect. 7.4.

7.3 The Effect of Turbulence on Astronomical Images

Optical Image Formation

The complex amplitude A of a wave ψ diffracted at an aperture P with area Π is given by Huygens' principle, which states that each point in the aperture can be considered as the center of an emerging spherical wave. In the far field (i.e., in the case of Fraunhofer diffraction), the spherical waves are equivalent to plane waves, and we can write down the expression for the amplitude as a function of position α in the focal plane:

$$A(\alpha) = \frac{1}{\sqrt{\Pi}} \int dx\, \psi(x) P(x) \exp(-2\pi i \alpha x / \lambda) . \qquad (112)$$

Here we describe the aperture P by a complex function $P(x)$. In the simple case of a fully transmissive and aberration-free aperture $P(x) \equiv 1$ inside the aperture, and $P(x) \equiv 0$ outside. Introducing the new variable $u \equiv x/\lambda$ we can write this as a Fourier relation:

$$A(\alpha) = \frac{1}{\sqrt{\Pi}} FT\left[\psi(u) P(u)\right] . \qquad (113)$$

The normalization in (112) and (113) has been chosen such that the illumination S in the focal plane is given by the square of the wave amplitude:

$$S(\alpha) = |A(\alpha)|^2 = \frac{1}{\Pi} \left| FT\left[\psi(u) P(u)\right] \right|^2 . \qquad (114)$$

Applying the Wiener-Khinchin Theorem (245) to this equation we get

$$S(f) = \frac{1}{\Pi} \int du\, \psi(u) \psi^*(u+f) P(u) P^*(u+f) . \qquad (115)$$

This equation describes the spatial frequency content $S(f)$ of images taken through the turbulent atmosphere, if ψ is identified with the wavefront after passing through the turbulence. Taking long exposures (in practice this means exposures of at least a few seconds) means averaging over many different realizations of the state of the atmosphere:

$$\langle S(f) \rangle = \frac{1}{\Pi} \int du \, \langle \psi(u)\psi^*(u+f) \rangle P(u) P^*(u+f)$$
$$= B(f) \cdot T(f) \,. \tag{116}$$

Here we have introduced the *telescope transfer function*

$$T(f) = \frac{1}{\Pi} \int du \, P(u) P^*(u+f) \,. \tag{117}$$

Equation (116) contains the important result that for long exposures the optical transfer function is the product of the telescope transfer function and the atmospheric transfer function, which is equal to the wavefront coherence function $B(f)$.

Diffraction-Limited Images and Seeing-Limited Images

The resolving power $\mathcal{R}$ of an optical system can very generally be defined by the integral over the optical transfer function. For the atmosphere–telescope system this means:

$$\mathcal{R} \equiv \int df \, S(f) = \int df \, B(f) T(f) \,. \tag{118}$$

In the absence of turbulence, $B(f) \equiv 1$, and we obtain the *diffraction-limited* resolving power of a telescope with diameter D:

$$\mathcal{R}_{\text{tel}} = \int df \, T(f) = \frac{1}{\Pi} \int \int dudf \, P(u) P^*(u+f)$$
$$= \frac{1}{\Pi} \left| \int du \, P(u) \right|^2 = \frac{\pi}{4} \left(\frac{D}{\lambda} \right)^2 \,. \tag{119}$$

The last equality assumes a circular aperture and shows the relation of $\mathcal{R}$ to the more familiar Rayleigh criterion $1.22 \cdot \lambda/D$. Working with $\mathcal{R}$ instead of using the Rayleigh criterion has the advantage that $\mathcal{R}$ is a well-defined quantity for arbitrary aperture shapes and in the presence of aberrations.

For strong turbulence and rather large telescope diameters, $T = 1$ in the region where B is significantly different from zero, and we get the *seeing-limited* resolving power:

$$\mathcal{R}_{\text{atm}} = \int df \, B(f) = \int df \, \exp\left[-3.44 \left(\frac{\lambda f}{r_0} \right)^{5/3} \right]$$
$$= \frac{6\pi}{5} \Gamma(\tfrac{6}{5}) \left[3.44 \left(\frac{\lambda}{r_0} \right)^{5/3} \right]^{-6/5} = \frac{\pi}{4} \left(\frac{r_0}{\lambda} \right)^2 \,. \tag{120}$$

Here we have used (111) with $r = \lambda f$ for the wavefront coherence function $B(f)$.

7.4 Fried Parameter and Strehl Ratio

The Significance of the Fried Parameter r_0

A comparison of (119) and (120) elucidates the significance of the Fried parameter for image formation, and reveals the reason for the peculiar choice of the numerical constant 0.423 in (110): *The resolution of seeing-limited images obtained through an atmosphere with turbulence characterized by a Fried parameter r_0 is the same as the resolution of diffraction-limited images taken with a telescope of diameter r_0.* Observations with telescopes much larger than r_0 are seeing-limited, whereas observations with telescopes smaller than r_0 are essentially diffraction-limited. It can also be shown that the mean-square phase variation over an aperture of diameter r_0 is about $1\,\text{rad}^2$ (more precisely, $\sigma_\phi^2 = 1.03\,\text{rad}^2$). These results can be captured in an extremely simplified picture that describes the atmospheric turbulence by r_0-sized "patches" of constant phase, and random phases between the individual patches. While this picture can be useful for some rough estimates, one should keep in mind that Kolmogorov turbulence has a continuous spectrum ranging from l_0 to L_0, as described by (100).

The scaling of r_0 with wavelength and zenith angle implied by (110) has far-reaching practical consequences. Since

$$r_0 \propto \lambda^{6/5}\,, \tag{121}$$

it is much easier to achieve diffraction-limited performance at longer wavelengths. For example, the number of degrees of freedom (the number of actuators on the deformable mirror and the number of subapertures in the wavefront sensor) in an adaptive optics system must be of order $(D/r_0)^2 \propto \lambda^{-12/5}$. An interferometer works well only if the wavefronts from the individual telescopes are coherent (i.e., have phase variances not larger than about $1\,\text{rad}^2$); therefore the maximum useful aperture area of an interferometer is $\propto \lambda^{12/5}$ (unless the wavefronts are corrected with adaptive optics). Equation (121) implies that the width of seeing-limited images, $\theta \approx 1.2 \cdot \lambda/r_0 \propto \lambda^{-1/5}$, varies only slowly with λ; it is somewhat better at longer wavelengths. In addition, we see from (110) that $r_0 \propto (\sec z)^{-3/5}$; the seeing gets worse with increasing zenith angle.

From this discussion it should be clear that the value of r_0 – given by the integral over C_N^2 – is a crucial parameter for high-resolution observations. At good sites, such as Mauna Kea or Cerro Paranal, r_0 is typically of order 20 cm at 500 nm, which corresponds to an image FWHM of $0\rlap{.}''6$. The scaling of r_0 with λ (121) implies that in the mid-infrared ($\lambda \gtrsim 10\,\mu\text{m}$) even the 10 m Keck Telescopes are nearly diffraction-limited, whereas a 1.8 m telescope has $D/r_0 \sim 2$ at $\lambda = 2\,\mu\text{m}$ and $D/r_0 \sim 5$ at $\lambda = 800\,\text{nm}$. It should be noted that at any given site r_0 varies dramatically from night to night; at any given time it may be a factor of 2 better than the median or a factor of 5 worse. In addition, the seeing fluctuates on all time scales down to minutes and seconds; this has to be taken into account in calibration procedures and in the design of servo loops for adaptive optics systems and of fringe trackers for interferometers.

Strehl Ratio

The quality of an aberrated imaging system, or of the wavefront after propagation through turbulence, is often measured by the *Strehl ratio S*. This quantity is defined as the on-axis intensity in the image of a point source divided by the peak intensity in a hypothetical diffraction-limited image taken through the same aperture. For a circular aperture with an aberration function $\psi(\rho,\theta)$, which describes the wavefront distortion (in units of µm or nm) as a function of the spherical coordinates (ρ,θ), the Strehl ratio is given by:

$$S = \frac{1}{\pi^2} \left| \int_0^1 \int_0^{2\pi} \rho\,d\rho\,d\theta\, e^{ik\psi(\rho,\theta)} \right|^2. \tag{122}$$

From this equation it is immediately clear that $0 \leq S \leq 1$, that $S = 1$ for $\psi = $ const., that $S \ll 1$ for strongly varying ψ, and that for any given (varying) ψ the Strehl ratio tends to be larger for longer wavelengths (smaller k). In the case of atmospheric turbulence, only the statistical properties of ψ are known. If the r.m.s. phase error $\sigma_\phi \equiv k\,\sigma_\psi$ is smaller than about 2 rad, S can be approximated by the so-called *extended Maréchal approximation*:

$$S = e^{-\sigma_\phi^2}. \tag{123}$$

We have seen above ((111) and discussion of the significance of r_0) that

$$\sigma_\phi^2 = 1.03 \left(\frac{D}{r_0}\right)^{5/3}. \tag{124}$$

Equations (123) and (124) show that the Strehl ratio for images obtained with a telescope of diameter $D = r_0$ is $S = 0.36$; for $D \gtrsim r_0$ the Strehl ratio decreases precipitously with telescope diameter. Equivalently, S decreases sharply with decreasing wavelength, since $r_0 \propto \lambda^{6/5}$.

If $S \gtrsim 0.1$ in an imaging application, deconvolution algorithms can usually be applied to obtain diffraction-limited images, but the dynamic range and signal-to-noise ratio are worse than for $S \sim 1$. For example, because of spherical aberration, the Hubble Space Telescope has $S \approx 0.1$ without corrective optics. Before the installation of COSTAR and WFPC2 in the first servicing mission, the imaging performance of HST was severely affected by the flawed optics, although diffraction-limited images could be obtained with image restoration software. In an interferometer, the maximum fringe contrast is roughly proportional to the Strehl ratio if no corrective measures (adaptive optics or mode filtering with pinholes or single-mode fibers) are taken. Planet detection with imaging requires an extremely high dynamic range, which usually means that a Strehl ratio close to 1 is desired.

7.5 Temporal Evolution of Atmospheric Turbulence

Taylor Hypothesis and τ_0

So far we have discussed the spatial structure of atmospheric turbulence and its effects on image formation. Now we turn to the question of temporal changes of the turbulence pattern. A convenient approximation assumes that the time scale for these changes is much longer than the time it takes the wind to blow the turbulence past the telescope aperture. According to this *Taylor hypothesis of frozen turbulence*, the variations of the turbulence caused by a single layer can therefore be modeled by a "frozen" pattern that is transported across the aperture by the wind in that layer. If multiple layers contribute to the total turbulence, the time evolution is more complicated, but the temporal behavior of the turbulence can still be characterized by a time constant

$$\tau_0 \equiv r_0/v , \quad (125)$$

where v is the wind speed in the dominant layer. With typical wind speeds of order $20\,\mathrm{ms}^{-1}$, $\tau_0 \approx 10\,\mathrm{ms}$ for $r_0 = 20\,\mathrm{cm}$. The wavelength scaling of τ_0 is obviously the same as that of r_0, i.e., $\tau_0 \propto \lambda^{6/5}$.

7.6 Temporal Structure Function and Power Spectra

It is sometimes necessary to quantify the temporal behavior of phase fluctuations at a given point in space. If Taylor's hypothesis is valid, we can of course convert the spatial structure function (111) into a temporal structure function:

$$D_\phi(t) = 6.88 \left(\frac{t}{\tau_0}\right)^{5/3} . \quad (126)$$

A calculation similar to the one leading to (100) can be carried out to compute the temporal phase power spectrum

$$\Phi_\phi(f) = 0.077\,\tau_0^{-5/3} f^{-8/3} . \quad (127)$$

This equation tells us which residual phase errors we have to expect if we try to correct atmospheric turbulence with a servo loop of a given bandwidth (e.g., in an adaptive optics system or an interferometric fringe tracker). For example, if we could correct the turbulence perfectly up to a limiting frequency f_0, and not at all at higher frequencies, we would obtain a phase variance that can be computed by integrating (127) from f_0 to ∞. For a more realistic calculation, we have to multiply the phase power spectrum with the response function of the servo loop.

The Long-Exposure and Short-Exposure Limits

Observations with exposure time $t \gg \tau_0$ average over the atmospheric random process; these are the *long exposures* for which (116) and (120) are applicable. In contrast, *short exposures* with $t \ll \tau_0$ produce images through a single instantaneous realization of the atmosphere; these *speckle images* contain information at high spatial frequencies up to the diffraction limit, which can be extracted from series of such images with computer processing (e.g., bispectrum analysis). The parameter τ_0 is also of great importance for the design of adaptive optics systems and interferometers. All control loops that have to reject atmospheric fluctuations – AO control loops, angle trackers, fringe trackers – must have bandwidths larger than $1/\tau_0$. Together r_0 and τ_0 set fundamental limits to the sensitivity of these wavefront control loops: a certain number of photons must arrive per r_0-sized patch during the time τ_0 for the wavefront sensor (or fringe sensor) to work. This implies that the sensitivity scales with $r_0^2 \cdot \tau_0 \propto \lambda^{18/5}$ (for equal photon flux per bandpass).

7.7 Angular Anisoplanatism

The light from two stars separated by an angle θ on the sky passes through different patches of the atmosphere and therefore experiences different phase variations. This *angular anisoplanatism* limits the field corrected by adaptive optics systems and causes phase decorrelation for off-axis objects in interferometers. To calculate the effect of anisoplanatism, we trace back the rays to two stars separated by an angle θ from the telescope pupil. They coincide at the pupil, and their separation $r(d)$ at a distance d is $\theta \cdot d$. At zenith angle z, the distance is related to the height h in the atmosphere by $d = h \sec z$. To calculate the phase variance between the two rays, we insert this relation in

$$\langle |\phi(0) - \phi(r)|^2 \rangle = D_\phi(r) = 2.914\, k^2 \sec z\, \delta h\, C_N^2\, r^{5/3} \qquad (128)$$

(see (phasestruct)), integrate over the height h, and obtain:

$$\langle \sigma_\phi^2 \rangle = 2.914\, k^2 (\sec z) \int dh\, C_N^2(h)\, (\theta h \sec z)^{5/3}$$
$$= 2.914\, k^2 (\sec z)^{8/3} \theta^{5/3} \int dh\, C_N^2(h) h^{5/3} \qquad (129)$$
$$= \left(\frac{\theta}{\theta_0} \right)^{5/3},$$

where we have introduced the *isoplanatic angle* θ_0, for which the variance of the relative phase is $1\,\mathrm{rad}^2$:

$$\theta_0 \equiv \left[2.914\, k^2 (\sec z)^{8/3} \int dh\, C_N^2(h) h^{5/3} \right]^{-3/5}. \qquad (130)$$

By comparing the definitions for the Fried parameter r_0 and for θ_0, (110) and (130), we see that

$$\theta_0 = 0.314 \, (\cos z) \, \frac{r_0}{H} \;, \tag{131}$$

where

$$H \equiv \left(\frac{\int dh \, C_N^2(h) h^{5/3}}{\int dh \, C_N^2(h)} \right)^{3/5} \tag{132}$$

is the *mean effective turbulence height*. Equations (isodef) and (131) show that the isoplanatic angle is affected mostly by high-altitude turbulence; the anisoplanatism associated with ground layers and dome seeing is very weak. Moreover, we see that θ_0 scales with $\lambda^{6/5}$, but it depends more strongly on zenith angle than r_0. For $r_0 = 20$ cm and an effective turbulence height of 7 km, (131) gives $\theta_0 = 1.8$ arcsec. For two stars separated by more than θ_0 the short-exposure point spread functions (or point spread functions generated by adaptive optics) are different. In contrast the long-exposure point spread functions, which represent averages over many realizations of the atmospheric turbulence, are nearly identical even over angles much larger than θ_0.

It should be pointed out that these calculations of anisoplanatism give results that are somewhat too pessimistic. The reason is that a large fraction of the phase variance between the two rays considered is a piston term (i.e., a difference in phase that is constant across the aperture), which doesn't lead to image motion or blurring.[29] Moreover, anisoplanatism is less severe for low spatial frequencies, which most adaptive optics systems correct much better than high spatial frequencies. The degradation of the Strehl ratio with off-axis angle is therefore not quite as bad as suggested by inserting (129) in (123).

7.8 Scintillation

The geometric optics approximation of light propagation that was used in Sect. 7.2 is only valid for propagation pathlengths shorter than the *Fresnel propagation length* $d_F \equiv r_0^2/\lambda$. In other words, the *Fresnel scale*

$$r_F \equiv \sqrt{\lambda L} = \sqrt{\lambda h \sec z} \;, \tag{133}$$

where L is the distance to the dominant layer of turbulence, must be smaller than the Fried scale r_0. For $r_0 = 20$ cm and $\lambda = 500$ nm, $d_F = 80$ km. This is significantly larger than the height of the layers contributing much to the C_N^2 integrals, and the geometric approximation is a good first-order approach at good sites for visible and infrared wavelengths, as long as the zenith angle is

---

[29] Note, however, that piston terms have to be taken into account in interferometry, where they are responsible for fluctuations in the relative delay between the two stars.

not too large. ($d_F \propto \lambda^{7/5}$ for Kolmogorov turbulence; therefore the geometric approximation is even better at longer wavelengths.) However, if the propagation length is comparable to d_F or longer, the rays diffracted at the turbulence cells interfere with each other, which causes intensity fluctuations in addition to the phase variations. This phenomenon is called *scintillation*; it is an important error source in high-precision photometry unless the exposure times are sufficiently long to average over the fluctuations. Since scintillation is an interference phenomenon, it is highly chromatic. This effect can be easily observed with the naked eye: bright stars close to the horizon twinkle strongly and change color on time scales of seconds.

Although scintillation is weak for most applications of adaptive optics and interferometry, it has to be taken into account when high Strehl ratios are desired. High-performance adaptive optics systems designed for the direct detection of extrasolar planets have to correct the wavefront errors so well that intensity fluctuations become important. In interferometers that use fringe detection schemes based on temporal pathlength modulation and synchronous photon detection, scintillation noise has to be considered when very small fringe amplitudes are to be measured.

The effects of scintillation can be quantified by determining the relative intensity fluctuations $\delta I/I$; for small amplitudes $\delta I/I = \delta \ln I$. A calculation similar to the one in Sect. 7.2 gives the variance of the log intensity fluctuations:

$$\sigma_{\ln I}^2 = 2.24 \, k^{7/6} (\sec z)^{11/6} \int dh \, C_N^2(h) h^{5/6} \,. \qquad (134)$$

This expression is valid only for small apertures with diameter $D \ll r_F$. For larger apertures, scintillation is reduced by averaging over multiple independent subapertures. This changes not only the amplitude of the intensity fluctuations, but also the functional dependence on zenith angle, wavelength and turbulence height. The expression

$$\sigma_{\ln I}^2 \propto D^{-7/3} (\sec z)^3 \int dh \, C_N^2(h) h^2 \,, \qquad (135)$$

which is valid for $D \gg r_F$ and $z \lesssim 60°$, shows the expected strong decrease of the scintillation amplitude with aperture size; note that it is independent of the observing wavelength. For larger zenith angles the assumption $\delta \ln I \ll 1$ is no longer valid, the fluctuations increase less strongly with $\sec z$ than predicted by (135), and eventually saturate.

7.9 Turbulence and Wind Profiles

We have seen in the preceding sections that the most important statistical properties of seeing can be characterized by a few numbers: the Fried parameter r_0, the coherence time τ_0, the isoplanatic angle θ_0, and the scintillation index $\sigma_{\ln I}$. For the design and performance evaluation of high-angular-resolution instruments it is of great importance to have reliable statistical information on these parameters. Therefore extensive seeing monitoring campaigns are normally conducted before decisions are made about the site selection for large telescopes and interferometers, or about the construction of expensive adaptive optics systems. Having access to the output of a continuously running seeing monitor which gives the instantaneous value of r_0 (and ideally also of the other seeing parameters) is also very convenient for debugging and for optimizing the performance of high-resolution instruments.

From equations (110), (125), (129), and (134) it is obvious that all seeing parameters can easily be calculated from moments

$$\mu_m \equiv \int dh\, C_N^2(h) h^m \tag{136}$$

of the turbulence profile $C_N^2(h)$, and (in the case of τ_0) from moments

$$v_m \equiv \int dh\, C_N^2(h) v^m(h) \tag{137}$$

of the wind profile $v(h)$. More complicated analyses such as performance estimates of adaptive optics systems with laser guide stars and of multi-conjugate AO systems also rely on knowledge of $C_N^2(h)$ and $v(h)$. In-situ measurements of these profiles with balloon flights and remote measurements with SCIDAR[30] or related methods are therefore needed to fully characterize the atmospheric turbulence. Figure 41 shows profiles measured on Cerro Paranal, the site of the European Southern Observatory's Very Large Telescope observatory. The decrease of C_N^2 with height is typical for most sites; frequently wind shear at altitudes near 10 km creates additional layers of enhanced turbulence. The highest wind speeds normally occur at heights between 9 and 12 km. Extensive sets of observed turbulence and wind profiles, combined with the analytic methods sketched in this section and numerical simulations, form a firm basis for the evaluation of astronomical sites, and for the design of interferometers and adaptive optics systems.

8 Introduction to Optical Interferometry

The angular resolution required for detection of extrasolar planets drives us to very large telescope sizes. Scaling this approach to mid-infrared wavelengths is certainly impractical; we are thus compelled to consider interferometry as a means to achieve the required resolution. This chapter introduces the basic concepts of optical and infrared interferometry, beam combination schemes, and fringe detection methods. We will then take a look at ways to transfer phase knowledge from one baseline to another, or from one wavelength to another, and discuss the limitations that instrumental and atmospheric errors

---

[30] The SCIDAR technique is based on auto-correlating pupil images of double stars.

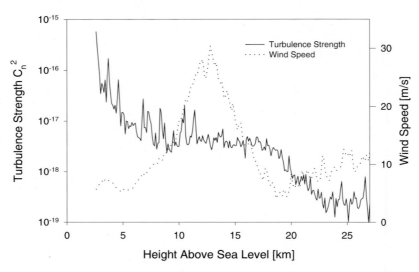

Fig. 41. Turbulence and wind profiles measured on Cerro Paranal, Chile. The turbulence is strongest close to the ground (2,635 m above sea level). The wind speed is highest at an altitude of ∼10 to 15 km. Wind shear often leads to additional layers of strong turbulence at high altitude (only weakly present in this data set)

impose on these techniques. These concepts will be applied to planet detection in the subsequent chapters. For a more detailed tutorial about optical interferometry the reader is referred to Lawson (2000); many details and references to the literature can also be found in the review by Quirrenbach (2001).

8.1 Schematic Design of an Optical Interferometer

Long-baseline interferometry is the coherent combination of light received with separate telescopes. This is shown schematically in Fig. 42. For the viewing direction indicated in this figure, light from a distant star arrives first at the telescope to the right, and a little later at the telescope to the left. The pathlength difference or *delay D* is given by the relation

$$D = \vec{B} \cdot \hat{s} \; , \tag{138}$$

where $\vec{B}$ is the *baseline vector* joining the two telescopes, and $\hat{s}$ the unit vector in the direction toward the star. D is of the order of the baseline length, i.e., up to tens or even hundreds of meters. This is much larger than the coherence length of the stellar light, which is given by $\lambda^2/\Delta\lambda$, where λ is the observing

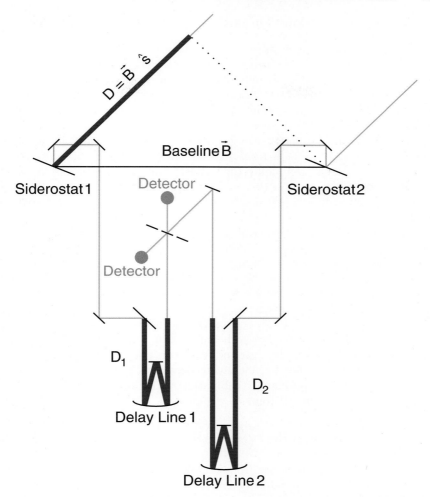

Fig. 42. Schematic drawing of the light path through a two-element interferometer. The external delay $D = \vec{B} \cdot \hat{s}$ is compensated by the two delay lines. The pathlengths D_1, D_2 through the delay lines are monitored with laser interferometers. The zero-order interference maximum occurs when the delay line positions are such that the internal delay $D_{\text{int}} = D_2 - D_1$ is equal to D

wavelength and $\Delta\lambda$ the bandwidth of the filter used for the observations. To observe interference fringes, it is therefore necessary to compensate the external delay D with an opposite internal delay.

$$V = \frac{I_{\max} - I_{\min}}{I_{\max} + I_{\min}}. \tag{139}$$

8.2 Beam Combination Concepts

The various beam combination schemes that can be employed in astronomical interferometry may be classified according to several criteria (?): the beam étendue (single-mode or multi-mode), the beam direction (co-axial or multi-axial), the combination plane (image plane or pupil plane), and the relation between input and output pupils (Michelson or Fizeau configuration, see below). For N telescopes in an array, there are $N(N-1)/2$ baselines. The $N(N-1)/2$ visibilities can either be measured by pairwise beam combination, or by bringing the light from all telescopes together on one detector. In the latter "all-on-one" techniques the fringes from the different baselines have to be encoded either spatially (by using a non-redundant output pupil) or temporally (by using different dither frequencies for the beams from individual telescopes).

Unlike in radio astronomy, where the radiation is detected and amplified before correlation, in an optical interferometer the beam combination occurs before detection. For pairwise beam combination the light from each telescope has to be divided in $(N-1)$ beams; in "all-on-one" schemes the visibility measurement for each baseline is affected by noise contributed by the $(N-2)$ other telescopes. This means that a baseline that is part of an N-element array is always less sensitive than an equivalent two-telescope interferometer. The detailed trade-offs between different beam combination schemes depend on the predominant noise source (background, detector, photon noise), detector cost and availability, and other technical considerations. Armstrong et al. (1998) discuss the case of pupil-plane combination with temporal encoding of the fringes, for which in the photon-rich regime

$$\mathrm{SNR} \propto \left(\frac{n_{\mathrm{phot}} N}{N_{\mathrm{out}}}\right)^{1/2} \frac{V}{N_{\mathrm{corr}}} , \qquad (140)$$

where n_{phot} is the photon rate from each telescope, N the number of array elements (equal to the number of input beams to the combiner), N_{out} the number of output beams from the combiner, and N_{corr} the number of input beams combined to produce each output beam. For pairwise combination, $N_{\mathrm{corr}} = 2$ and $N_{\mathrm{out}} = N(N-1)$. (Note that combining beams at beamsplitting surfaces produces two output beams.) For "all-on-one" combination $N_{\mathrm{corr}} = N$ and $N_{\mathrm{out}} = 2$. In both cases $\mathrm{SNR} \propto N^{-1/2}$, which demonstrates that multi-element arrays are indeed less sensitive than single-baseline instruments. Equation (140) gives a $\sim\sqrt{2}$ advantage of the "all-on-one" technique over pairwise beam combination, but the required temporal encoding of the fringes is difficult to realize technically.

Michelson and Fizeau Interferometers

In a Fizeau interferometer the output pupil is an exact replica of the input pupil, scaled only by a constant factor. This is also known as homothetic mapping between input and output pupil. In contrast, in a Michelson interferometer there is no homothetic relation between the input and output pupils.[31] This means that the object–image relationship can no longer be described as a convolution, because the rearrangement of the apertures rearranges the high-spatial frequency part of the object spectrum in the Fourier plane (Tallon and Tallon-Bosc 1992). This has an important consequence for off-axis objects: the image position does not coincide with the white-light fringe position (see Fig. 1 in Tallon and Tallon-Bosc 1992). For a finite spectral bandwidth this means that the fringe contrast decreases with field angle and the field-of-view is limited; the maximum size of an image from a Michelson interferometer is $\sim R \equiv \lambda/\Delta\lambda$ resolution elements in diameter. This effect is known as "bandwidth smearing" in radio astronomy (see Sect. 8.4). If a Michelson interferometer is used with image plane beam combination, and the visibilities are estimated by integration over each fringe peak, the field-of-view is additionally restricted to the size of one Airy disk of the individual telescopes.

8.3 Source Coherence and Interferometer Response

The *source coherence function* is defined as

$$\gamma(\xi_1, \xi_2, \tau) = \langle E(\xi_1, t) E^*(\xi_2, t - \tau) \rangle , \tag{141}$$

where E is the radiation field, and ξ the direction cosine on the sky. For $\xi_1 = \xi_2 = \xi$, γ is the time autocorrelation function of the radiation from direction ξ; for $\tau = 0$, this is the time-averaged brightness from that direction $\langle |E(\xi)|^2 \rangle$. An extended source is *spatially incoherent* if $\gamma = 0$ for $\xi_1 \neq \xi_2$; in this case

$$\gamma(\xi_1, \xi_2, \tau) = \gamma(\xi_1, \tau) \cdot \delta(\xi_1 - \xi_2) . \tag{142}$$

An interferometer with antennae at positions u_1, u_2 measures essentially the Fourier transform of γ:

$$\Gamma(u_1, u_2, \tau) = \int_{-\infty}^{\infty} \int_{-\infty}^{\infty} \gamma(\xi_1, \xi_2, \tau) \, e^{-2\pi i (\xi_1 u_1 - \xi_2 u_2)} d\xi_1 d\xi_2 . \tag{143}$$

In the spatially incoherent case, the interferometer response depends only on the *difference vector* of the antenna positions:

$$\Gamma(u, \tau) = \int_{-\infty}^{\infty} \gamma(\xi, \tau) \, e^{-2\pi i \xi u} d\xi . \tag{144}$$

---

[31] I take this as the definition of Fizeau and Michelson interferometers. Sometimes these terms are also used to mean "image plane" and "pupil plane" interferometer, respectively. In my nomenclature, it is possible to build an image plane Michelson interferometer.

The interferometer output at zero delay is called *complex visibility*; the complex visibility is the Fourier transform of the source brightness distribution:

$$V = \Gamma(u, 0) = \int_{-\infty}^{\infty} \left\langle |E(\xi)|^2 \right\rangle e^{-2\pi i \xi u} d\xi. \tag{145}$$

Each observation on one baseline measures one Fourier component of the sky brightness distribution.

8.4 Bandwidth and Interferometric Field-of-View

For monochromatic light, the interferometer response is:

$$F = \cos\left(\frac{2\pi B}{\lambda} \sin\theta\right) = \cos\left(\frac{2\pi B \xi}{\lambda}\right). \tag{146}$$

For a rectangular bandpass with width $\Delta\nu$, the response is:

$$\begin{aligned} F(\nu_0) &= \frac{1}{\Delta\nu} \int_{\nu_0-\Delta\nu/2}^{\nu_0+\Delta\nu/2} \cos\left(\frac{2\pi B \xi \nu}{c}\right) d\nu \\ &= \cos\left(\frac{2\pi B \xi \nu_0}{c}\right) \cdot \frac{\sin(\pi B \xi \Delta\nu/c)}{\pi B \xi \Delta\nu/c} \\ &= \cos\left(\frac{2\pi B \xi \nu_0}{c}\right) \cdot \mathrm{sinc}\left(\frac{\pi B \xi \Delta\nu}{c}\right), \end{aligned} \tag{147}$$

where we have again used the function $\mathrm{sinc}(x) \equiv \sin(x)/x$. This *bandwidth smearing* limits the field-of-view of the interferometer; the maximum size of the field is of order $R = \nu/\Delta\nu$ resolution elements:

$$\xi_{\max} \approx R \cdot \lambda/B. \tag{148}$$

8.5 Fringe Detection

Delay Modulation and ABCD Detection

In each of the spectral channels, arriving photons are counted synchronously with the delay modulation in bins corresponding to $\lambda/4$. (Since the physical stroke is equal to λ only in the channel with the longest wavelength, dead time has to be added in the electronics at the end of the stroke in the other three channels.) From the four bin counts A, B, C, and D (see in Fig. 43, the square of the visibility V^2 can be estimated using

$$V^2 = \frac{\pi^2}{2} \cdot \frac{\langle X^2 + Y^2 - N \rangle}{\langle N - N_{\mathrm{dark}} \rangle^2}, \tag{149}$$

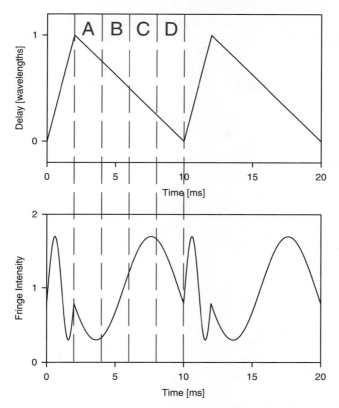

Fig. 43. ABCD fringe scanning scheme. The delay is modulated by a sawtooth pattern with amplitude one wavelength *(top)*. The time for the fringe scan is divided in four intervals of equal length, A, B, C, and D. The detector readout is synchronized with these four intervals, and the intensity integrated during each interval is measured. The fringe amplitude and phase can be derived from the four bin counts (see (149) and (150))

where $X = C - A$ and $Y = D - B$ are the real and imaginary parts of the visibility, $N = A + B + C + D$ is the total number of photons counted, and N_{dark} is the background count rate determined separately on blank sky. This estimator for V^2 is not biased by photon noise (Shao et al. 1988). The visibility phase is estimated from

$$\phi = \arctan\left(\frac{Y}{X}\right) - \frac{\pi}{4}. \qquad (150)$$

For delay modulation and synchronous detection of the photon count rate in n bins per wavelength of modulation, (149) can be generalized to

$$V^2 = \left(\frac{4\pi^2}{n^2 \sin^2(\pi/n)}\right) \frac{\langle X^2 + Y^2 - \sigma_N^2 \rangle}{\langle N - N_{\text{dark}} \rangle^2}, \qquad (151)$$

where X and Y are again the real and imaginary parts of the visibility constructed from the bin counts, N the total number of counts in all bins, σ_N^2 the variance of N due to noise (Benson et al. 1998).

Coherent and Incoherent Visibility Integration

The data are averaged using a combination of coherent and incoherent integrations.[32] By choosing a coherent integration time T, an observation of total duration $M \cdot T$ is divided into M intervals, which are averaged incoherently. The variance of the V^2-estimator (149) is then given by

$$\sigma^2 = \frac{\pi^4}{4MN^2} + \frac{\pi^2 V^2}{MN}, \qquad (152)$$

where N is the number of photons detected per coherent integration time (Colavita 1985). The signal-to-noise-ratio of V^2 is therefore:

$$\mathrm{SNR}(V^2) = \frac{2}{\pi^2} \cdot \frac{\sqrt{M} N V^2}{\sqrt{1 + \frac{4}{\pi^2} N V^2}}. \qquad (153)$$

If $NV^2 \gg 1$, the second term in (152) dominates, and the variance depends only on the total number of photons detected, MN. If, however, $NV^2 \ll 1$, the first term is the dominant one, and the variance for a given total duration of the observation (i.e., constant total number of photons MN) decreases with increasing coherent integration time, $\sigma^2 \propto N^{-1} \propto T^{-1}$; this implies that the signal-to-noise ratio of V^2 is $\propto T^{1/2}$ (for constant $M \cdot T$). We will call the two cases the "photon-rich" and "photon-starved" regimes, although NV^2, and not N, is the critical quantity.

The important results captured in (152) and (153) have a simple intuitive interpretation. If the coherent integration time is sufficiently long, we get a good estimate of the amplitude *and phase* of the complex visibility. We can then stop the coherent integration, write out V^2 for a data sample, and average over these samples later without losing sensitivity. This is the photon-rich regime. If we are forced to stop the coherent integration (e.g., because of variations in the atmospheric or instrumental phase) before we get a meaningful phase measurement, we can still estimate V^2 for each data sample, but averaging over these estimates gives the poorer signal-to-noise characteristic of the photon-starved regime.

---

[32] Coherent integration means that we sort each photon arriving during the integration time in one of the bins A, B, C, D, and use (149) to get an estimate of V^2. Incoherent integration means that we average over many estimates of V^2. The intuitive meaning is that the coherent integration is used to estimate both amplitude *and phase* of the visibility, whereas the incoherent integration averages over the *modulus* of the visibility.

While these considerations show that it is advantageous to choose T large enough to get into the photon-rich regime, values larger than a fraction of the atmospheric coherence time will lead to serious phase changes and therefore to unacceptable degradation of the visibility. In the MkIII "standard" data reduction for measurements of stellar diameters and binary stars, $T = 4$ ms is adopted, which gives a coherence loss of a few percent for seeing conditions typical for Mt. Wilson.

Visibility Calibration

The different interferometers use somewhat different observing strategies and calibration procedures, but they are all based on measurements of stars with known diameters. In the case of the MkIII, several calibrator stars were normally included in the observing list for each night. They were used to determine the "system visibility" V_{sys}^2, i.e., the value of V^2 observed for unresolved stars, as a function of seeing, zenith angle, time, and angle of incidence on the siderostat mirrors. For the seeing calibration, a seeing index S is calculated for each observation from the residual delay that the fringe tracker was unable to remove (Mozurkewich et al. 1991). After removing the relatively strong dependence of V^2 on S, calibration with respect to the other variables normally gave only a slight further improvement. (This situation is changed for phase-referenced data, where an additional strong decrease of V^2 with zenith angle has to be taken into account, see Sect. 8.6). The raw values of V^2 determined from (149) are then divided by V_{sys}^2 to obtain calibrated data V_{cal}^2 for further analysis. Both the internal noise, with contributions from photon noise and from short-term fluctuations, and the calibration uncertainty contribute to the error of V_{cal}^2. The two terms can be added in quadrature to obtain formal error bars.

Fringe Tracking Trade-Offs

The optimization of the fringe-tracking sensitivity is critically important for every interferometer. An important consideration in this regard concerns the trade-off between white-light fringe tracking and techniques based on dispersed fringes (le Poole and Quirrenbach 2002). White-light fringe tracking gives the highest sensitivity, because it allows for the simplest optical design (and thus for the highest optical throughput), and because it uses all available light with the smallest possible number of pixel-reads. On the other hand, white-light fringe tracking is also most sensitive to fringe mis-identification and fringe jumps. The best way to overcome this is by considering spectrally dispersed fringes. A compromise between these two conflicting requirements is to send part of the light to the white-light tracker, and part to a dispersed fringe sensor to locate the central fringe at a lower sampling rate. It is obvious that this works well only if fringe jumps are sufficiently rare, which in turn sets a limit on the minimum necessary signal-to-noise ratio and thus on the required brightness of the reference star. Detectors with intrinsic spectral resolution would be very advantageous to solve this dilemma; they could combine the advantages of white-light fringe tracking (efficiency and sensitivity) with a simultaneous capability to measure the group delay, and thus to ensure proper identification of the central fringe. Detector types that are currently under development include Superconducting Tunneling Junctions (STJs) and Superconducting Edge Sensors (Perryman and Peacock 2000; Romani et al. 1999; Bruijn et al. 2000). A fringe tracker based on an adaptation of these detector types for near-infrared observations with a spectral resolving power of $R \approx 20$ would produce an almost optimum solution for fringe tracking on faint sources.

Single-Mode Fibers and Modal Filtering

Single-mode optical fibers can be used for many of the functions required in an optical interferometer: beam transport, beam combination (in **X** couplers), modulation of the optical path difference (by physically stretching a fiber, Shaklan 1990), polarization control, and modal filtering. The last capability is particularly attractive; single-mode fibers can eliminate the decrease in fringe visibility caused by atmospheric turbulence and thus alleviate the calibration difficulties of ground-based interferometers. The coupling efficiency into single-mode fibers depends on the wavefront shape; it is roughly equal to the Strehl ratio of the input beam (Shaklan and Roddier 1988). The output of the fiber is a perfectly flat wavefront. Introducing a single-mode fiber in each of the interferometer arms thus converts atmospheric wavefront aberrations into intensity fluctuations. Splitting off some of the light from each telescope before beam combination in a **Y** coupler allows monitoring of I_1 and I_2 (Coudé du Foresto et al. 1997). The corrected interferogram

$$I_{\text{cor}} \equiv \frac{I_{\text{out}} - I_1 - I_2}{2\sqrt{I_1 I_2}} = \Gamma(u, 0) \tag{154}$$

is then independent of atmospheric wavefront degradation. By definition, the étendue of a single-mode fiber is λ^2, i.e., the field-of-view is limited to one Airy disk.

Phase-Referenced Visibility Averaging

The wide-band tracking channel in the MkIII Interferometer provides a phase reference, which can be used to extend the coherent integration time T beyond the limit imposed by the atmospheric turbulence. This method provides a means of obtaining substantially better signal-to-noise in the photon-starved regime, or even to make a transition into the photon-rich regime. The phase-referenced quantities X_r, Y_r, V_r, and ϕ_r are defined by

$$X_r + iY_r = V_r\, e^{i\phi_r} = V_s\, e^{i(\phi_s - \frac{\lambda_t}{\lambda_s}\phi_t)} , \tag{155}$$

where λ_s, V_s, ϕ_s are the wavelength, visibility, and phase in the signal channel, and λ_t, ϕ_t the wavelength and phase in the tracking channel. In practice, V_r^2 is computed from (149) using X_r and Y_r instead of X and Y; this procedure retains the advantage of using an unbiased estimator.

Equation (155) assumes that the atmospheric phase at λ_s is given by $(\lambda_t/\lambda_s)\phi_t$. If this were the case exactly, there would be no coherence losses, and the integration time could be arbitrarily long. A number of systematic effects (discussed in more detail in Sect. 8.6, see also Quirrenbach et al. 1994) can lead to a decorrelation of the phases between the signal and tracking channels, however. They introduce additional phase noise, which reduces the system visibility and limits the maximum integration time. The dependence of the system visibility on seeing and zenith angle is also made steeper, which increases the uncertainty of the calibration. In practice, therefore, phase-referenced averaging involves trading off some calibration accuracy for the gain in signal-to-noise.

8.6 Phase Decorrelation Mechanisms

Phase Errors and Coherence Losses

We will now discuss a number of mechanisms that lead to phase errors and therefore to coherence losses and to a reduction of the phase-referenced visibility. These effects can be broadly divided into two classes, namely those mechanisms that are due to errors in the determination of the phase in the reference channel, and those that are due to differential atmospheric propagation effects. While some of the former processes are instrument-dependent and can be reduced (or even avoided) by improved interferometer and fringe-detector designs, the latter class sets fundamental limits to the application of phase-referencing methods from the ground. We will again use phase-referenced visibility averaging with the MkIII Interferometer to give some specific numerical examples (see also Quirrenbach et al. 1994).

If the variance of the referenced phase ϕ_r associated with a decorrelation mechanism is $\sigma_{\phi,r}^2$, it will reduce V_r^2 by a factor η, which can be computed from

$$\eta = e^{-\sigma_{\phi,r}^2} . \qquad (156)$$

For assessing the individual mechanisms, it is not only important to compare the numerical values of the associated phase variances, but also to note their dependencies on observing conditions (e.g. seeing, zenith angle) and particularly on stellar parameters (e.g. colors). While the standard calibration procedure will correct for a uniform reduction of V^2, and to some extent for variations with observing conditions, effects that differ from star to star can introduce systematic errors that are difficult to detect. A priori limits on these effects are therefore necessary for practical applications of phase-referenced visibility averaging.

Photon Noise in the Tracking Channel

The finite number of photons detected during each coherent integration interval (4 ms in the MkIII case) sets a fundamental limit to the precision of the reference phase determination. The variance of ϕ_r due to photon noise in the tracking channel is

$$\sigma_{\phi,r}^2 = \left(\frac{\lambda_t}{\lambda_s}\right)^2 \sigma_{\phi,t,\text{phot}}^2 = \left(\frac{\lambda_t}{\lambda_s}\right)^2 \cdot \frac{2}{N_t V_t^2} \;, \tag{157}$$

where N_t and V_t are the number of the photons counted and the visibility in the tracking channel. $\sigma_{\phi,r}^2$ depends on the brightness and color of the star, and even on the baseline length (through V_t^2). However, for the fringe tracker to work reliably under average seeing conditions, $N_t V_t^2 \approx 70$ is needed for the 4 ms sampling interval, giving $\eta \approx 0.98$ for $\lambda_t = 700$ nm, $\lambda_s = 800$ nm, and $\eta \approx 0.95$ for $\lambda_t = 700$ nm, $\lambda_s = 500$ nm. Thus the visibility reduction is slight even for stars that are close to the sensitivity limit of the fringe tracker, and negligible for stars that are substantially brighter. It is also possible to introduce the signal-to-noise in the tracking channel as an additional independent variable in the calibration process, if very high accuracy is required.

Color and Visibility Dependence of the Effective Tracking Wavelength

To achieve high sensitivity (and to keep the errors due to photon noise small), the bandpass in the fringe-tracking channel should be made as wide as possible. The effective wavelength to be used in (155) is then given by

$$\lambda_t = \frac{\int d\lambda\, \lambda\, W_t(\lambda) N(\lambda) V(\lambda)}{\int d\lambda\, W_t(\lambda) N(\lambda) V(\lambda)} \;, \tag{158}$$

where $N(\lambda)$ is the number of photons emitted by a star as a function of wavelength, $V(\lambda)$ the visibility, and $W_t(\lambda)$ the combined response of atmosphere, instrument, and detector. If the wavelength used in (155) differs from the true effective wavelength by $\delta\lambda_t$, the resultant variance of the reference phase is

$$\sigma_{\phi,r}^2 = \left(\frac{\delta\lambda_t}{\lambda_s}\right)^2 \cdot \langle\phi_t^2\rangle \;. \tag{159}$$

As evident from (158), the true effective wavelength depends on stellar colors and diameters, and on the baseline length. If for simplicity one uses $\lambda_t = 700$ nm for all stars in the data reduction, $\delta\lambda_t \lesssim 25$ nm for the parameters of the MkIII Interferometer. With the additional assumption that the residual atmospheric phase r.m.s. not tracked by the fringe tracker $\sqrt{\langle\phi_t^2\rangle} \lesssim 2$ rad, $\eta \gtrsim 0.99$ is derived from (159).

Stroke Mismatch

In pathlength modulation schemes like that used by the MkIII, any difference between the stroke of the 500 Hz pathlength modulation and the wavelength λ will also lead to errors in the phase estimation, since then the bins A, B, C, and D do not correspond exactly to $\lambda/4$. (This correspondence is assumed implicitly in (150).) For each channel, the gating of the electronic counters for A, B, C, and D has to be set by the on-line control system to match one quarter of the nominal wavelength. In this way, an effective stroke s is created for each channel. Defining

$$\varepsilon = \frac{2\pi}{\lambda} \cdot (s - \lambda) \quad \text{and} \quad \delta = \frac{\cos \varepsilon/4}{1 + \sin \varepsilon/4}, \tag{160}$$

it has been shown by Colavita (1985) that

$$\tan \phi_{\text{est}} = \delta \cdot \tan \phi_{\text{true}}, \tag{161}$$

where ϕ_{est} is the phase estimated from (150), and ϕ_{true} is the true phase. For a complete treatment of the effect of the stroke mismatch, these equations have to be integrated over λ, with a suitable weighting function representing the bandpass of the tracking channel. To first order, however, it can be assumed that the phase error is given by (160) and (161), evaluated at $\lambda = \lambda_t$. For $s_t - \lambda_t \leq 25\,\text{nm}$, a phase error $\phi_{\text{est}} - \phi_{\text{true}} \leq 2°$ is then obtained. Errors of this order can be safely ignored for most visibility averaging applications, but may be important for phase-referenced imaging and spectroscopy.

Fringe Jumps

An ideal fringe tracker would follow the atmospheric pathlength fluctuations to a fraction of λ_t, and ϕ_t would always be well within the interval $(-\pi, \pi]$. In practice, however, temporary excursions from the central fringe that are larger than $\lambda/2$ may occur, and the phase has to be "unwrapped" by the phase-referencing algorithm. This is done by imposing the requirement that the phase in successive data segments (4 ms intervals for the MkIII) should be continuous. While this process normally works well, occasional misidentifications are possible. It is obvious from (155) that a 360° error in ϕ_t will lead to a phase jump in ϕ_r.

If the average number of these jumps during the coherent integration time T is small, the coherence loss is not dramatic. This requirement sets an upper limit to T. Since the probability of unwrapping errors depends only on the seeing and on the signal-to-noise in the tracking channel, it can be accounted for in the calibration procedure. In a series of tests with the MkIII, it turned out that the degradation of the phase-referenced visibility V_r due to fringe jumps was not serious for integration times up to 2 s, for average seeing conditions on Mt. Wilson.

Dispersion

While (155) assumes that the atmospheric pathlength fluctuations are independent of wavelength, they are actually larger in the blue spectral range than in the red, because of dispersion. The two-color dispersion coefficient D is defined by

$$D = \frac{n(\lambda_t) - 1}{n(\lambda_s) - n(\lambda_t)} \;, \qquad (162)$$

where $n(\lambda)$ is the refractive index of air at λ. Typical values for $\lambda_t = 700$ nm and $\lambda_s = 450, 500, 550$, and 800 nm are $D = 59, 87, 137$, and -364, respectively. If the total "unwrapped" phase in the tracking channel is denoted Φ_t, a phase error $(\lambda_t/\lambda_s)(\Phi_t/D)$ is introduced by the dispersion. Since the largest phase excursions occur on long time scales, this sets a limit to the coherence time. For Kolmogorov turbulence, the coherence time $t_{0,r}$ of ϕ_r is given by

$$t_{0,r} = |D|^{6/5} t_{0,s} \;, \qquad (163)$$

where $t_{0,s}$ is the atmospheric coherence time in the data channel (Colavita 1994). Under average conditions on Mt. Wilson, $t_{0,s}$ is of order 6 to 8 ms at 500 nm. For integration times up to about 2 s, the coherence losses due to dispersion are therefore tolerable for visibility averaging, and they can be taken into account by the calibration procedure.

It is obviously possible to deal with dispersion explicitly by using

$$\tilde{\phi}_r = \phi_s - \frac{\lambda_t}{\lambda_s} \phi_t - \frac{\lambda_t}{\lambda_s} \cdot \frac{\Phi_t}{D} \qquad (164)$$

instead of ϕ_r as defined in (155). While this approach can reduce the phase errors by a factor ~ 10, a residual effect due to water vapor fluctuations remains, because their dispersion is different from the values applicable to dry air.

Anisoplanatism

If the reference phase is measured on a star at an angular separation θ from the target object, there will be some decorrelation because the light from the two sources passes through different turbulence cells, as discussed in Sect. 7.7. In interferometric applications, the independent contributions from the two arms of the interferometer have to be taken into account. Under the assumption of a Kolmogorov turbulence spectrum, the interferometric isoplanatic angle therefore is

$$\theta_i = 2^{-3/5} \theta_0 = \left[5.82 \, k^2 (\sec z)^{8/3} \int_0^\infty dh \, C_N^2(h) h^{5/3} \right]^{-3/5} \;, \qquad (165)$$

where $k = 2\pi/\lambda$ is the wavenumber (assumed here to be equal for the target and reference channels), and z the zenith angle. While this expression holds for

Table 11. Astrometric signature from different planet / parent star combinations

planet	orbit [AU]	star	amplitude $\cdot\, d$ [μas $\cdot$ pc]
Earth	1	G2	3
Jupiter	5	G2	4,800
Uranus	20	G2	880
"hot" Jupiter	0.1	G2	96
Brown Dwarf	0.1	M5	7,500
Jupiter	5	A5	2,200
Jupiter	5	M5	24,000

Note that the numbers given in the last column have to be divided by the distance to give the astrometric signature. 15 $M_{\rm jup}$ was used for the mass of the brown dwarf

- The detection bias of astrometry with orbital radius is opposite to that of the radial-velocity method, favoring planets at larger separations from their parent stars.

It should be pointed out that for circular orbits the observed astrometric signal is an ellipse with semi-major axis θ independent of the orbital inclination; the mass of the planet can therefore be derived directly from (170) if the mass of the parent star is known. The situation is a bit more complicated for non-circular orbits, but even in that case can the orbital inclination be determined from the astrometric data with techniques analogous to those used for fitting orbits of visual binaries (e.g., Binnendijk 1960).

In Table 12, the numbers from Table 11 have been converted to the maximum distance at which the planet is detectable, for different assumptions about the astrometric precision.

9.2 Upper Mass Limits and Astrometric Detection of Gl 876 B

Looking at planets that are already known from radial-velocity surveys is an obvious first application of the astrometric technique, because of its ability to determine the planet's mass without $\sin i$ ambiguity. It is clear from that this is a challenging task with a precision of slightly better than a milliarcsecond, which is currently achievable with Hipparcos data or with the Hubble Fine Guidance Sensors. Observations of Gl 876 with the latter instrument resulted in the detection of the astrometric wobble due to its companion Gl 876 b, and thus mark the first secure astrometric detection of an extrasolar planet (Benedict et al. 2002). The parameters of this system are $\phi = 0.25 \pm 0.06$ mas, $i = 84° \pm 6°$, and $m_p = 1.9 \pm 0.5\, M_{\rm jup}$.

In many cases, interesting upper limits on the companion mass can be derived from astrometric observations even if the signature is below the detection

Table 12. Maximum distance to which different planets can be observed with a ground-based interferometer in astrometric mode, with either 50 µas or 10 µas precision

precision	planet	orbit [AU]	star	max. distance [pc]
50 µas	Jupiter	5	G2	48
	Uranus	20	G2	9
	Brown Dwarf	0.1	M5	75
	Jupiter	5	A5	22
	Jupiter	5	M5	240
10 µas	Jupiter	5	$1\,M_\odot$	240
	Uranus	20	G2	44
	Jupiter	5	A5	110
	10 Earths	1	G2	1.5
1 µas	Jupiter	0.1	G2	48
	Earth	1	G2	1.5

1 µas can probably be attained only from space. It is assumed that a 4σ peak-to-peak variation is required for secure detection of a planet

limit. A good example is the case of ι Dra b (Frink et al. 2002). The radial-velocity observations give $P = 536$ days, $e = 0.70$, and $m_p \sin i = 8.9\,M_{\rm jup}$ for this object. The non-detection in the Hipparcos data places a 3σ upper limit of $45\,M_{\rm jup}$ on the mass of ι Dra B, and thus firmly establishes its sub-stellar nature.

9.3 Astrometric Measurements with an Interferometer

Improving the astrometric precision by one to two orders of magnitude is required to make astrometry a truly powerful and versatile tool for planet detection. Dramatic progress is indeed expected from the development of interferometric techniques, which have the potential to achieve ∼10 µas from the ground, and ∼1 µas from space.

Principles of Interferometric Astrometry

Astrometric observations by interferometry are based on measurements of the delay $D = D_{\rm int} + (\lambda/2\pi)\phi$, where $D_{\rm int} = D_2 - D_1$ is the internal delay measured by a metrology system (see Fig. 42), and ϕ the observed fringe phase. Here ϕ has to be unwrapped, i.e., not restricted to the interval $[0, 2\pi)$. In other words, one has to determine which of the sinusoidal fringes was observed. This can, for example, be done with dispersed-fringe techniques (see Sect. 8.5). D is related to the baseline $\vec{B}$ by

$$D = \vec{B} \cdot \hat{s} = B \cos \theta , \qquad (171)$$

where $\hat{s}$ is a unit vector in the direction toward the star, and θ the angle between $\vec{B}$ and $\hat{s}$. Each data point is thus a one-dimensional measurement of the position of the star θ, provided that the length and direction of the baseline are accurately known. The second coordinate can be measured with a separate baseline at a roughly orthogonal orientation.

In a ground-based interferometer the endpoints of the baseline (i.e., the positions of the telescopes or siderostats) can be related to the solid ground of the site and thus tied to the Earth's rotation. One can either rely on the stability of the telescope mount, or, if greater precision is needed, use a truss of laser interferometers to monitor changes of the positions of the siderostat pivot points with respect to "optical anchors" attached directly to bedrock (Armstrong et al. 1998). Repeated observations of a set of stars throughout a night can then be used to determine the baseline vector $\vec{B}$ in the rotating reference frame of the Earth. With a sufficient number of observations for each star one obtains more observables (delays) than unknowns (stellar and baseline coordinates), so that one can solve for $\vec{B}$ from a set of over-constrained equations (Thompson et al. 1986).

In space, no convenient stable platform like the Earth is available. The Space Interferometry Mission (SIM, Sect. 9.6) therefore uses two additional interferometers that look at two stars to stabilize the spacecraft attitude; these guide interferometers are essentially extremely precise star trackers (Milman and Turyshev 2000). Since the spacecraft structure is not sufficiently stiff on the sub-µm scale, an "optical truss" is again formed by laser interferometers and used to monitor the exact position of all important optical elements, including the baseline length and the relative orientation of the main and guide interferometers.

Astrometric Precision

The photon noise limit for the precision σ of an astrometric measurement is given by the expression

$$\sigma = \frac{1}{\text{SNR}} \cdot \frac{\lambda}{2\pi B} . \qquad (172)$$

Since high signal-to-noise ratios can be obtained for bright stars, σ can be orders of magnitude smaller than the resolution λ/B of the interferometer. For example, the resolution of SIM ($B = 10$ m) is about 10 mas, but the astrometric precision of measurements with SIM should approach 1 µas.

To reach this photon noise limit, it is of course necessary to control all other statistical and systematic errors very precisely. Keeping track of all instrumental contributions to the final astrometric error is quite an arduous task; it is usually accomplished with a formal error budget. This is essentially a tree in which error sources are organized in a hierarchical structure. Each box in this error budget can be further subdivided in sub-boxes down to the noise of individual metrology detectors, surface errors of individual mirrors, or vibration spectra of the mounts of individual optical elements. Once this error tree has been established, it is possible to perform "top-down" or "bottom-up" analyses of the error budget. In a top-down error budget, one starts with a desired measurement accuracy, and divides the allowable error to individual systems, sub-systems, and so on to derive performance specifications for all components in the instrument. In a bottom-up analysis, one starts with known data or manufacturers' specifications for all components to arrive at a prediction of the final system performance. In the end, hopefully, both analyses agree, and one ends up with a plan for building an instrument that will perform to specifications. During the commissioning and operation of the instrument, the error tree is an important tool for debugging and implementation of improvements.

Radio and Millimeter Interferometry

The application of interferometric methods to astrometry is not limited to the visible and infrared wavelength ranges, of course. The technique of very long baseline interferometry (VLBI), in which telescopes on different continents are coupled coherently, has indeed produced astrometric data of such quality that the International Celestial Reference Frame has been based on radio sources. VLBI can also reach sub-milliarcsecond precision for the measurement of the photocenter position of radio stars (Lestrade 2000a). Unfortunately, the relation between the center of mass and the region from which the radio emission emanates, is frequently poorly understood, which makes the interpretation of the data difficult. Nevertheless, pilot VLBI observations of a number of dMe stars have been performed with the ultimate goal to search for planets around them (Guirado et al. 2002). The sub-millimeter array ALMA currently under construction in the Atacama desert in Chile will have sufficient sensitivity to detect the photospheres of nearby stars at 345 GHz and could thus also search for the astrometric signature of planets around them (Lestrade 2000b). However, the intrinsic faintness of stellar photospheres far in the Rayleigh-Jeans regime of their thermal emission limits the scope of such projects in comparison to the more promising astrometric programs at visible and near-infrared wavelengths.

9.4 Atmospheric Limitations of Astrometry

The Narrow-Angle Atmospheric Regime

The Earth's atmosphere imposes serious limitations on the precision that can be achieved with astrometric measurements from the ground. An obvious difficulty is atmospheric refraction (Gubler and Tytler 1998), but the effects

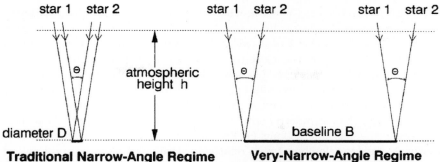

Fig. 46. Schematic of the atmospheric paths for narrow-angle astrometry with short and long baselines B (or telescope diameter D). In each panel, rays from two stars to the two telescopes at the ends of the baseline are shown. The atmosphere is represented by a single layer at height h. In this layer, the two rays originating from the same star to the two telescopes are separated by B; the two rays originating from the two stars to the same telescope are separated by θh. In the left panel the baseline B is short, $\theta h \gg B$, in the right panel the baseline is long, $\theta h \ll B$. From Shao and Colavita (1992)

of seeing are even more disturbing. In Sect. 7 we have discussed the blurring of optical images due to atmospheric turbulence. The first-order terms (frequently referred to as *tip* and *tilt*) of this blurring are global wavefront gradients, which correspond to a motion of the centroid of the stellar light in the two coordinates. Because most of the power of atmospheric turbulence is in these low-order modes (e.g. Hardy 1998), the amplitude of this image motion is similar to the width of the stellar images, i.e., $\approx \lambda/r_0 \approx 0\rlap{.}''5...1''$ (Sect. 7.4). One can obviously reduce this error by taking many exposures and thus averaging over many independent realizations of the atmospheric turbulence, but achieving a precision of a milliarcsecond or even better in this way is clearly not possible.

It helps, however, to make differential measurements over small angles on the sky, i.e., to measure the position of the target star with respect to that of a nearby reference (see Fig. 46). If the reference star is sufficiently close on the sky, the rays from the two stars are affected in almost the same way by the atmospheric turbulence, and the error in the relative position between the two stars is greatly reduced. From Fig. 46 it should be intuitively clear that the length of the baseline[33] also plays an important role.

---

[33] An analogous analysis can also be carried out for single telescopes, where the telescope diameter plays the role of the baseline length. It is indeed possible to perform precise narrow-angle astrometry with large telescopes (Pravdo and Shaklan 1996).

If the baseline is short ($B \ll \theta h$, where θ is the angle between the two stars, and h the effective height of the atmospheric turbulence, see the left panel of Fig. 46), all rays from star 1 pass close to each other through the atmosphere, and all rays from star 2 pass close to each other; the separation between the two ray bundles is large in comparison. This means that a localized patch of atmospheric turbulence could affect all rays from star 1, but leave those from star 2 unaffected. Therefore the image motions of the two stars are only weakly correlated, and the astrometric error is independent of the baseline length. In contrast, if the baseline is long ($B \gg \theta h$, right panel of Fig. 46), the rays from both stars to telescope 1 pass relatively close to each other, and those to telescope 2 pass close to each other. A localized patch of turbulence will thus tend to affect rays from star 1 and 2 in nearly the same way, which leads to a stronger correlation of the image motions, and therefore to an astrometric error that decreases with increasing B.

With a calculation similar to those demonstrated in Sect. 7 it can be shown that the variance σ_θ^2 of measurements of the angle θ is given by (Shao and Colavita 1992)

$$\sigma_\theta^2 \approx \frac{16\pi^2}{B^2 t} \int_0^\infty dh\, v^{-1}(h) \int_0^\infty d\kappa\, \Phi(\kappa, h) \left[1 - \cos(B\kappa)\right] \cdot \left[1 - \cos(\theta h \kappa)\right] \, , \quad (173)$$

provided that the integration time $t \gg \max(B, \theta h)/v$. Here $v(h)$ is the wind speed at altitude h, and $\Phi(\kappa, h)$ denotes the three-dimensional spatial power spectrum of the refractive index (see Footnote 28 on p. 132). It may at first seem surprising that stronger winds should give a smaller measurement error, but within the frozen-turbulence picture (see Sect. 7.5) a higher wind speed means that one averages faster over independent realizations of the stochastic refractive index fluctuations. Inserting a Kolmogorov power spectrum in (173) one obtains the two limiting cases

$$\sigma_\theta^2 \approx \begin{cases} 5.25\, B^{-4/3} \theta^2 t^{-1} \int_0^\infty dh\, C_N^2(h) h^2 v^{-1}(h) & \text{for } \theta \ll B/h, \quad t \gg B/v \\ 5.25\, \theta^{2/3} t^{-1} \int_0^\infty dh\, C_N^2(h) h^{2/3} v^{-1}(h) & \text{for } \theta \gg B/h, \quad t \gg \theta h/v \end{cases}$$
(174)

for long and short baselines, respectively. In particular we see that for sufficiently small angles θ the important scaling relations $\sigma_\theta \propto \theta$ and $\sigma_\theta \propto B^{-2/3}$ hold for the astrometric error σ_θ. This error is plotted as a function of θ for three different baseline lengths in Fig. 47; the knees in the curves mark the transition between the two limiting cases in (174). We see that for a good site such as Mauna Kea astrometric measurements with a precision of $\sim 10\,\mu$as are possible over angles of $\sim 10''$. It is also apparent from the factor h^2 under the integral in this equation that the astrometric error is dominated by the turbulence at high altitudes. The low level of high-altitude turbulence at the South Pole would therefore make an astrometric interferometer at a site on the high Antarctic plateau an attractive possibility (Lloyd et al. 2002).

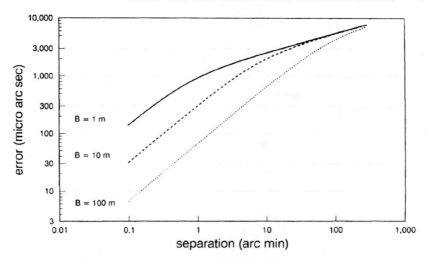

Fig. 47. Error of relative astrometric measurements for different values of the baseline length B, using a turbulence profile as measured for Mauna Kea, and assuming an integration time of 1 h. From Shao and Colavita (1992)

Influence of the Outer Scale of Atmospheric Turbulence

The calculation leading to (174), which is based on a Kolmogorov power law for the turbulence spectrum, actually gives somewhat pessimistic results. The reason is that in the Kolmogorov theory there must be an outer scale, beyond which the power in the turbulence spectrum flattens out (see Sect. 7.1). The mathematical treatment of this regime is much more involved than that of the inertial range. Two different parameterizations are in common use, namely the von Karman spectrum

$$\Phi(|\vec{\kappa}|) \propto \left(\kappa^2 + L_0^{-2}\right)^{-11/6} \tag{175}$$

and the Greenwood–Tarazano prescription

$$\Phi(|\vec{\kappa}|) \propto \left(\kappa^2 + \kappa/L_0\right)^{-11/6} . \tag{176}$$

It is obvious that in the limit of small spatial scales (large κ) both functional forms asymptotically approach the Kolmogorov spectrum $\Phi(|\vec{\kappa}|) \propto \kappa^{-11/3}$ (see Footnote 28 on p. 132).

Reliable measurements of the outer scale are very sparse, so that little is known about numerical values for L_0, much less about temporal variability of L_0 or which of the functional forms (175) or (176) is to be preferred (see Quirrenbach 2002b and references therein). Plausible typical values are in the range $L_0 \approx 10 \ldots 100$ m, but the uncertainty is very large. The best one can therefore currently do is calculate the astrometric errors for a range of outer scales – and base the design of astrometric instruments on the more conservative assumption of a Kolmogorov spectrum with infinite outer scale.

9.5 Dual-Star Interferometry

Simultaneous Observations of Two Stars

For faint objects one would like to emulate the phase calibration procedure widely used in radio astronomy in which the atmospheric phase is determined from a bright source near the target. In radio interferometry one can slew the telescope between target and reference in intervals of several minutes, but because of the short atmospheric coherence time at visible and near-infrared wavelengths, here the target and the reference have to be observed truly simultaneously. Off-source fringe tracking is therefore possible only in interferometers with a field much wider than feasible in a Michelson instrument; either a wide-field (e.g., Fizeau) setup or a *dual-star system* is required. The discussion of atmospheric limitations of ground-based interferometric astrometry (Sect. 9.4) also assumed that the target and reference are observed simultaneously, again calling for a wide-field or dual-star system.

In a dual-star interferometer, each telescope accepts two small fields and sends two separate beams through the delay lines (see Fig. 48). The delay difference between the two fields is taken out with an additional short-stroke differential delay line; an internal laser metrology system is used to monitor the delay difference (which is equal to the phase difference multiplied with $\lambda/2\pi$, of course). For astrometric observations, this delay difference ΔD is the observable of interest, because it is directly related to the coordinate difference between the target (subscript t) and reference stars (subscript r); from (171) it follows immediately that

$$\Delta D \equiv D_t - D_r = \vec{B} \cdot (\hat{s}_t - \hat{s}_r) = B(\cos\theta_t - \cos\theta_r). \qquad (177)$$

To get robust two-dimensional position measurements, observations of the target with respect to several references and with a number of baseline orientations are required.

Narrow-Angle Astrometry

Measurements of the delay difference between two stars give *relative* astrometric information; this means that the position information is not obtained in a global reference frame, but only with respect to the nearby comparison stars, which define a local reference frame on a small patch of sky. We have seen that this approach greatly reduces the atmospheric errors, and some instrumental requirements are also relaxed (see below). The downside is that the information that can be obtained in this way is more restricted, because the local frame may have a motion and rotation of its own. This obviously makes it impossible to measure proper motions. Moreover, all parallax ellipses have the same orientation and axial ratio, which allows only "relative parallaxes" to be measured.

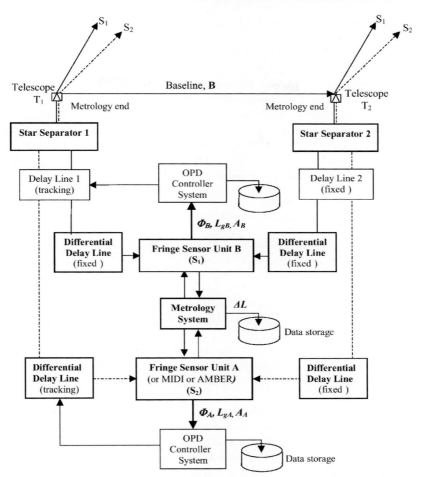

Fig. 48. Schematic setup of a dual-star interferometer. The star separators (also called dual-star modules) send the light from two stars down the main delay lines and separate differential delay lines into two beam combiners. The differential delay ΔL is measured by an internal metrology system, which measures the pathlengths backwards from the beam combiners to common metrology fiducials. The observed fringe phases $\Phi_{A,B}$, amplitudes $A_{A,B}$ and group delays $L_{gA,B}$ are recorded along with ΔL.

Narrow-angle astrometry is generally sufficient for planet detection, but there are a few caveats. First of all, if only one reference star is used, there is an ambiguity to which star an astrometric wobble has to be attributed. For example, if one chooses a distant star as reference, and this star happens to be an unrecognized binary, the resulting variation of the position difference could be mistaken for planetary signature of the much closer target (see (170)).

This can of course be avoided by using multiple reference stars, at the cost of an increase in observing time and somewhat reduced sensitivity. (Some of the signal-to-noise is used to sort out which of the references are binaries and which are not.) A somewhat more subtle effect is caused by the rotation of the local reference frame, due to uncertainties in the proper motions of the reference stars.[34] This rotation couples to the proper motion of the target star, and produces a spurious "Coriolis" acceleration, which could be mistaken for the signature of a planet in an orbit with a period longer than the time span covered by the observations. The detection of planets in long-period orbits with narrow-angle astrometry therefore requires accurate knowledge of the proper motions of the reference stars.

Anisoplanatism and Sky Coverage

The dual-star technique is being developed for interferometric astrometry and for phase-referenced visibility averaging or phase-referenced imaging. The most important problem encountered by these off-source phase-referencing techniques is anisoplanatism. The phase noise associated with anisoplanatism reduces the phase-referenced visibility dramatically if the distance to the reference source exceeds the isoplanatic angle. The need to find a reference object within the isoplanatic patch (165) is a severe limitation for off-source phase-referencing; the chances to find a suitably bright star for a randomly chosen target are typically one in a hundred or worse.[35] This is a substantial problem for interferometric observations of faint targets such as extragalactic objects or microlensing events.

The interferometric reference source can also be used for adaptive optics wavefront sensing, if such a system is available. In this case the whole entrance pupil of the interferometer is made fully co-phased and the sensitivity of the interferometer is essentially identical to the sensitivity of a single telescope with the same diameter. It is thus important to realize that bright objects are needed to co-phase an interferometer, but very faint sources can be observed in a limited field around these reference sources.

These principles can also be applied to astrometric observations; where we need to find astrometric reference stars near the intended target. In the case of astrometric planet searches, the target is a nearby and therefore bright star, which can be used for fringe tracking. The point is that now the astrometric

---

[34] For illustration purposes, assume that the local frame is defined by two stars. The star in the North has an Eastward proper motion, the star in the South a Westward one; these proper motions are not known to the observer. The reference frame in which these stars are at rest will have a counter-clockwise rotation.

[35] The probability depends strongly on the observing wavelength (because of the $\lambda^{6/5}$ scaling of the isoplanatic angle, (165)) and on the Galactic latitude. The stellar environment also plays a role, e.g. for observations in clusters or toward dark clouds.

references can be quite faint, because phase-referenced fringe tracking can be applied to them: the *astrometric target* is the *interferometric reference*. It is thus possible to measure the fringe phase on the astrometric reference stars with a co-phased instrument; the only limitation on their brightness is the photon noise contribution to the astrometric error budget. For a half-hour integration with 1.8 m apertures and an intended astrometric error of $10\ldots 20\,\mu$as this limit is about $m_K = 16\ldots 17$. The probability that suitable reference stars can be found will be further discussed in Sect. 9.8.

Instrumental Requirements

The fundamental instrumental requirements can be derived directly from (177), which can be written as

$$\Delta D \equiv D_t - D_r = \vec{B} \cdot (\hat{s}_t - \hat{s}_r) \equiv \vec{B} \cdot \Delta \vec{s}. \qquad (178)$$

The propagation of systematic errors in measurements of the differential delay $\delta \Delta D$ and of the baseline vector δB to errors in the derived position difference $\delta \Delta s$ can be estimated from the total differential

$$\delta \Delta s \approx \frac{\delta \Delta D}{B} + \frac{\Delta D}{B^2} \delta B = \frac{\delta \Delta D}{B} + \Delta s \frac{\delta B}{B}. \qquad (179)$$

This formula allows us to draw two important conclusions. First, the systematic astrometric error is inversely proportional to the baseline length. Together with the $B^{-2/3}$ scaling of the atmospheric differential delay r.m.s. (174) this clearly favors longer baselines, up to the limit where the target star gets resolved by the interferometer. The second important conclusion from (179) is that the relative error of the baseline measurement gets multiplied with Δs; this means that the requirement on the knowledge of the baseline vector is sufficiently relaxed to make calibration schemes possible that rely primarily on the stability of the telescope mount. For a 10 µas (50 prad) contribution to the error budget for a measurement over a 20″ angle, with an interferometer with a 100 m baseline, the metrology system must measure $\delta \Delta D$ with a 5 nm precision; the baseline vector has to be known to $\delta B \approx 50\,\mu$m (Quirrenbach et al. 1998). These are very demanding specifications, but within reach of the Very Large Telescope Interferometer and the Keck Interferometer.

9.6 Interferometric Astrometry from Space

The Space Interferometry Mission (SIM) and GAIA

While many interesting astrometric projects on extrasolar planets can be carried out from the ground (Sect. 9.7), overcoming the fundamental limitations imposed by atmospheric anisoplanatism requires going to space. Orbits that leave the immediate vicinity of the Earth (e.g., L2 orbits or drift-away orbits) provide the added bonus of a quiet environment with stable heat flux and low vibration levels. NASA's Space Interferometry Mission (SIM), to be launched in 2010, will exploit these advantages to perform a diverse astrometric observing program (NASA 1999a). SIM is essentially a single-baseline interferometer with 30 cm telescopes on a baseline of length 10 m. SIM is a pointed mission, i.e., targets can be observed whenever there is a scientific need (subject only to scheduling and Solar exclusion angle constraints), and the integration time can be matched to the desired signal-to-noise ratio. The limiting magnitude of SIM with "reasonable" observing times (of order one hour per visit) is thus $m_V \approx 19\ldots 20$. The main scientific driver of SIM is observing extrasolar planets, but a wide range projects addressing Galactic and extragalactic astrophysics will also be carried out (e.g., Quirrenbach 2002a).

The European Space Agency is planning to launch an astrometric satellite of their own, called GAIA, in roughly the same time frame as SIM. GAIA's architecture builds on the successful Hipparcos mission (Lindegren and Perryman 1996). Unlike SIM, GAIA will be a continuously scanning survey instrument with a large field of view, which results in an enormous number of observed stars. The main thrust of GAIA will thus be in the area of Galactic structure (Gilmore et al. 2000; Perryman et al. 2001), but it can also detect extrasolar planets (Lattanzi et al. 2000; Sozzetti et al. 2001). SIM and GAIA will be complementary to each other in this area. SIM will provide at least one order of magnitude better accuracy and the ability to obtain well-sampled orbits (especially of multiple systems) with suitably timed observations, whereas GAIA will potentially discover massive planets around a larger number of stars.

SIM Observing Modes

For any fixed spacecraft orientation, SIM will be able to access stars in a field with a diameter of ∼15°. Within each such "tile", a few stars will be selected before the mission to define an astrometric reference grid. A basic observing block will consist of observations of the target object(s) interleaved with observations of the grid stars in the tile; the measured delay differences will thus yield one-dimensional positions of the target(s) relative to the grid stars. The second coordinate can be obtained by rotating the spacecraft around the line of sight by an angle close to 90°. During the course of the mission, the grid stars will be visited regularly, about four to five times per year. By observing overlapping tiles, a full-sky reference grid can be constructed in the same manner as overlapping plates have been used to assemble all-sky astrometric catalogs. The inclusion of quasars in the grid will ensure that this reference system will represent an inertial frame. The expected performance of SIM in this "global astrometry" mode is 4 µas precision on the derived astrometric parameters (position, parallax) at the end of the nominal 5-year mission.

Many terms in the error budget of SIM measurements depend on the angle between the target and the astrometric references. It is thus possible to course of a several decade-long program that the astrometric accuracy gets severely degraded. It is obviously possible to increase the number of potential targets by requiring only one or two references, but then the risk gets high that the references turn out to be members of multiple systems, which can lead to unusable data, or worse, to false "planet" detections if the motion of the reference is ascribed to the target.

It is also possible to search for planets in double stars (Quirrenbach 2000a). The Washington Double Star Catalog (Worley and Douglass 1997) contains 745 F, G, and K main sequence stars with $\delta \leq 20°$ and $V \leq 10$ in pairs with separation between $5''$ and $20''$; 23 of these are G main sequence stars with $V \leq 7.5$. Most of these are members of wide physical binaries, and searching for planets in these systems is scientifically interesting (see Sect. 3.4) and technically somewhat less challenging than searches around single stars. The downside of this approach is the difficulty of determining to which of the two system components any detected planet belongs.

The bottom line is that the availability of nearby astrometric references is an important criterion for the selection of suitable targets for ground-based astrometric observations. Optimizing the sensitivity of the faint star channel is clearly very important, because this enhances the chances of finding astrometric references. The general considerations in this regard are similar to the optimum design of a fringe tracker (Sect. 8.5). In addition, one should use as many photons as possible; simultaneous operation in the H and K bands is therefore highly advantageous.

Reference Stars for the Space Interferometry Mission

The success of SIM also depends critically on the selection of suitable grid and reference stars. About $2,000\ldots 3,000$ stars distributed evenly over the sky are needed for the grid. These stars must be astrometrically stable (i.e., they must not show any motion other than parallax and linear proper motion) on the level of a few µas. For each narrow-angle target, at least three reference stars stable to better than 1 µas must be available within a $\sim 1°$ diameter field. Finding stars that meet these stability requirements is by no means a trivial task. They have to be identified and characterized before the launch of SIM, because the potential presence of even a fairly small number of unstable references can only be compensated by a dramatic increase in redundancy of the observations (Frink et al. 2001), which is very costly in terms of SIM observing time.

A key criterion for selecting suitable grid and reference stars for SIM is their distance d, because the wobble induced by planets or other unseen companions scales with $1/d$ (170). The best class of reference stars are therefore K giants, which are numerous even at high galactic latitudes, and intrinsically bright. Samples of candidate SIM reference stars can either be selected from existing astrometric catalogs (Frink et al. 2000a,b), or identified in a specialized survey (Patterson et al. 1999; Rhee et al. 2001). The case for distant K giants as good grid stars rests on a three-fold argument (Frink et al. 2001):

- The wobble due to planetary companions is sufficiently small.
- We know from the radial-velocity (RV) surveys that brown dwarfs are rare as companions to G dwarfs (the "brown dwarf desert", see Sect. 3.2 or e.g. Marcy and Butler 2000), which are the progenitors of K giants. Brown-dwarf companions to K giants will therefore also be rare.
- Stellar companions can be detected efficiently before the launch of SIM by an RV survey. This is a non-trivial statement as photospheric activity could corrupt precise RV measurements. A survey with the Hamilton Echelle Spectrograph at Lick Observatory has shown, however, that many K giants are sufficiently stable; about 2/3 of all K giants are drawn from a distribution with a mean of $\sim 20\,\mathrm{m\,s}^{-1}$ (see Fig. 49). This allows the detection of most stellar companions with only two or three RV data points.

It should thus be possible to define the SIM grid, and to gain confidence in its integrity, well before launch. The selection of narrow-angle reference stars is a more difficult problem because of the increased accuracy requirement and more limited search area for suitable stars. Observational programs aimed at identifying candidate reference stars for high-priority narrow-angle targets

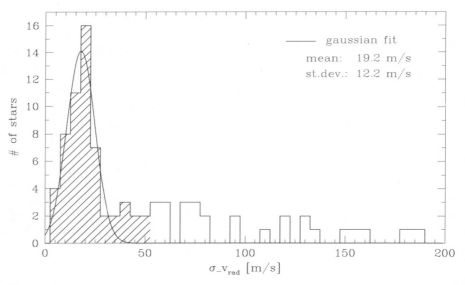

Fig. 49. Histogram of radial velocity scatter (i.e., dispersion of repeated RV measurements) observed in a sample of K giants (updated from Frink et al. 2001). About 2/3 of the observed stars have radial velocity scatter of $19.2 \pm 12.2\,\mathrm{m\,s}^{-1}$ and are good candidates for the SIM grid. Stars showing larger RV scatter could have companions and would not be included in the grid

have recently been started. Without this preparation it would not be possible to carry out an efficient planet detection program with SIM.

9.9 The Differential-Phase Method

Another interesting application of phase referencing, which is related to astrometry, consists of making differential phase measurements between different wavelengths (Akeson and Swain 1999; Quirrenbach 2000a). The near-infrared spectra of giant extrasolar planets should be characterized by extremely deep absorption bands of water and methane (e.g. Burrows et al. 1997a, see also Sect. 6.6). This opens the possibility of using wavelength-dependent astrometric information for the detection of extrasolar planets, and even to obtain their spectra.

The photocenter of a star–planet system is slightly different between two wavelengths, one of which falls in a region free of molecular bands, where the planet is relatively bright, and the other inside a band, where the planet is much fainter (see Fig. 50). Actually, the shift of the photocenter is proportional to the planet/star brightness ratio and can thus be used as a proxy for the planet spectrum (Quirrenbach 2000a). The shift of the photocenter gives rise to a corresponding wavelength dependence of the interferometer phase, which can be measured if the signal-to-noise ratio is sufficient and systematic effects are kept small. In the case of "hot Jupiters" like 51 Peg b, which is quite favorable because the planet is close to the star and therefore hot and bright, the expected effect on the interferometer phase is ≈ 0.5 mrad on the longest baselines of the VLTI (see Fig. 51). To measure such a small phase difference, a signal-to-noise ratio of $\approx 3,000$ is needed, and it remains to be seen whether

Fig. 50. The shift of the star–planet photocenter with wavelength gives rise to an interferometric phase shift that can be exploited to obtain a spectrum of the planet

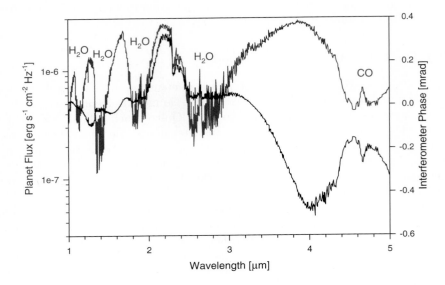

Fig. 51. Model spectrum of the planet 51 Peg b (red, from Sudarsky et al. 2003), and interferometer phase predicted for a 100 m baseline aligned with the star–planet separation vector (blue). The planetary spectrum is dominated by absorption bands of water and carbon monoxide. At short wavelengths ($\leq 2\,\mu\mathrm{m}$) it is very difficult to detect the planet because of the very high contrast between the star and the planet. The phase changes significantly across the K band near $2.2\,\mu\mathrm{m}$ due to water bands shortward and longward of the window in the Earth's atmosphere that defines this observing band. For the specific baseline length chosen, the phase goes through zero near $3.3\,\mu\mathrm{m}$

the systematic instrumental and atmospheric effects can be overcome at this level. For this technique, the dispersion in the air in the delay lines is a serious difficulty (Daigne and Lestrade 1999; Meisner and le Poole 2002). This problem can be overcome either by the use of a evacuated delay lines, or by making double differential measurements with respect to a nearby reference star

$$\phi_{dd} \equiv \phi_t(\lambda_1) - \phi_t(\lambda_2) - [\phi_r(\lambda_1) - \phi_r(\lambda_2)] \;, \tag{180}$$

where the subscripts t and r stand for "target" and "reference", respectively (le Poole and Quirrenbach 2002). If the reference star has no companion, its photocenter position is independent of wavelength, and the term $[\phi_r(\lambda_1) - \phi_r(\lambda_2)]$ is equal to the phase offset caused by dispersion, which is thus subtracted from the phase difference of the target in (180). But even with this double differential technique the spectroscopy of extrasolar planets will be a very challenging project.

10 Nulling Interferometry

The principal problem of direct planet detection is the large contrast between the planet and the parent star. Bracewell (1978) proposed to overcome this difficulty by using an interferometer to suppress the starlight. The key to the success of this method is the creation of an *achromatic null*, which ensures that the light arriving on axis interferes destructively at all wavelengths within the observing bandpass. Nulling interferometry in the mid-IR from the ground is a promising approach to the detection of exozodiacal dust disks. From space, the contrast and signal-to-noise ratio can be made sufficient for low-resolution spectroscopy of Earth-like planets. This is one of the leading architectures that have been proposed for the DARWIN and Terrestrial Planet Finder missions.

10.1 Principles of Nulling

Starlight Rejection in a Michelson Interferometer

For monochromatic light, the output intensity of a standard Michelson interferometer with an ideal 50% beam combiner varies with the phase ϕ as

$$I_{\text{out}} = I_{\text{in}}(1 + V \cos \phi), \quad (181)$$

where V is the fringe visibility and $I_{\text{in}} \equiv I_1 + I_2$ the sum of the intensities in the two interferometer input arms. This means that the intensity oscillates between $I_{\min} = (1-V)I_{\text{in}}$ and $I_{\max} = (1+V)I_{\text{in}}$ when the delay D is varied, in agreement with (139). If $V = 1$, we can set the delay line such that $\phi = 2\pi D/\lambda = 180°$ and get completely destructive interference, $I_{\text{out}} = 0$. For a non-zero bandwidth, we have to integrate the right-hand side of (181) over frequency; for $V = 1$ and a rectangular bandpass we thus get

$$I_{\text{out}} = \frac{1}{\Delta\nu} I_{\text{in}} \int_{\nu_0 - \Delta\nu/2}^{\nu_0 + \Delta\nu/2} \left[1 + \cos(2\pi D \nu/c)\right] d\nu$$

$$= I_{\text{in}} \left[1 + \cos\left(\frac{2\pi D \nu_0}{c}\right) \cdot \text{sinc}\left(\frac{\pi D \Delta\nu}{c}\right)\right], \quad (182)$$

in analogy to (147). If $\Delta\nu/\nu_0 \approx \Delta\lambda/\lambda \ll 1$, the cosine term in (182) varies much faster than the sinc term; the minima of the right-hand side therefore occur close to the delays for which the cosine term assumes the value -1. The condition for the first and deepest of these minima is $D = c/(2\nu_0)$, and we obtain

$$\frac{I_{\min}}{I_{\max}} \approx \frac{1}{2}\left[1 - \text{sinc}\left(\frac{\pi}{2}\frac{\Delta\lambda}{\lambda}\right)\right]. \quad (183)$$

For bandwidths of 10%, 30%, and 50%, the depth of the first minimum is 0.2%, 2%, and 5%, respectively. It is thus possible to reject most of the starlight by simply offsetting the delay line to the first interference minimum, but rejection factors of more than a few hundred are possible only with a very small bandwidth.[37] An achromatic method to generate destructive interference is therefore needed.

The Principle of Achromatic Nulling

If we can introduce a phase shift in the interferometer, which is exactly π rad at all wavelengths, the output signal from an interferometer is changed to

$$I_\text{out} = I_\text{in}(1 - V \cos \phi). \tag{184}$$

Now the intensity is zero if $V = 1$ and $\phi = 0$, i.e., the light from an on-axis point source is completely rejected at zero delay. This is the principle of *achromatic nulling*. A nulling interferometer can be used to detect extrasolar planets in the following way (see Fig. 52): The parent star is placed on the interferometer line-of-sight, and a fringe tracker assures that $\phi = 0$, so that no light from the star is received. The planet, on the other hand, is located at an off-axis angle which is comparable to the interferometer resolution λ/B. The light from the planet therefore has a significantly non-zero phase, leading to a detectable interferometer output according to (184). The nulling interferometer thus acts as an ideal coronographic mask, with complete rejection of the starlight and only moderate attenuation of the planetary signal.

In practice, the null is never perfect, of course, because there are always wavefront corrugations, phase fluctuations, and internal contrast losses, which reduce the visibility. This means that $I_\text{out} \neq 0$ even in the absence of planets. It is therefore necessary to modulate the signal from the planet, for instance by rotating the interferometer around its axis, which leads to a periodic modulation of the projected baseline length and therefore of ϕ and I_out. In this way it is possible to separate the constant term due to the starlight leak (or a uniform face-on dust disk) from the AC term due to the planet (see Fig. 53).

To characterize the quality of a nulling instrument and to quantify the leakage of unwanted photons, we introduce the *null depth*

$$N \equiv I_\text{min}/I_\text{max}. \tag{185}$$

For low-resolution spectroscopy of Earth-like planets with space-based interferometers an extremely deep null ($N \lesssim 10^{-6}$) is required; for the detection of exozodiacal dust disks $N \approx 10^{-3} \ldots 10^{-4}$ is sufficient.

---

[37] This conclusion is strictly valid only for single-baseline nulling interferometers. Mieremet and Braat (2002a) have shown that interferometric arrays with multiple telescopes can produce a deep broad-band null if appropriate delays are introduced in the interferometer arms.

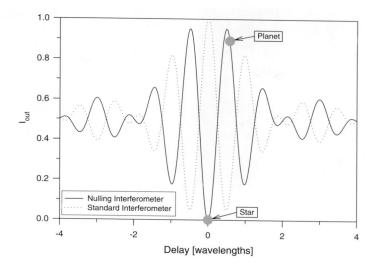

Fig. 52. Fringe pattern for a standard and a nulling interferometer. The fractional bandwidth $\Delta\lambda/\lambda = 0.5$. The pattern of the nulling interferometer has a central zero; the depth of the first minima of the standard interferometer are 5%. Fringe tracking ensures that the delay for the star is zero. The resolution of the interferometer is matched to the star–planet separation such that the planet is close to a transmission maximum

Symmetry and Stability Requirements

To produce a deep null, the two arms of the interferometer must be made symmetric with very high precision (apart from the π phase shift, of course). Any of the following imperfections can ruin the performance of the nuller: residual phase fluctuations, differences between the two arms in dispersion or polarization properties, a rotation between the two beams, and a mismatch between the two intensities (e.g., Serabyn 2000; Wallner et al. 2001). We denote the residual phase (i.e., the difference between the actual phase and the "best compromise" for all wavelengths within the bandpass and the two polarization states) by $\Delta\phi$, the rms phase difference due to dispersion mismatch, averaged over the bandpass, by $\Delta\phi_\lambda$, the phase difference between the two polarization states by $\Delta\phi_{s-p}$, the relative rotation angle between the two beams by α, and the normalized intensity mismatch by $\delta I/I \approx \delta \ln I$. A fairly straightforward analysis shows that the null depth is then given by (Serabyn 2000)

$$N = \frac{1}{4}\left[(\Delta\phi)^2 + (\Delta\phi_\lambda)^2 + \frac{1}{4}(\Delta\phi_{s-p})^2 + \alpha^2 + (\delta \ln I)^2\right]. \qquad (186)$$

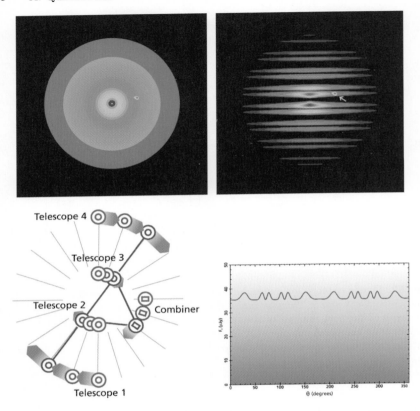

Fig. 53. *Top left*: model face-on planetary system with one planet and exozodiacal dust. *Top right*: the same system, multiplied with the response function of a linear four-element nulling interferometer. *Bottom left*: positions of the four telescopes and the beam combiner during rotation of the interferometer. *Bottom right*: output of the nuller for one full rotation. From NASA (1999b)

For illustration, we show how N can be calculated in the case that intensity mismatch is the only imperfection in the interferometer. $I_{\min}$ and $I_{\max}$ are then given by

$$\begin{aligned}
I_{\min} &= (I + \delta I) + (I - \delta I) - 2\sqrt{(I + \delta I)(I - \delta I)} \\
&= 2\left(I - \sqrt{I^2 - (\delta I)^2}\right) \\
&= 2I\left(1 - \sqrt{1 - \left(\frac{\delta I}{I}\right)^2}\right) \approx \frac{(\delta I)^2}{I} ,
\end{aligned} \quad (187)$$

and

$$I_{\max} = (I + \delta I) + (I - \delta I) + 2\sqrt{(I + \delta I)(I - \delta I)} \approx 4I , \quad (188)$$

from which we get

$$N \equiv I_{\min}/I_{\max} \approx \frac{1}{4}\left(\frac{\delta I}{I}\right)^2 = \frac{1}{4}(\delta \ln I)^2 \, , \qquad (189)$$

in agreement with the more general expression given in (186). To achieve a good null depth it is obviously necessary to control the optical path difference between the two interferometer arms, and to balance their intensities. If the mean values for $\Delta\phi$ and $\delta \ln I$ are kept at zero, the time-averaged null depth can be written as

$$\overline{N} = \frac{1}{4}\left[\sigma_\phi^2 + (\Delta\phi_\lambda)^2 + \frac{1}{4}(\Delta\phi_{s-p})^2 + \alpha^2 + \sigma_{\ln I}^2\right] \, , \qquad (190)$$

where σ_ϕ^2 is the residual phase variance and $\sigma_{\ln I}^2$ the variance of the residual intensity fluctuations. If we assume that the purely instrumental effects (dispersion, polarization, beam rotation) are relatively stable, the fluctuations of the null depth around its mean value are dominated by the fluctuations of the phase and intensity. It can then be shown that the variations of the null depth are given by (Serabyn 2000)

$$\sigma_N^2 = \frac{1}{8}\left(\sigma_\phi^4 + \sigma_{\ln I}^4\right) \, . \qquad (191)$$

If a certain depth of the null is desired, (190) and (191) can be used to construct an error budget for the various contributions to $\overline{N}$. A slight complication arises from the fact that the phase and intensity variations determine not only the mean null depth, but also the level of the fluctuations around this mean value. For example, if the null depth is dominated by phase fluctuations, a $+2\sigma$ excursion of N from its mean value means that the instantaneous null depth is

$$N_{2\sigma} = \overline{N} + 2\sigma_N = \left(\frac{1}{4} + \frac{2}{\sqrt{8}}\right)\sigma_\phi^2 \approx \sigma_\phi^2 \, . \qquad (192)$$

This means that specifying an error budget for $+2\sigma$ fluctuations requires placing a four times more stringent requirement on the phase variance than constructing an error budget for the mean null depth. The same argument obviously also applies to intensity variations. In the error budget one can allocate a maximum value N_i to each one of the contributing effects; the sum of these values must be smaller than the desired null depth. We thus get the set of requirements

$$\sigma_\phi < \sqrt{N_1} \qquad (193)$$

$$\sigma_{\ln I} < \sqrt{N_2} \qquad (194)$$

$$\alpha < 2\sqrt{N_3} \qquad (195)$$

$$\Delta\phi_\lambda < 2\sqrt{N_4} \qquad (196)$$

$$\Delta\phi_{s-p} < 4\sqrt{N_5} \qquad (197)$$

$$\sum_{i=1}^{5} N_i < \overline{N}. \qquad (198)$$

Wavefront Aberrations and Modal Filtering

In the previous section (186) and (190) we have implicitly assumed that at each wavelength and for each polarization there is a unique phase difference between the two interferometer arms. This is not necessarily the case, however, as aberrations and atmospheric turbulence distort the wavefronts and thus create phase variations across each of the apertures. Denoting the phase variances across the two pupils with $\sigma_{w_{1,2}}^2$, we can write in analogy to (190)

$$\overline{N} = \frac{1}{4}\left(\sigma_{w_1}^2 + \sigma_{w_2}^2\right) = \frac{1}{2}\sigma_w^2, \qquad (199)$$

where the second equality holds if the contributions from the two telescopes are equal. We will see in Sect. 10.3 that (199) places a requirement on the wavefront quality that is impossible to achieve even with state-of-the-art adaptive optics systems, if a null depth of $\sim 10^{-3}$ is desired at $10\,\mu\mathrm{m}$ in a ground-based interferometer. Similarly, it appears infeasible to produce optics that would allow a space interferometer to obtain a 10^{-6} null in this way.

It is thus necessary to introduce a modal filter in the interferometer, as described in Sect. 8.5 (Clark and Roychoudhuri 1979; Mennesson et al. 2002). Since the coupling efficiency into a single-mode fiber is roughly given by the Strehl number of the input beam, we can use (123) to calculate the intensity fluctuations

$$\delta I \equiv \frac{1}{2}(I_1 - I_2) \approx \frac{1}{2}\left(e^{-\sigma_{w_1}^2} - e^{-\sigma_{w_2}^2}\right) I_0 \approx \frac{1}{2}\left(\sigma_{w_2}^2 - \sigma_{w_1}^2\right) I_0. \qquad (200)$$

The second approximation is valid only if $\sigma_w^2 \ll 1$, which is the case for nulling interferometers in space. Inserting this result in (186) gives

$$N = \frac{1}{4}\left(\frac{\delta I}{I}\right)^2 = \frac{1}{16}\left(\sigma_{w_2}^2 - \sigma_{w_1}^2\right)^2. \qquad (201)$$

It is thus the *difference* between the wavefront qualities in the two beams that matters for the null depth, but in most practical circumstances this difference is directly related to the wavefront quality itself. It is also worth pointing out that (201) implies that $N \propto \lambda^{-4}$, which is a very steep scaling with the observing wavelength.

Pointing Requirements

If the telescopes are not pointed exactly at the target star, the phase across the pupil is not constant, and we expect an associated null leakage. If there is a pointing error $\delta\theta$ in the x-direction, the phase varies across the pupil according to

$$\phi(x) = \frac{2\pi x}{\lambda}\delta\theta = \frac{2\pi r \cos\psi}{\lambda}\delta\theta , \qquad (202)$$

where we have introduced polar coordinates (r, ψ). The phase variance is thus

$$\begin{aligned}\sigma_w^2 &= \frac{1}{\pi R^2}\int_0^{2\pi} d\psi \int_0^R r\,dr\,\phi^2 \\ &= \frac{1}{\pi R^2}\frac{4\pi^2(\delta\theta)^2}{\lambda^2}\int_0^{2\pi} d\psi \int_0^R dr\,r^3 \cos^2\psi \qquad (203) \\ &= \left(\frac{\pi D}{2\lambda}\right)^2 (\delta\theta)^2 .\end{aligned}$$

If there is no modal filtering, the resulting null depth can be computed by inserting (203) in (199)

$$\overline{N} = \frac{1}{8}\left(\frac{\pi D \sigma_\theta}{\lambda}\right)^2 . \qquad (204)$$

In the case of modal filtering we have to insert (203) in (201) and obtain

$$\overline{N} = \frac{1}{256}\left(\frac{\pi D}{\lambda}\right)^4 \left\langle \left[(\delta\theta_1)^2 - (\delta\theta_2)^2\right]^2 \right\rangle . \qquad (205)$$

To simplify this expression we note that for Gaussian variables χ with zero mean and standard deviation σ_χ

$$\langle \chi^4 \rangle = 3\sigma_\chi^4 . \qquad (206)$$

Therefore

$$\left\langle \left[(\delta\theta_1)^2 - (\delta\theta_2)^2\right]^2 \right\rangle = \langle(\delta\theta_1)^4\rangle - 2\langle(\delta\theta_1)^2(\delta\theta_2)^2\rangle + \langle(\delta\theta_2)^4\rangle = 4\sigma_\theta^4 . \qquad (207)$$

Inserting this result in (205) we finally obtain the desired expression for the null depth

$$\overline{N} = \frac{1}{64}\left(\frac{\pi D \sigma_\theta}{\lambda}\right)^4 . \qquad (208)$$

To achieve a deep null it is therefore necessary to stabilize the telescope pointing at a rather small fraction of the width of an Airy disk, which is of order λ/D.

Leakage from the Stellar Disk

Up to this point we have computed limitations of the null depth due to a variety of instrumental effects, under the assumption that we want to reject light from a point source. This assumption is violated if the star is partially resolved by the interferometer; in that case the light from the stellar limb arriving at an off-axis angle is not completely rejected even by a perfect nulling device. To first order (ignoring limb darkening) the star can be modeled as a uniform disk of angular diameter $\theta_{\rm dia}$ and normalized intensity

$$I(\Omega) = \left(\frac{\pi \theta_{\rm dia}^2}{4}\right)^{-1}. \tag{209}$$

The fringe phase for light emanating from a point on the stellar surface with polar coordinates (θ_r, ψ) is

$$\phi = \frac{2\pi B \sin(\theta_r \cos \psi)}{\lambda} \approx \frac{2\pi B \, \theta_r \cos \psi}{\lambda}. \tag{210}$$

The null depth can easily be calculated by integrating ϕ^2 over the stellar surface, which results in

$$\begin{aligned}
N &= \frac{1}{4} \int d\Omega \, \phi^2(\theta_r, \psi) I(\Omega) \\
&= \frac{1}{4} \int_0^{\theta_{\rm dia}/2} \int_0^{2\pi} \theta_r d\theta_r d\psi \, \phi^2(\theta_r, \psi) I(\Omega) \\
&= \frac{1}{\pi} \left(\frac{2\pi B}{\theta_{\rm dia}\lambda}\right)^2 \int_0^{\theta_{\rm dia}/2} \int_0^{2\pi} \theta_r d\theta_r d\psi \, \theta_r^2 \cos^2 \psi \\
&= \frac{1}{\pi} \left(\frac{2\pi B}{\theta_{\rm dia}\lambda}\right)^2 \int_0^{2\pi} d\psi \cos^2 \psi \int_0^{\theta_{\rm dia}/2} d\theta_r \theta_r^3 \\
&= \frac{\pi^2}{16} \left(\frac{B \theta_{\rm dia}}{\lambda}\right)^2.
\end{aligned} \tag{211}$$

This expression relates the null depth to the ratio of the angular diameter and the resolution of the interferometer λ/B. To get a quick overview of the importance of this stellar leak for stars of different types, it is convenient to derive a scaling relation with stellar magnitude and effective temperature (Quirrenbach 2000b). For observations in the infrared ($\lambda \gtrsim 2.2\,\mu$m) we can use the Rayleigh–Jeans approximation of blackbody radiation

$$\mathcal{F} \propto \theta_{\rm dia}^2 \cdot T_{\rm eff} \tag{212}$$

to rewrite (211) as

$$N = 0.26 \left(\frac{B}{100\,{\rm m}}\right)^2 \left(\frac{\lambda}{2.2\,\mu{\rm m}}\right)^{-2} \left(\frac{T_{\rm eff}}{10,000\,{\rm K}}\right)^{-1} \cdot 10^{-0.4\,m_{\rm K}}, \tag{213}$$

where m_K is the K-band magnitude of the star. The leakage from the stellar disk is tolerable for ground-based nulling if the baseline is not too long, but it may limit the null depth for the very brightest stars. For space interferometers with a requirement of $N \approx 10^{-6}$ the stellar leakage seems to be an insurmountable obstacle, and indeed it is necessary to employ more complicated nulling schemes involving multiple baselines. These will be discussed in Sect. 10.4.

10.2 Implementation of Achromatic Phase Shifts

There are two fundamentally different approaches to implement achromatic phase shifts. The first is the introduction of a "geometric" phase shift, in most cases by 180°. Techniques based on this principle are truly achromatic in that they manipulate the phase, not the delay. The alternative is the use of dispersive elements to create "pseudo-achromatic" phase shifts; the goal here is find a combination of different materials that approximates a constant phase shift over a broad band. Both methods have shown great potential, and they will now be discussed in turn.

Geometrical Field Reversal

The basic principle of creating a field flip by purely geometrical means is illustrated in Fig. 54. The beams in the two arms of the interferometer are sent

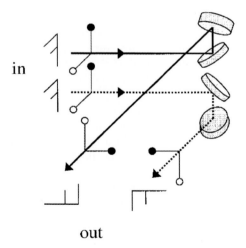

Fig. 54. Effect of a mirror-reflected pair of right-angle periscopes. Each beam encounters two mirrors at the locations of the 90° folds. Both the apertures and the fields undergo a relative rotation of 180°, as shown schematically by the clock hands and the letters "F". Each polarization component undergoes one s-plane and one p-plane reflection. From Serabyn and Colavita (2001)

through right-angle periscopes that are mirror-images of each other. This leads to an inversion of the apertures and of the relative direction of the electric fields in the two beams, which is equivalent to a 180° phase shift. Since this phase shift is achieved by reflections, it is strictly achromatic. In every other respect the field inverter is fully symmetric, e.g., each polarization component undergoes one s-plane and one p-plane reflection at a 45° incidence angle. This symmetry facilitates the task of meeting the stringent tolerances discussed in Sect. 10.1.

After passage through the pair of periscopes, the light from the two interferometer arms is sent to a pupil plane beam combiner. It is important again to make the combiner as symmetric as possible, to keep δI and $\Delta\phi_{s-p}$ as small as possible. A single pass through the beam splitter as drawn in Fig. 42 is not suitable, because it is impossible to manufacture a beam splitter for which the reflectivity and transmissivity are exactly equal to each other over a large wavelength range and for both polarizations. It is thus necessary to design a beam combiner in which each beam undergoes one reflection and one transmission before emerging at the nulled output. A further subtlety arises from the point that beam splitters are multi-layer dielectric films on a transparent substrate. In the presence of slight internal absorption, the reflectivity of the beam splitter is different between the front and back sides, whereas the transmissivity does not depend on the direction.[38] Figure 55 shows an example of a fully symmetric beam combiner, based on a classical Mach-Zehnder interferometer. Note that each of the beams emanating at the balanced outputs (shown as solid heave arrows) undergoes two reflections at a mirror, one reflection at the front side of the beam splitter (r), and one transmission through the beam splitter. The transmissions are in different directions (front side first (t), and back side first (t'), respectively), but this doesn't matter since $t = t'$. The combination of the field inverter (Fig. 54) and the beam combiner (Fig. 55) is thus fully symmetric by design; the null depth is only limited by imperfections of the coatings, and the quality and alignment of the optical elements.

A variation on the theme of geometric field reversal is the *rotational shearing interferometer*. Here the functions of field reversal and beam combination are not separated, but are both done in a modified Michelson interferometer (Serabyn 1999; Serabyn et al. 1999). The field reversal can for example be performed by rooftop (i.e., V-shaped) mirrors in the two interferometer arms. One of the rooftops is oriented vertically, the other horizontally, so that the two beams are rotated by 180° with respect to each other (see Fig. 56).[39] Although this arrangement is not fully symmetric (one of the beams undergoes

---

[38] This is the "Left-and-Right Incidence Theorem", see e.g. Knittl (1976).

[39] Take two identical copies of a written page. Flip one of them around a horizontal axis, the other around a vertical axis. The two pages now have a relative rotation of 180°.

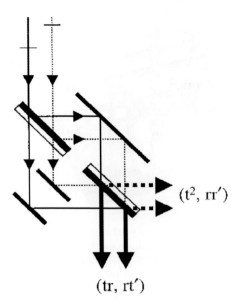

Fig. 55. Symmetric beam combiner derived from a classical Mach-Zehnder interferometer. At zero optical path difference, constructive interference occurs at the balanced outputs (shown as solid heavy arrows). In conjunction with a prior field flip these balanced outputs become nulled outputs at zero OPD. The pair of markers on the input beams indicate the wavefront offset needed for pathlength matching at the outputs. From Serabyn and Colavita (2001)

a front-side (r) reflection, the other back-side (r') reflection), it has been used to demonstrate a $\overline{N} = 7 \cdot 10^{-5}$ null in the laboratory with broadband visible light (Wallace et al. 2000).

Asymmetric Passage through Focus and Use of Berry's Phase

An alternative way of introducing a geometric phase shift of 180° is the introduction of a focus in one arm of the interferometer (Gay and Rabbia 1996). This method should in principle work well, but because of its intrinsic asymmetry it is probably more difficult to achieve a very deep null in this way than by symmetric field inversion.

Yet another geometric method is based on an effect known as Pancharatnam's phase or Berry's phase (Pancharatnam 1956). The essence of this phenomenon is that converting two beams from identical initial polarization states to identical final polarization states via different intermediate states (with a suitable arrangement of polarizers, quarter-wave plates etc.) in general results

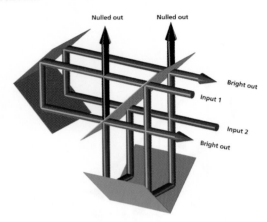

Fig. 56. Schematic layout of the beam paths in a rooftop-based rotational shearing interferometer. As a result of the double-pass beam splitter, the two input beams to be combined (red and blue) yield four output beams, two of which (black) are nulled at zero optical path difference and two of which add constructively (violet). From Serabyn (1999)

in a phase shift between the two beams.[40] Nulling with Pancharatnam's phase has been demonstrated experimentally in the laboratory; the most important technical challenge is obtaining fully achromatic wave plates (Baba et al. 2001).

Phase Shift with Dispersive Elements

If we had a plane-parallel plate of thickness s, made of a material with strictly linear dispersion, $n(\lambda) = n_0 + a\lambda$, we could use it to create an achromatic phase shift. Inserting the plate in one of the interferometer arms, and compensating for the fixed delay $D_0 \equiv (n_0 - 1) \cdot s$, the resulting phase would be

$$\phi = \frac{2\pi}{\lambda} \cdot a\lambda \cdot s = 2\pi a s , \qquad (214)$$

which is obviously independent of λ. Unfortunately real optical materials do not have linear dispersion; it is therefore not possible to obtain a truly achromatic null in this way. For a plate with refractive index $n(\lambda)$, the general expression for the phase is

$$\phi(\lambda) = \frac{2\pi}{\lambda} \cdot \left[D + \big(n(\lambda) - 1\big) \cdot s \right] , \qquad (215)$$

---

[40] The acquired phase difference is equal to $-1/2$ times the area enclosed by the two paths of the polarization state on the Poincaré sphere (for definition see e.g. Born and Wolf 1997). Pancharatnam's phase is closely analogous to the well-known Aharonov–Bohm effect, according to which two electron beams acquire a relative phase proportional to the magnetic flux they enclose (Berry 1987).

where D is some extra pathlength introduced by the delay line. If we pick two wavelengths λ_1, λ_2 and a desired phase shift $\tilde{\phi}$, (215) becomes a system of two equations with two free variables D and s, which can thus be chosen such that $\phi(\lambda_1) = \phi(\lambda_2) = \tilde{\phi}$.[41] The residual phase error $\phi(\lambda) - \tilde{\phi}$ at $\lambda \neq \lambda_1, \lambda_2$ is then due to the second order curvature of the dispersion $n(\lambda)$. This concept can easily be generalized to an arbitrary number N of plates made of different materials, for which

$$\phi(\lambda) = \frac{2\pi}{\lambda} \cdot \left[D + \sum_{i=1}^{N} (n_i(\lambda) - 1) \cdot s_i \right], \quad (216)$$

which allows for perfect phase adjustment at $N+1$ distinct wavelengths, $\phi(\lambda_j) = \tilde{\phi}$ for $j = 1 \ldots N+1$. One can choose the λ_j optimally spaced across the bandpass of interest, and thus get $\phi(\lambda) \approx \tilde{\phi}$ with very small errors, because now the dispersion can be balanced up to order N in the Taylor expansion of the $n_i(\lambda)$. Many different glasses are available at visible wavelengths to implement this approach, but the choices in the mid-infrared are much more limited. Nevertheless, just two materials (ZnSe and ZnS) are sufficient to achieve a nearly achromatic phase shift consistent with a 10^{-5} null across a $7\,\mu\text{m} \ldots 19\,\mu\text{m}$ bandpass (Morgan et al. 2000). A nulling interferometer working in the $8\,\mu\text{m} \ldots 13\,\mu\text{m}$ band has been tested on the Multiple Mirror Telescope (Hinz et al. 1998). In this instrument a single ZnSe plate was used for the phase shift. This would ideally have allowed to reach a null depth $N \approx 10^{-4}$, but the observed null depth was limited to only 0.06 by atmospheric turbulence.

One obvious difficulty with the plane-parallel plate setup are manufacturing tolerances; one needs plates of precisely prescribed thickness. It is possible, however, to adjust the effective thickness by introducing a slight tilt of the plate (Mieremet et al. 2000).[42] The delay $D(\lambda)$ introduced by a tilted plate can be calculated by referring to Fig. 57; we get

$$\begin{aligned} D(\lambda) &= n(\lambda)\,\overline{AB} + \overline{BC} - \overline{AD} \\ &= \frac{n(\lambda)s}{\cos\beta} + s\sin\alpha(\tan\alpha - \tan\beta) - \frac{s}{\cos\alpha} \\ &= \frac{n(\lambda)s}{\cos\beta}\left(1 - \sin^2\beta\right) + \frac{s}{\cos\alpha}\left(\sin^2\alpha - 1\right) \quad (217) \\ &= s\sqrt{n^2(\lambda) - n^2(\lambda)\sin^2\beta} - s\cos\alpha \\ &= \left(\sqrt{n^2(\lambda) - \sin^2\alpha} - \cos\alpha\right) s. \end{aligned}$$

---

[41] Note that we can make D or s negative by introducing the extra delay or the plate in the other arm of the interferometer. For symmetry reasons, one should in any case put a plate in each arm; in that case s is the thickness difference between these plates.

[42] Note that there is a typo in (10) of Mieremet et al. (2000). It should read $W = n_1 \cos(\alpha - \beta) d_1 / \cos(\alpha)$.

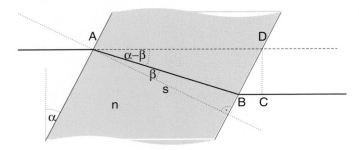

Fig. 57. Light path through a tilted plate with refractive index n. The angles α and β are related by Snell's Law $\sin\alpha = n\sin\beta$

Expansion of this expression in powers of the tilt angle α gives for $\alpha \ll 1$

$$D = (n-1)\left(1 + \frac{1}{2}\frac{\alpha^2}{n}\right)s \;,\quad \frac{dD}{d\lambda} = \left(1 + \frac{1}{2}\frac{\alpha^2}{n^2}\right)s\frac{dn}{d\lambda}; \qquad (218)$$

this means that tilting the plate affects not only its effective thickness but also the effective dispersion.

The phase shifter can also be implemented with a pair of prisms in place of the plane-parallel plate (Bokhove et al. 2002). The pathlength through a pair of prisms (see Fig. 58) is

$$\begin{aligned}
D(\lambda) &= n(\lambda)\left(\overline{AB} + \overline{CD}\right) + \overline{BC} \\
&= n(\lambda)\left(s - h\tan\alpha\right) + g\frac{\cos\alpha}{\cos\beta} \\
&= n(\lambda)\left(s - g\frac{\cos\alpha}{\cos\beta}\sin(\beta-\alpha)\tan\alpha\right) + g\frac{\cos\alpha}{\cos\beta} \\
&= n(\lambda)\left(s + g\frac{\sin\alpha}{\cos\beta}(\sin\alpha\cos\beta - \cos\alpha\sin\beta)\right) + g\frac{\cos\alpha}{\cos\beta} \qquad (219) \\
&= n(\lambda)\left(s + g\sin^2\alpha\right) - g\frac{\cos\alpha\sin^2\beta}{\cos\beta} + g\frac{\cos\alpha}{\cos\beta} \\
&= n(\lambda)\left(s + g\sin^2\alpha\right) + g\cos\alpha\cos\beta \\
&= n(\lambda)\left(s + g\sin^2\alpha\right) + g\cos\alpha\sqrt{1 - n^2(\lambda)\sin^2\alpha}\,.
\end{aligned}$$

The pair of prisms is thus equivalent to two plane-parallel plates; one with thickness s and index $n(\lambda)$, and one with thickness g and index

$$n_e \equiv n(\lambda)\sin^2\alpha + \cos\alpha\sqrt{1 - n^2(\lambda)\sin^2\alpha}\,. \qquad (220)$$

This is a remarkable result, because we now have effectively twice as many materials at our disposal to flatten the phase across the bandpass, a significant

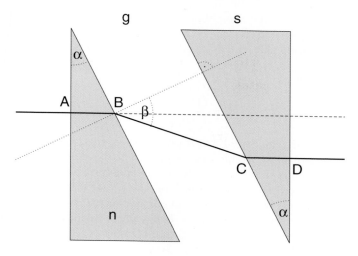

Fig. 58. Light path through a pair of prisms with refractive index n. The angles α and β are related by Snell's Law $n \sin \alpha = \sin \beta$

advantage especially in the mid-infrared, where only very few optical materials are available. Dispersive elements appear thus equally suited for nulling interferometry as the geometric methods discussed above, because they can generate phase shifts that are close enough to achromatic. In addition, dispersive elements can produce any desired phase shift (not only π rad), which is required in some classes of multi-element nulling arrays (see Sect. 10.4). An additional degree of freedom can sometimes be used to minimize achromatic effects, namely the possibility to use a (nominal) phase shift of 2π instead of 0 (Mieremet and Braat 2002b). For example, a three-telescope configuration with relative phases of $(0, \pi, 0)$ is equivalent to one with phases of $(0, \pi, 2\pi)$, but the achromatic errors are different between the two cases.

10.3 Nulling Interferometry from the Ground

Strehl Fluctuations, and Tip-Tilt Accuracy

We will now consider a simple Bracewell nulling interferometer on the ground, and we assume that the two telescopes are equipped with adaptive optics systems, which correct the beams from the two telescopes with a Strehl ratio S. From (199) and (123) we obtain

$$\overline{N} = \frac{1}{2}\sigma_w^2 \approx -\frac{1}{2}\ln S \approx \frac{1}{2}(1-S) \;, \tag{221}$$

which means that the AO systems have to provide $S = 0.998$ to achieve a 10^{-3} null. This is hardly feasible even in the thermal infrared, which means that modal filtering of the wavefront is necessary. We will therefore analyze performance requirements for the AO system to reach a specified null depth under the assumption that the nulling beam combiner contains a single-mode modal filter. Using (123), (190), and (200), we see that we can then write the average null depth as

$$\overline{N} = \frac{\langle (S_1 - S_2)^2 \rangle}{16 \langle S^2 \rangle} \, , \tag{222}$$

where S_1 and S_2 are the Strehl ratios produced by the two AO systems, and $\langle S^2 \rangle$ the mean squared Strehl ratio (assumed to be equal for the two AO systems). This means that the null depth depends primarily on the stability of the wavefront correction, not so much on the actual value of $\langle S^2 \rangle$. In practice, the stability and the quality of AO correction are closely linked to each other, however. For strictly stationary Kolmogorov turbulence it is possible to calculate the expected variability of the AO performance (Yura and Fried 1998), but in practice the non-stationary nature of turbulence is probably the limiting factor.[43] Slow variations of the seeing produce equal variations of the Strehl ratio in both telescopes, but fast variations must be uncorrelated over distances of ∼100 m. These fast Strehl fluctuations are difficult to measure with most AO systems, but it is plausible to assume that an AO system producing a mean Strehl ratio of 0.5 actually provides an instantaneous Strehl ratio fluctuating between 0.3 and 0.7. We can therefore estimate $\sqrt{\langle (S_1 - S_2)^2 \rangle} \approx 0.2$ and get $\overline{N} \approx 0.01$ from (222). This is not very satisfactory, but fortunately the null depth is $\propto \lambda^{-4}$ as pointed out above. If we can build an AO system that achieves the above performance in the K band (2.2 μm), it will provide a mean Strehl ratio of 0.97 and a null depth of $2 \cdot 10^{-5}$ at 10 μm (see Table 13).

The requirements on the quality of the tip-tilt correction follow immediately from (208). The angle tracker has to follow not only the atmospheric image motion, but also telescope vibrations induced by wind shake. The performance should be independent of the observing wavelength. If the residual tip-tilt fluctuation are of order 10 mas (which is realistic for the 10 m Keck telescopes), we get $\overline{N} \approx 3.6 \cdot 10^{-3}$ in the K band, about a factor of three smaller than the effect of Strehl ratio fluctuations. For the 8 m telescopes of the VLTI, $\overline{N} \approx 1.5 \cdot 10^{-3}$. Since again $\overline{N} \propto \lambda^{-4}$, the ratio between the effects of AO compensation and tip-tilt correction is the same for all wavelengths. We conclude that on an 8 m to 10 m telescope a tip-tilt servo loop that provides 10 mas rms angle tracking is well-matched to an AO system that delivers a Strehl ratio of 0.5 in the K band.

---

[43] In practice the seeing is observed to vary on all accessible time scales. "Lucky moments" are interrupted by spells of bad seeing, which may last only seconds. Real-time seeing monitors typically average over a few minutes to get representative figures, yet they frequently report variations of r_0 by sizeable amounts within less than one hour.

Table 13. Expected null depth for the near-infrared bands

band	wavelength	$\langle S \rangle$	$\overline{N}_{AO}$	$\overline{N}_{OPD}$	$\overline{N}_{scint}^0$	$\overline{N}_{scint}^{8m}$
K	2.2 μm	0.50	$1.0 \cdot 10^{-2}$	$3.8 \cdot 10^{-3}$	$4.4 \cdot 10^{-3}$	$7.4 \cdot 10^{-7}$
L	3.6 μm	0.77	$1.4 \cdot 10^{-3}$	$1.4 \cdot 10^{-3}$	$2.5 \cdot 10^{-3}$	$7.4 \cdot 10^{-7}$
M	5.0 μm	0.87	$3.9 \cdot 10^{-4}$	$7.4 \cdot 10^{-4}$	$1.7 \cdot 10^{-3}$	$7.4 \cdot 10^{-7}$
N	10.0 μm	0.97	$2.4 \cdot 10^{-5}$	$1.9 \cdot 10^{-4}$	$7.5 \cdot 10^{-4}$	$7.4 \cdot 10^{-7}$
Q	20.0 μm	0.99	$1.5 \cdot 10^{-6}$	$4.6 \cdot 10^{-5}$	$3.3 \cdot 10^{-4}$	$7.4 \cdot 10^{-7}$

The following assumptions have been made (for details see text). AO system: mean Strehl at K-band = 0.5, r.m.s. Strehl imbalance at K-band = 0.2. Atmosphere: f_G = 21.35 Hz at 500 nm, one layer at 10 km with $r_0 = 0.2$ m at 500 nm for scintillation calculation. Fringe tracker: 2 ms loop lag

Fringe Tracking

Nulling interferometry requires precise fringe tracking. The null depth due to residual uncorrected high-frequency fluctuations of the optical path difference can be calculated

$$\overline{N} = \frac{1}{4}\sigma_R^2 = \frac{1}{2}\kappa\left(\frac{f_G}{f_S}\right)^{5/3}. \tag{223}$$

A careful modeling of the dynamical behavior of the control loop is needed to determine the value of κ. To get a feeling for the numerical values involved, consider the case of a "pure delay" of 2 ms, so that $\kappa = 28.4$, and $f_S = 500$ Hz. Let's further assume that at 500 nm the Greenwood frequency $f_{G,500} = 21.35$ Hz; this corresponds to a wind velocity $v = 10$ m s^{-1} and Fried parameter $r_{0,500} = 20$ cm. Scaling f_G to 2.2 μm and inserting into (223), we obtain a K-band null depth of $3.8 \cdot 10^{-3}$. We see that the speed of the control loop must be much faster than the Greenwood frequency to obtain a good null depth; lags due to detector readout or processing time are particularly damaging because of the large value of κ associated with them. The wavelength scaling of high-frequency fringe tracking residuals is $\overline{N} \propto \lambda^{-2}$.

Photon noise in the fringe tracker is also a source of phase noise, which becomes important for faint stars. The null depth due to photon noise from (157) and (190) is

$$\overline{N} = \frac{1}{4}\left(\frac{\lambda_t}{\lambda_n}\right)^2 \sigma_{\phi,t,\mathrm{phot}}^2 = \left(\frac{\lambda_t}{\lambda_n}\right)^2 \cdot \frac{1}{2N_t V_t^2}, \tag{224}$$

where λ_t is the wavelength at which the fringe tracking is performed, and λ_n the wavelength at which the nuller operates. The factor $(\lambda_t/\lambda_n)^2$ in (224) favors fringe tracking at a short wavelength. This means that one has to be careful about additional errors due to dispersion between λ_t and λ_n. One possibility is using two nested control loops. This scheme uses a fast servo at λ_t, which tracks the rapid atmospheric phase fluctuations, and an "outer" loop with a sensor that measures the residual phase at λ_n, and determines tracking offsets between the two wavelengths with a much slower update rate.

Another implication of (224) is the degradation of the quality of the null when the star is partially resolved at the tracking wavelength. If we require $V_t^2 \gtrsim 0.03$, the stellar diameter has to be $\lesssim \lambda_t/B$. The numerical value for a 50 m baseline at K band is 10 mas, which is exceeded only by a relatively small number of bright cool giants and Mira stars. On much longer baselines or at much shorter λ_t, however, many of the nearby main-sequence stars will have low visibilities. It thus appears reasonable to perform the fringe tracking at K band. In this band a 0 mag star observed with two 8 m telescopes with a 10% total efficiency generates $\sim 2 \cdot 10^{10}$ photons per second. Scaling this value to $K = 12$ gives 320 photons per millisecond. If the AO systems achieve a Strehl ratio of 0.5, half of these photons are rejected by the modal filter. The corresponding K-band null depth from (224) is $3.1 \cdot 10^{-3}$; it scales with $\overline{N} \propto \lambda^{-2}$. Consequently, the nulling performance should not depend much on the stellar brightness down to $K \approx 12$, if the fringe sensor remains photon-noise limited down to that magnitude.

Scintillation Effects

While the atmospheric phase fluctuations will be corrected to a large extent by an adaptive optics system, it may be difficult to implement an intensity control that works on similar time scales. In that case atmospheric scintillation creates a rapidly variable imbalance between the two beams. The consequent limitation on the null depth can be calculated combining (110) and (134)

$$\sigma_{\ln I}^2 = 1.14 \left(\frac{\sqrt{\lambda h \sec z}}{r_0(h)} \right)^{5/3} \qquad (225)$$

and by inserting it in (190); this gives[44]

$$\overline{N} = 0.14 \left(\frac{\sqrt{\lambda h \sec z}}{r_0(h)} \right)^{5/3}. \qquad (226)$$

A layer at 10 km above the observatory with $r_0 = 20$ cm at 500 nm will thus produce a K-band null depth of $4.4 \cdot 10^{-3}$. The improvement with wavelength, $\overline{N} \propto \lambda^{-7/6}$, is the slowest of all effects considered.

---

[44] A little care has to be taken about factors of 2 here. A fluctuation δI in one beam increases the average intensity I; the two beams thus have intensities $I + \delta I/2$ and $I - \delta I/2$. So the variance in (190) is 1/4 of the variance in (225). Multiplying this by 2 for the contributions from the two telescopes we get the numerical factor in (226).

In the derivation of (226) we have made use of (134), which is valid for the intensity fluctuations at any given point of the wavefront. This means that (226) is applicable if the interferometer consists of two very small telescopes, or if a classical co-axial beam combiner is used.[45] If, on the other hand, a single-mode fiber is used in the beam combiner, the contributions from different parts of the pupil are mixed, and it is the intensity averaged over the telescope aperture that matters for the null depth. The scintillation pattern varies spatially on the Fresnel scale $r_F \equiv \sqrt{\lambda h \sec z}$. For telescopes much larger than r_F most of the fluctuations represent high-order modes that are rejected by the modal filter; only variations of the average intensity couple efficiently into a single-mode fiber. According to Ryan (2002), aperture averaging can be approximately described by multiplying the right-hand side of (134) or (225) by a factor

$$A \approx \left[1 + 1.1 \left(\frac{D^2}{\lambda h \sec z}\right)^{7/6}\right]^{-1}. \tag{227}$$

This leads to a null depth of $7.4 \cdot 10^{-7}$ independent of λ for $D = 8\,\mathrm{m}$, so that scintillation should be negligible for nulling with the VLTI UTs or the Keck 10 m telescopes.

Subtraction of the Thermal Background

Infrared nulling observations from the ground have to cope with the thermal background radiation of the atmosphere and instrument. This can in principle be done by the standard chopping and nodding techniques that are widely used in infrared astronomy (e.g., McLean 1997). This is technically not easy, however, because it means that the chopping devices (usually the telescope secondaries), adaptive optics systems, and the fringe tracker have to be synchronized. The adaptive optics and fringe tracking loops have to be opened during the off-source part of the chopping cycle; at the beginning of the on-source part, the adaptive optics loops have to be closed, the fringes have to be re-acquired and the servo has to find the zero optical path difference position before data taking can resume. Since canceling the atmospheric fluctuations requires chopping frequencies of several Hz, this all has to be done very quickly to keep the resulting overhead tolerable.

A more efficient way of dealing with the thermal background is creating an internal modulation of the interferometric signal (see Fig. 59). For this purpose the apertures of the two telescopes are split in halves, so that four beams are sent to the beam combination laboratory. The right halves and the

---

[45] Recall that in co-axial beam combiner a 50% beam splitter is used to superpose the pupils from the two telescopes. Light from each point of the first pupil therefore interferes only with light from the corresponding point of the second pupil.

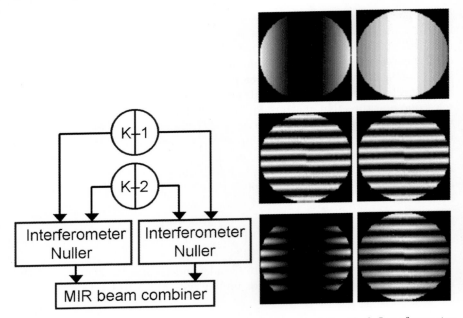

Fig. 59. Schematic setup (left) and transmission maps of the Keck Interferometer nulling experiment. Each of the two telescope apertures is split into two halves; the right halves and the left halves form two nulling interferometers. The outputs of these nulling interferometers are combined on a third nuller. Modulating the optical path difference of this aperture nuller by $\lambda/2$ alternates between the transmission maps shown in the top panels. Multiplying by the transmission maps of the two nulling interferometers (center panels) gives the resulting transmission maps shown in the bottom panels. A point source on the optical axis is rejected at all times, while the light from an extended source (e.g., dust disk) or off-axis companion produces a modulated signal on top of the constant thermal background

left halves form two nulling interferometers with baselines B, which reject the light of an on-axis star. The angular scale of these nulls is therefore λ/B. The two outputs are combined on a third nuller, which interferes the light from the sub-aperture pairs and therefore creates a null with angular scale λ/D. Modulating the optical path difference of this aperture nuller by $\lambda/2$ creates the desired modulated signal. Whereas the light from an on-axis point source is rejected at all times by the λ/B nuller, and the incoherent background is not affected by the modulation, any source at a separation between λ/B and λ/D (or of corresponding size) will alternatively be passed an rejected. For stars at $d = 20\,\mathrm{pc}$, these two numbers correspond to 0.4 AU and 4 AU for an interferometer with two 10 m telescopes on a 100 m baseline, operated at 10 µm. This setup is therefore well-suited for the detection of dust disks and planets around nearby stars.

10.4 Design of Nulling Arrays

Quadratic and Higher-Order Nulling

We have seen above (213) that the non-zero diameter of the stellar disk causes a leak, which severely limits the null depth that can be obtained on bright stars. The reason is that the null depth is proportional to the square of the off-axis angle (see (186) and (210)). The only way around this problem is creating a higher-order null, which is possible if the light from more than two telescopes is combined with proper phase shifts. Let us consider an array consisting of n telescopes with diameters D_k, located at positions with polar coordinates (L_k, δ_k). If a phase shift ϕ_k is introduced in the beam of telescope k before beam combination, the complex amplitude for a point source with polar coordinates (θ, ψ) is given by

$$A_{\text{out}} = \sum_{k=1}^{n} D_k \cdot e^{2\pi i (L_k \theta / \lambda) \cos(\delta_k - \psi)} \cdot e^{i\phi_k} . \qquad (228)$$

We are interested in the scaling of A_{out} for small off-axis angles; therefore we expand the exponential into a Taylor series in θ. With the abbreviation $x_k \equiv 2\pi(L_k \theta / \lambda)$ we obtain

$$e^{ix_k \cos(\delta_k - \psi)} = 1 + ix_k \cos(\delta_k - \psi) - \tfrac{1}{2} x_k^2 \cos^2(\delta_k - \psi) + \mathcal{O}(x_k^3) . \qquad (229)$$

Back-substituting this expansion into (228) we see that the condition for on-axis nulling ($A_{\text{out}} = 0$ for $\theta = 0$) is

$$\sum_{k=1}^{n} D_k \cdot e^{i\phi_k} = 0 . \qquad (230)$$

The complex amplitude is then $\propto \theta$ and the intensity leaking through the nuller $\propto \theta^2$. Setting the second term of the expansion (228) and (229) to zero we get

$$\sum_{k=1}^{n} D_k \cdot x_k \cdot \cos(\delta_k - \psi) \cdot e^{i\phi_k} = 0 . \qquad (231)$$

If (230) and (231) are satisfied simultaneously for all values of ψ, the intensity varies $\propto \theta^4$. This result can easily be generalized by adding equations from higher orders of the expansion; the additional condition to achieve a θ^6 null is

$$\sum_{k=1}^{n} D_k \cdot x_k^2 \cdot \cos^2(\delta_k - \psi) \cdot e^{i\phi_k} = 0 . \qquad (232)$$

Using the standard formula for the cosine of a difference, we see that requiring (231) to be satisfied for all values of ψ is equivalent to the two conditions

$$\sum_{k=1}^{n} D_k \cdot x_k \cdot \cos \delta_k \cdot e^{i\phi_k} = 0, \qquad (233)$$

$$\sum_{k=1}^{n} D_k \cdot x_k \cdot \sin \delta_k \cdot e^{i\phi_k} = 0. \qquad (234)$$

Similarly, (232) is equivalent to

$$\sum_{k=1}^{n} D_k \cdot x_k^2 \cdot \cos^2 \delta_k \cdot e^{i\phi_k} = 0, \qquad (235)$$

$$\sum_{k=1}^{n} D_k \cdot x_k^2 \cdot \sin^2 \delta_k \cdot e^{i\phi_k} = 0, \qquad (236)$$

$$\sum_{k=1}^{n} D_k \cdot x_k^2 \cdot \sin 2\delta_k \cdot e^{i\phi_k} = 0. \qquad (237)$$

Equations (230), (233), and (235) provide a systematic framework for the design of arrays that perform quadratic, fourth-order, or sixth-order nulling. An example for a linear array that generates a sixth-order null is the OASES concept (Angel and Woolf 1997). The transmission of this configuration is compared to that of a two-element Bracewell interferometer in Fig. 60.

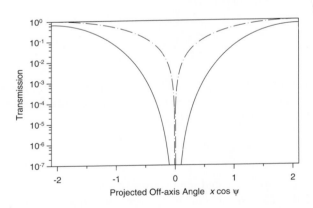

Fig. 60. Transmission of the linear OASES interferometer concept consisting of four telescopes with diameters $(1, 2, 2, 1)$ located at positions $(-2, -1, 1, 2)$, and combined with phases $(0, \pi, 0, \pi)$. The horizontal axis is the component of the off-axis angle parallel to the array. The θ^6 null of OASES (full line) is compared to the θ^2 null of a two-telescope Bracewell interferometer 3/8 as long, for which the first maximum in the transmission occurs at the same off-axis angle (*dash-dotted line*)

External and Internal Modulation

The next important question to consider is the way of extracting useful information from the nulling array. A pupil plane interferometer does not produce images, but rather a photon count from a single-pixel detector as a function of time. The output of the beam combiner for light arriving from the direction (θ, ψ) is given by (228); the total amplitude received can therefore be computed by multiplying this expression with the electric field at (θ, ψ), and integrating over the plane of the sky. This process is illustrated in the top panels of Fig. 53; the observed photon count is the total integrated intensity contained in the top right panel. This single number obviously does not provide us with the information we are seeking, namely the position and brightness of the planet(s) associated with the target star. The simplest way of increasing the amount of information from the nulling interferometer is to rotate it around the line-of-sight toward the star, as illustrated in the bottom panels of Fig. 53. In the course of the rotation the planet is traversed by regions of high and low response, which leads to a modulation of the observed intensity. The information about the position and brightness of the planet is encoded in this signal, which is plotted as a function of time for a full rotation in the bottom right panel.

The rotational modulation method has a significant difficulty, namely the requirement that the instrumental response has to remain stable over a full rotation period, which is typically of order several hours. Small drifts of the detector sensitivity or background, of the telescope pointing, or other components of the interferometer can mask the presence of a planet or lead to spurious detections. It is therefore highly desirable to modulate the output at a higher rate (≈ 1 Hz). In a "standard" Michelson stellar interferometer one can modulate the delay and measure the visibility with a lock-in detection scheme as described in Sect. 8.5. This is generally not possible for a nulling interferometer, because delay scanning changes the ϕ_k and thus violates the nulling conditions. It is obvious from (230), however, that four identical telescopes combined with phases $(0, \pi, \alpha,$ and $\alpha + \pi)$ will produce a central null irrespective of the value of α. One can, for example, quickly alternate between $\alpha = \pi/2$ and $\alpha = -\pi/2$, which generally changes the interferometer response pattern on the sky (see Fig. 61). This technique of modulating the planetary signal is called "internal chopping".

We can think of this configuration as constructed from two Bracewell pairs with phases $(0, \pi)$ and $(\alpha, \alpha+\pi)$, respectively. This concept can be generalized to arbitrary nulling configurations. If the outputs from N arrays, each of which produces a θ^μ null, are combined with relative phases $\alpha_1, \ldots, \alpha_N$, the output produces again a θ^μ null, and the α_i can be varied to produce a modulation of the signals from off-axis sources. This recipe can be used to construct complicated nulling arrays with desirable properties from simpler building blocks. We should also note that compressing or stretching an array in one direction changes the width of the null in that direction, but preserves the order of the null.

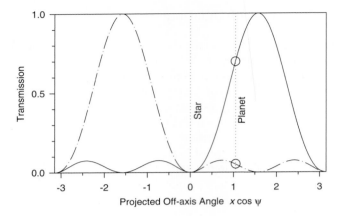

Fig. 61. Transmission of a linear double-Bracewell interferometer consisting of four telescopes of identical diameter with unit spacing. The horizontal axis is the component of the off-axis angle parallel to the array. The *full curve* is the transmission for phases $(0, -\pi/2, \pi, \pi/2)$, the *dash-dotted curve* for $(0, \pi/2, \pi, -\pi/2)$. Chopping between these states modulates the planetary signal between the two values indicated by *circles*

Nulling Array Geometries

Various array geometries have been proposed for the direct detection of extrasolar planets in the mid-infrared, each one with certain technical advantages and disadvantages. If all telescopes are mounted on a single long structure, linear configurations are certainly easier to assemble and maneuver. On the other hand, two-dimensional arrays in which each telescope is mounted on its own free-flying spacecraft have more versatility. In that case, an "ideal" nulling array should perhaps fulfill the following criteria:

- All telescopes are at equal distance from the beam combiner spacecraft. If this is not the case, the beam transport is more complicated, and diffraction effects will be different between the interferometer arms.
- All telescopes and the beam combiner are in one plane (perpendicular to the line of sight to the target star). This allows a thermal design that minimizes radiative coupling between the illuminated and cold parts of the spacecraft.
- The null is of fourth or sixth order. We have seen above that second-order starlight suppression is not sufficient.
- The only phase shift needed is π. While it is possible to introduce shifts by other amounts with dispersive elements, they are more difficult to control precisely.
- All telescopes have the same diameter. This is certainly cheaper than launching telescopes with different sizes.

- Internal phase modulation is possible. This relaxes the required stability of the optics and detectors.
- The transmission function should not have any rotational symmetries, which lead to ambiguities in the position of any detected planets. (For example, linear configurations with mirror-symmetric transmission functions have a 180° degree ambiguity.) The determination of planet orbits from multiple observations is much easier if the positions are unambiguous, especially in multi-planet systems.

As we will see, it is possible to design a arrays that have all of these properties. The price to be paid is in the number of telescopes and in the complexity of the beam combination, when a high-order null and chopping capability are required. Since the more complicated arrays are constructed from simpler building blocks, it is useful to look at the various arrangements, progressing from the elementary to the more complex. A summary is also given in Table 14. All concepts assume that the whole interferometer is oriented perpendicular to the line-of-sight to the target star, i.e., there has to be a spacecraft maneuver to reorient the array for each observed star. The array can then be rotated around the line-of-sight during the observation, either by a full 360° or a smaller angle, as shown in Fig. 53.

Bracewell Pair.

This is the most basic nulling interferometer, consisting of two telescopes with equal size and a π phase shift between them. The resulting θ^2 null is insufficient for the direct detection of Earth-like planets.

Table 14. Configurations for space-borne nulling interferometers

name	N_{tel}	configuration	order	chopping	ambiguities
Bracewell	2	Single Baseline	θ^2	no	yes
Double Bracewell	4	Linear 1:1:1:1	θ^2	yes	no
OASES	4	Linear 1:2:2:1	θ^6	no	yes
Angel's Cross	4	Cross-shaped	θ^4	no	yes
DAC	3	Linear 1:$\sqrt{2}$:1	θ^4	no	yes
Double DAC	4	Linear 1:$\sqrt{3}$:$\sqrt{3}$:1	θ^4	yes	no
Mariotti 3-DAC	6	Triangular	θ^4	yes	no
DARWIN 5-Telescope	5	Compressed Pentagon	θ^4	no	no
Robin Laurance	6	Hexagon	θ^4	yes	no

For explanations of the individual configurations see text. Note that variants of many of the concepts exist; only the main version for each one is included in the table

Double Bracewell Interferometer.

This is the simplest nulling interferometer that allows for internal chopping. It consists of four identical telescopes in a linear arrangement with equal spacing. The first and third telescope form one Bracewell pair, the second and fourth the other. Switching the relative phase between the two Bracewell pairs phases between $\pi/2$ and $-\pi/2$ is illustrated in Fig. 61. The double Bracewell interferometer inherits the θ^2 null from the single Bracewell pair.

OASES.

The OASES concept is a linear array that provides θ^6 nulling (Angel and Woolf 1997). It consists of two Bracewell pairs; one of them has twice the baseline but only half the aperture diameter of the other. The phase between the two Bracewell pairs is π. The OASES configuration can thus be described by $D_k = (1, 2, 2, 1)$, $L_k = (-2, -1, 1, 2)$, and $\phi_k = (0, \pi, 0, \pi)$.

Angel's Cross.

This is a two-dimensional configuration, in which four telescopes of equal diameter are placed at the corners of a rhombus, so that their arrangement resembles a cross with pairwise equal bar lengths (Angel 1990). The telescopes located opposite of each other on the same bar have the same phase; the phase difference between the two bars is π. Angel's cross produces a θ^4 null.

Degenerate Angel's Cross (DAC).

This configuration is derived from Angel's cross by collapsing it along one of the bars, i.e., the two telescopes of one of the bars are replaced by a single telescope with twice the aperture area. Taking a linear configuration with equal spacings, telescope diameters $(1, \sqrt{2}, 1)$, and phases $(0, \pi, 0)$, and dividing the light from the central telescope in two equal beams, we get an array of four telescopes with equal diameters at positions $(-1, 0, 0, 1)$ and with phases $(0, \pi, \pi, 0)$. This is just the projection of an Angel's cross on one coordinate axis.[46] The order of the null in the DAC configuration is also θ^4.

Double Degenerate Angel's Cross.

If we construct two DACs, we can combine their outputs with a time-variable phase and thus obtain an array with a θ^4 null and internal chopping. A double

---

[46] There seems to be a paradox here. According to our formalism (230), one should expect diameters of $(1, 2, 1)$, not $(1, \sqrt{2}, 1)$. Actually, both versions are "correct". Consistency with our formalism requires that if we choose $(1, 2, 1)$, we must send the light from the central telescope in one entrance of the beam combiner; if we choose $(1, \sqrt{2}, 1)$ we must divide the light from the central telescope and send the two beams into two different entrances. It is not too difficult to convince oneself that the light in the nulled output is the same in both cases; the extra photons in the $(1, 2, 1)$ case end up in the non-nulled outputs.

DAC can either be realized with six independent telescopes (Woolf and Angel 1997), or with a linear array of four equally spaced telescopes with diameters $(1, \sqrt{3}, \sqrt{3}, 1)$. The latter approach is based on the idea of overlaying separate nulling arrays such that they share one or more telescopes. The aperture area of a shared telescopes must be equal to the sum of the areas of the constituent telescopes at that position. In the case of the double DAC, splitting the light from the large telescopes with a 2:1 intensity ratio forms two arrays that are equivalent to diameters $(1, \sqrt{2}, 1, 0)$, and $(0, 1, \sqrt{2}, 1)$, respectively; these are the two constituent DACs.

Mariotti 3-DAC.

This configuration consists of three DACs, which form the three sides of an equilateral triangle. Each of the telescopes at the vertices of the triangle is shared between two DACs, so that all telescopes are of equal size. This fact, together with the θ^4 null, chopping capability, redundancy, and two-dimensional coverage of the uv plane make the Mariotti 3-DAC quite attractive for a space nulling array.

DARWIN 5-Telescope Configuration.

A five-telescope solution to (230) and (233) is given by $x_k = \text{const.}$, $\delta_k = 2(k-1)\pi/5$, and $\phi_k = 4(k-1)\pi/5$, i.e., the telescopes are located on a regular pentagon, and the phase difference between neighboring telescopes is 144° (Mennesson and Mariotti 1997). The DARWIN 5-telescope configuration is derived from this solution by compressing the array along one axis by a factor ~ 2. Upon rotation of the array (internal chopping is not possible), the transmission function provides fairly strong modulation of planetary signals but only weak modulation of the light from a symmetric exozodiacal disk. Unlike most other non-chopping configurations, the DARWIN 5-telescope concept does not suffer from ambiguities of the planet position; the five-fold symmetry of the regular pentagon is broken by the compression along one axis.

Robin Laurance Interferometers.

This is a class of nulling arrays that consists of building blocks, which can be called generalized Angel's crosses (GACs); a GAC is here defined simply as a nulling interferometer that satisfies (230) and (233) and thus produces a θ^4 null. One can easily verify that assigning telescope diameters $D_k = (3, 2, 0, 1, 0, 2)$ and phases $\phi_k = (0, \pi, \times, 0, \times, \pi)$ to the six vertices of a regular hexagon defines a GAC. Three such GACs rotated by 120° with respect to each other and overlayed on the hexagon as shown in Fig. 62 constitute a Robin Laurance interferometer (Karlsson and Mennesson 2000). This configuration satisfies all of the requirements on an "ideal" nulling array listed above. The most significant drawbacks are its complexity and the need to divide the light from three of the telescopes asymmetrically with a 4:4:1 intensity ratio. Many similar overlays of GACs on regular hexagons and on pentagons exist. It is

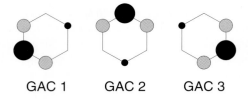

Fig. 62. Robin Laurance interferometer. Three generalized Angel's crosses (GACs) are overlayed on a hexagonal telescope array. The designation of this specific configuration RL3(3,2,0,1,0,2) gives the number of GACs, and the amplitudes contributing to each GAC, starting with the largest and counting along the hexagon's vertices. Within each GAC, the large and small telescope are combined with phase 0 (black), the two intermediate-size telescopes with phase π (light gray). The contributions from the three GACs add up to 9 units for each telescope aperture area

therefore possible to devise layouts that are technically simpler (for example solutions in which all beams have the same intensities), and to optimize various aspects of their performance (Karlsson and Mennesson 2000).

Location of the Beam Combiner

In the preceding section the various array geometries have been described in terms of the location and size of the telescopes, and of the applied phase shifts. It is also necessary, of course, to provide for a beam combiner. In the case of the Robin Laurance configuration, for example, the beam combiner should clearly be located at the center of the hexagon. The pathlength from each telescope is then equal, and the beam relay to the central hub comparatively simple. If there is no location from which all telescopes have the same distance (e.g., in the OASES and Mariotti 3-DAC configurations), the beam can be passed through several spacecraft rather than sent directly to the beam combiner. An example is shown in the bottom left panel of Fig. 53. Sending the beams from telescopes 1 and 3 to the combiner via telescope 2 ensures equality of the paths. It is thus normally necessary to provide a separate beam combiner spacecraft in addition to those carrying the telescopes.

Image Reconstruction

Like any other interferometer, a nulling array produces a non-intuitive output when a complicated object is observed. Whereas in a "standard" two-element interferometer the signal and the source structure are related by a Fourier transform (145), we can derive an analogous generalized equation for a nulling array by multiplying (228) with the electric field distribution $E(\theta, \psi)$ on the sky, and integrating over θ and ψ. Because more than two telescopes contribute to the observed signal, there is no simple Fourier relation between source structure and data, but general requirements on the sampling of the uv plane, and the use of image reconstruction algorithms, apply in a similar way.

Figure 63 shows an example of these principles applied to a simulated observation of a planetary system with a linear nulling interferometer, which

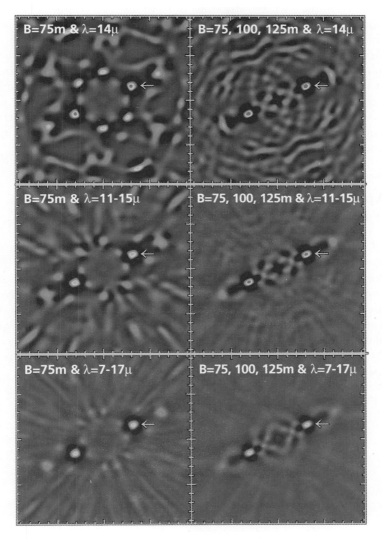

Fig. 63. Reconstruction of an observation of a terrestrial planet at a distance of 10 pc with a linear nulling interferometer. The level of exozodiacal emission was assumed to be equal to that in the Solar System; the inclination was taken to be $i = 30°$. The panels show how the reconstructed image improves as more baselines (left to right) and more wavelengths (top to bottom) are added. From NASA (1999b)

was rotated around the viewing direction to synthesize a reasonable uv plane coverage as illustrated in Fig. 53. The left-hand panels assume that the telescopes of the interferometer are fixed with respect to each other, whereas the right-hand panels assume that the size of the array can be varied. The top panels use only monochromatic light for the reconstruction, whereas the center and bottom panels take advantage of wavelength synthesis. These images demonstrate clearly how the quality of the reconstruction improves as more baselines and more wavelengths are added. A few artifacts remain, however, even in the best case. Most significantly, there is a mirror image of the planet due to the symmetric transmission pattern of the array, and an aliasing artifact at twice the orbital distance. These effects can be avoided by asymmetric array configurations or asymmetric internal chopping techniques. It is clear that a careful choice of the array geometry and of the image reconstruction technique are necessary to optimize the ability of a nulling interferometer to perform meaningful observations of systems with multiple planets.

11 Appendix: Useful Definitions and Results from Fourier Theory

For reference, this appendix lists a few useful results from Fourier theory without proofs. In the notation adopted, $g \Longleftrightarrow G$ means "G is the Fourier transform of g", and it is understood that small and capital letters designate Fourier transforms pairs, i.e., $g \Longleftrightarrow G$ and $h \Longleftrightarrow H$. H^* is the complex conjugate of H. Introductions into Fourier theory and more details can be found in many textbooks (e.g. Bracewell 1965). The results are frequently formulated for the one-dimensional Fourier pair time and frequency ($t \longleftrightarrow f$), but they can equally be applied to the three-dimensional variables position and spatial frequency ($x \longleftrightarrow \kappa$).

The *convolution* $g * h$ and *correlation* $\mathrm{Corr}(g, h)$ of two functions g and h are defined by:

$$g * h \equiv \int_{-\infty}^{\infty} d\tau\, g(t - \tau) h(\tau) \qquad (238)$$

and

$$\mathrm{Corr}(g, h) \equiv \int_{-\infty}^{\infty} d\tau\, g(t + \tau) h(\tau)\,. \qquad (239)$$

A special case of the latter is the correlation of a function with itself, the *covariance*:

$$B_g \equiv \mathrm{Corr}(g, g)\,. \qquad (240)$$

For complex functions, the *coherence function* is defined by:

$$B_g \equiv \mathrm{Corr}(g, g^*). \tag{241}$$

The customary use of the same symbol B for covariance and coherence function is somewhat unfortunate, but should not be too confusing. The power spectral density $\Phi(f)$ is defined as

$$\Phi(f) \equiv |G(f)|^2. \tag{242}$$

The famous *Convolution Theorem* and *Correlation Theorem* are:

$$g * h \iff G(f)H(f) \tag{243}$$

and

$$\mathrm{Corr}(g, h) \iff G(f)H^*(f). \tag{244}$$

The special case of the Correlation Theorem for the covariance is the *Wiener-Khinchin Theorem*:

$$B_g = \mathrm{Corr}(g, g) \iff |G(f)|^2 = \Phi(f). \tag{245}$$

The *structure function* D_g of a function g is defined by:

$$D_g(t_1, t_2) \equiv \left\langle |g(t_1) - g(t_2)|^2 \right\rangle. \tag{246}$$

If g describes a homogeneous and isotropic random process, D_g depends only on $t = |t_1 - t_2|$. By expanding the square in (246), we see that in this case

$$D_g(t) = 2\bigl(B_g(0) - B_g(t)\bigr). \tag{247}$$

Finally, *Parseval's Theorem* states that the total power in a time series is the same as the total power in the corresponding spectrum:

$$\mathrm{TotalPower} \equiv \int_{-\infty}^{\infty} dt\, |g(t)|^2 = \int_{-\infty}^{\infty} df\, |G(f)|^2. \tag{248}$$

Acknowledgment

I thank Laurance Doyle and Oswald Wallner for careful reading of parts of the manuscript.

References

Aikawa Y, Herbst E. 1999. Molecular evolution in protoplanetary disks. Two-dimensional distributions and column densities of gaseous molecules. *ApJ* 351:233–46

Akeson RL, Swain MR. 1999. Differential phase mode with the Keck Interferometer. In *Working on the fringe,* ed. S Unwin, R Stachnik, pp. 89–94. ASP Conf. Ser. Vol. 194. San Francisco, CA

Alard C, Lupton R. 1998. A method for optimal image subtraction. *ApJ* 503:325–31

Albrow MD, An J, Beaulieu JP, Caldwell JAR, DePoy DL, et al. 2001. Limits on the abundance of Galactic planets from 5 years of PLANET observations. *ApJ* 556:L113–16

Albrow MD, An J, Beaulieu JP, Caldwell JAR, DePoy DL, et al. 2002. A short, non-planetary, microlensing anomaly: observations and lightcurve analysis of MACHO 99-BLG-47. *ApJ* 572:1031–1040

Albrow MD, Beaulieu JP, Birch P, Caldwell JAR, Kane S, et al. 1998. The 1995 pilot campaign of PLANET: searching for microlensing anomalies through precise, rapid, round-the-clock monitoring. *ApJ* 509:687–702

Albrow MD, Beaulieu JP, Caldwell JAR, DePoy DL, Dominik M, et al. 2000a. Limits on stellar and planetary companions in microlensing event OGLE-1998-BUL-14. *ApJ* 535:176–89

Albrow MD, Beaulieu JP, Caldwell JAR, Dominik M, Gaudi BS, et al. 2000b. Detection of rotation in a binary microlens: PLANET photometry of MACHO 97-BLG-41. *ApJ* 534:894–906

Alcock C. 2000. The dark halo of the Milky Way. *Science* 287:74–79

Alcock C, Akerlof CW, Allsman RA, Axelrod TS, Bennett DP, et al. 1993. Possible gravitational microlensing of a star in the Large Magellanic Cloud. *Nature* 365:621–23

Alcock C, Allen WH, Allsman RA, Alves D, Axelrod TS, et al. 1997. MACHO alert 95-30: first real-time observation of extended source effects in graviational microlensing. *ApJ* 491:436–50

Alcock C, Allsman RA, Alves D, Axelrod TS, Baines D, et al. 2000a. Binary microlensing events from the MACHO project. *ApJ* 541:270–97

Alcock C, Allsman RA, Alves DR, Axelrod TS, Becker AC, et al. 2000b. The MACHO Project: microlensing optical depth toward the Galactic bulge from difference image analysis. *ApJ* 541:734–66

Allard F, Hauschildt PH, Alexander DR, Starrfield S. 1997. Model atmospheres of very low mass stars and brown dwarfs. *ARAA* 35:137–77

André P. 1994. Observations of protostars and protostellar stages. In *The cold Universe,* ed. T Montmerle, CJ Lada, IF Mirabel, J Trân Thanh Vân, pp. 179–92. Editions Frontieres, Gif-sur-Yvette, France

Angel JRP. 1990. Use of a 16 m telescope to detect Earthlike planets. In *The Next Generation Space Telescope,* ed. PY Bely, CJ Burrows, GD Illingworth, pp. 81–94. Baltimore, MD

Angel JRP, Woolf NJ. 1997. An imaging nulling interferometer to study extrasolar planets. *ApJ* 475:373–79

Armitage PJ. 2000. Suppression of giant planet formation in stellar clusters. *A&A* 362:968–72

Armitage PJ. 2003. A reduced efficiency of terrestrial planet formation following giant planet migration. *ApJ* 582:L47–50

Armstrong JT, Mozurkewich D, Rickard LJ, Hutter DJ, Benson JA, et al. 1998. The Navy Prototype Optical Interferometer. *ApJ* 496:550–71

Artymowicz P. 1997. Beta Pictoris: an early Solar System? *Ann Rev Earth Planet Sci* 25:175–219

Arzoumanian Z, Joshi K, Rasio FA, Thorsett SE. 1996. Orbital parameters of the PSR 1620−26 triple system. In *Pulsars: problems and progress*, ed. S Johnston, MA Walker, M Bailes, pp. 525–30. IAU Coll. 160, ASP Conf. Ser. Vol. 105. San Francisco, CA

Ashton CE, Lewis GF. 2001. Gravitational microlensing of planets: the influence of planetary phase and caustic orientation. *MNRAS* 325:305–11

Aubourg E, Bareyre P, Br'ehin S, Gros M, Lachieze-Rey M, et al. 1993. Evidence for gravitational microlensing by dark objects in the Galactic halo. *Nature* 365:623–25

Aumann HH, Gillett FC, Beichman CA, de John T, Houck JR, et al. 1984. Discovery of a shell around Alpha Lyrae. *ApJ* 278:L23–27

Baba N, Murakami N, Ishigaki T. 2001. Nulling interferometry by use of geometric phase. *Opt Lett* 26:1167–69

Backman DE, Paresce F. 1993. Main-sequence stars with circumstellar solid material: the Vega phenomenon. In *Protostars and planets III*, ed. EH Levy, JI Lunine, pp. 1253–1304. University of Arizona Press.

Baglin A, Auvergne M, Barge P, Buey JT, Catala C, et al. 2002. COROT: asteroseismology and planet finding. In *Proceedings of the first Eddington workshop on stellar structure and habitable planet finding*, ed. F Favata, IW Roxburgh, D Galadi, pp. 17–24. ESA SP-485, Noordwijk: ESA Publications Division

Bailes M, Lyne AG, Shemar SL. 1991. A planet orbiting the neutron star PSR1829−10. *Nature* 352:311–13

Bally J, O'Dell CR, McCaughrean MJ. 2000. Disks, microjets, windblown bubbles, and outflows in the Orion nebula. *AJ* 119:2919–59

Baraffe I, Chabrier G, Barman TS, Allard F, Hauschildt PH. 2003. Evolutionary models for cool brown dwarfs and extrasolar giant planets. The case of HD 209458. *A&A* 402:701–712

Baranne A. 1999. Spectrographs for the measurement of radial velocities. In *Precise stellar radial velocities*, ed. JB Hearnshaw, CD Scarfe, pp. 1–12. IAU Coll. 170, ASP Conf. Ser. Vol. 185. San Francisco, CA

Baranne A, Queloz D, Mayor M, Adrianzyk G, Knispel G, et al. 1996. ELODIE: A spectrograph for accurate radial velocity measurements. *A&AS* 119:373–90

Barbieri M, Gratton RG. 2002. Galactic orbits of stars with planets. *A&A* 384:879–83

Barbieri M, Marzari F, Scholl H. 2002. Formation of terrestrial planets in close binary systems: the case of α Centauri A. *A&A* 396:219–24

Barman TS, Hauschildt PH, Allard F. 2001. Irradiated planets. *ApJ* 556:885–95

Barman TS, Hauschildt PH, Schweitzer A, Stanch, PC, Baron E, Allard F. 2002. Non-LTE effects of Na I in the atmosphere of HD 209458 b. *ApJ* 569:L51–54

Barnes JW, O'Brien DP. 2002. Stability of satellites around close-in extrasolar giant planets. *ApJ* 575:1087–93

Barnes SA. 2001. An assessment of the rotation rates of the host stars of extrasolar planets. *ApJ* 561:1095–1106

Bastian TS, Dulk GA, Leblanc Y. 2000. A search for radio emission from extrasolar planets. *ApJ* 545:1058–63

Batten AH. 1973. *Binary and multiple systems of stars.* Oxford: Pergamon Press, 278 pp.

Beckwith SVW, Sargent AI. 1993. The occurence and properties of disks around young stars. In *Protostars and planets III,* ed. EH Levy, JI Lunine, pp. 521–41. University of Arizona Press.

Béjar VJS, Martín EL, Zapatero Osorio MR, Rebolo R, Barrado y Navascués D, et al. 2001. The substellar mass function in σ Orionis. *ApJ* 556:830–36

Benedict GF, McArthur BE, Forveille T, Delfosse X, Nelan E, et al. 2002. A mass for the extrasolar planet Gl 876 b determined from Hubble Space Telescope Fine Guidance Sensor 3 astrometry and high-precision radial velocities. *ApJ* 581:L115–18

Bennett DP, Alcock C, Allsman RA, Axelrod TS, Cook KH, et al. 1995. Recent developments in gravitational microlensing and the latest MACHO results: microlensing towards the Galactic bulge. In *Dark matter,* ed. S Holt, DP Bennett, pp. 77–90. AIP Conf. Proc. Vol. 336, New York

Bennett DP, Rhie SH. 1996. Detecting Earth-mass planets with gravitational microlensing. *ApJ* 472:660–64

Bennett DP, Rhie SH. 2000. The Galactic Exoplanet Survey Telescope: a proposed space-based microlensing survey for terrestrial extrasolar planets. In *Disks, planetesimals, and planets,* ed. F Garzón, C Eiroa, D de Winter, TJ Mahoney, pp. 542–49. ASP Conf. Ser. Vol. 219, San Francisco, CA

Bennett DP, Rhie SH, Becker AC, Butler N, Dann J, et al. 1999. Discovery of a planet orbiting a binary star system from gravitational microlensing. *Nature* 402:57–59

Benson JA, Mozurkewich D, Jefferies SM. 1998. Active optical fringe tracking at the NPOI. In *Astronomical interferometry,* ed. RD Reasenberg, pp. 493–96. SPIE Vol. 3350. Bellingham, WA

Berry MV. 1987. The adiabatic phase and Pancharatnam's phase for polarized light. *J Mod Opt* 34:1401–07

Bertout C. 1989. T Tauri stars: wild as dust. *ARAA* 27:351–95

Cumming A, Marcy GW, Butler RP, Vogt SS. 2002. The statistics of extrasolar planets: results from the Keck survey. In *Scientific Frontiers iin Research on Extrasolar Planets* ed. D Deming and S. Seager *ASP Conf Ser 294*: 27–30

Cuntz M, Shkolnik E. 2002. Chromospheres, flares and exoplanets. *Astron Nachr* 323:387–91

Daigne G, Lestrade JF. 1999. Astrometric optical interferometry with non-evacuated delay lines. *A&AS* 138:355–63

Davies MB, Sigurdsson S. 2001. Planets in 47 Tuc. *MNRAS* 324:612–16

Davis J, Lawson PR, Booth AJ, Tango WJ, Thorvaldson ED. 1995. Atmospheric path variations for baselines up to 80 m measured with the Sydney University Stellar Interferometer. *MNRAS* 273:L53–58

Deeg HJ, Doyle LR, Kozhevnikov VP, Martín EL, Oetiker B, et al. 1998. Near-term detectability of terrestrial extrasolar planets: TEP network observations of CM Draconis. *A&A* 338:479–90

Deeg HJ, Doyle LR, Kozhevnikov VP, et al. 2000. A search for Jovian-Mass planets around CM Draconis using eclipse minima timing. *A & A* 358:L5.

Delfosse X, Forveille T, Mayor M, Perrier C, Naef D, Queloz D. 1998. The closest extrasolar planet. A giant planet around the M4 dwarf GL 876. *A&A* 338:L67–70

Delplancke F, Górski KM, Richichi A. 2001. Resolving gravitational microlensing events with long-baseline optical interferometry. Prospects for the ESO Very Large Telescope Interferometer. *A&A* 375:701–10

De Pater I, Lissauer JJ. 2001. *Planetary sciences.* Cambridge: Cambridge University Press. 528 pp.

Dermott SF, Grogan K, Durda DD, Jayaraman S, Kehoe TJJ, et al. 2001. Orbital evolution of interplanetary dust. In *Interplanetary dust,* ed. E Grün, BÅS Gustafson, SF Dermott, H Fechtig, pp. 569–639. Springer-Verlag

Derue F, Afonso C, Alard C, Albert JN, Andersen J, et al. 2001. Observation of microlensing toward the Galactic spiral arms. EROS II 3 year survey. *A&A* 373:126–38

Díaz-Cordovés J, Claret A, Giménez A. 1995. Linear and non-linear limb-darkening coefficients for LTE model atmospheres. *A&AS* 110:329–50

Dick SJ. 1982. *Plurality of worlds: the extraterrestrial life debate from Democritus to Kant.* Cambridge: Cambridge University Press. 246 pp.

Dick SJ. 1998. *Life on other worlds: the twentieth century extraterrestrial life debate.* Cambridge: Cambridge University Press. 304 pp.

Dominik C, Laureijs RJ, Jourdain de Muizon M, Habing HJ. 1998. A Vega-like disk associated with the planetary system of ρ^1 Cnc. *A&A* 329:L53–56

Doyle LR, Deeg HJ, Kozhevnikov VP, Oetiker B, Martín EL, et al. 2000. Observational limits on terrestrial-sized inner planets around the CM Draconis system using the photometric transit method with a matched-filter algorithm. *ApJ* 535:338–49

Drake FD. 1962. *Intelligent life in space.* New York: Macmillan. 128 pp.

Dravins D, Lindegren L, Mezey E, Young AT. 1998. Atmospheric intensity scintillation of stars. III. Effects for Different Telescope Apertures. *PASP* 110:610–33

Dreher JW, Cullers DK. 1997. SETI figure of merit. In *Astronomical and biochemical origins and the search for life in the Universe*, ed. CB Cosmovici, S Bowyer, D Werthimer, pp. 711–17. Bologna: Editrice Compositori

Dreizler S, Hauschildt PH, Kley W, Rauch T, Schuh SL, et al. 2003. OGLE-TR-3: a possible new transiting planet. *A&A* 402:791–99

Dreizler S, Rauch T, Hauschildt P, Schuh S, Kley W, Werner K. 2002. Spectral types of planetary host star candidates: two new transiting planets? *Astron Nachrichten Suppl* 324:23–28

Dutrey A. 1999. Latest stages of star formation and circumstellar environment of young stellar objects. In *Planets outside the Solar System: theory and observations*, ed. JM Mariotti, D Alloin, pp. 13–49. NATO ASI Vol. 532, Dordrecht: Kluwer

Dutrey A, Guilloteau S, Guelin M. 1997. Chemistry of protosolar-like nebulae: the molecular content of the DM Tau and GG Tau disks. *A&A* 317:L55–58

Dutrey A, Guilloteau S, Prato L, Simon M, Duvert G. 1998. CO study of the GM Aurigae Keplerian disk. *A&A* 338:L63–66

Dyson FW, Eddington AS, Davidson C. 1920. A Determination of the deflection of light by the Sun's gravitational field, from observations made at the total eclipse of May 29, 1919. *Phil Trans Roy Soc Series A* 220:291–333

Eddington AS. 1920. *Space, time and gravitation.* Cambridge: Cambridge University Press.

Einstein A. 1936. Lens-like action of a star by the deviation of light in the gravitational field. *Science* 84:506–07

Eisenhauer F, Quirrenbach A, Zinnecker H, Genzel R. 1998. Stellar content of the galactic starburst template NGC 3603 from adaptive optics observations. *ApJ* 498:278–92

Eisner JA, Kulkarni SR. 2001a. Sensitivity of the radial-velocity technique in detecting outer planets. *ApJ* 550:871–83

Elliot JL. 1978. Direct imaging of extrasolar planets with stationary occultations viewed by a space telescope. *Icarus* 35:156–64

Encrenaz T. 2001. The formation of planets. In *Solar and extra-solar planetary systems*, ed. IP Williams, N Thomas, pp. 76–90. Lecture Notes in Physics, Springer, Berlin

Endl M, Kürster M, Els S, Hatzes AP, Cochran WD, et al. 2002. The planet search program at the ESO Coudé Echelle Spectrometer. III. The complete long camera survey results. *A&A* 392:671–90

Erdl H, Schneider P. 1993. Classification of the multiple deflection two point-mass gravitational lens models and application of catastrophe theory in lensing. 268:453–71

European Space Agency. 1997. *The Hipparcos and Tycho catalogues.* ESA SP-1200

Evans NJ. 1999. Physical conditions in regions of star formation. *ARAA* 37:311–62

Everett ME, Howell SB. 2001. A technique for ultrahigh-precision CCD photometry. *PASP* 113:1428–35

Farrell WM, Desch MD, Zarka P. 1999. On the possibility of coherent cyclotron emission from extrasolar planets. *JGR* 104:14025–32

Favata F. 2002. The Eddington baseline mission. In *Proceedings of the first Eddington workshop on stellar structure and habitable planet finding*, ed. F Favata, IW Roxburgh, D Galadi, pp. 3–10. ESA SP-485, Noordwijk: ESA Publications Division

Ferlet R, Vidal-Madjar A, Hobbs LM. 1987. The Beta Pictoris circumstellar disk. V. Time variations of the Ca II-K line. *A&A* 185:267–70

Fischer DA, Butler RP, Marcy GW, Vogt SS, Henry GW. 2003a. A sub-Saturn mass companion to HD 3651. *ApJ*, submitted

Fischer DA, Marcy GW, Butler RP, Laughlin G, Vogt SS. 2002a. A Second Planet Orbiting 47 Ursae Majoris. *ApJ* 564:1028–34

Fischer DA, Marcy GW, Butler RP, Vogt SS, Frink S, Apps K. 2001. Planetary companions to HD 12661, HD 92788, and HD 38529 and variations in Keplerian residuals of extrasolar planets. *ApJ* 551:1107–18

Fischer DA, Marcy GW, Butler RP, Vogt SS, Henry GW, et al. 2003b. A planetary companion to HD 40979 and additional planets orbiting HD 12661 and HD 38529. *ApJ* 586:1394–408

Fischer DA, Marcy GW, Butler RP, Vogt SS, Walp B, Apps K. 2002b. Planetary companions to HD 136118, HD 50554, and HD 106252. *PASP* 114:529–35

Ford EB, Joshi KJ, Rasio FA, Zbarsky B. 2000. Theoretical implications of the PSR B1620−26 triple system and its planet. *ApJ* 528:336–50

Freundling W, Lagrange AM, Vidal-Madjar A, Ferlet R, Forveille T. 1995. Gas around β Pictoris: an upper limit on the H I content. *A&A* 301:231–35

Fried DL. 1994. Atmospheric turbulence optical effects: understanding the adaptive-optics implications. In *Adaptive optics for astronomy*, ed. DM Alloin, JM Mariotti, pp. 25–57. NATO ASI Vol. 423, Dordrecht: Kluwer

Frink S, Mitchell DS, Quirrenbach A, Fischer DA, Marcy GW, Butler RP. 2002. Discovery of a substellar companion to the K2 III giant ι Draconis. *ApJ* 576:478–84

Frink S, Quirrenbach A, Fischer D, Röser S, Schilbach E. 2000a. K giants as astrometric reference stars for the Space Interferometry Mission. In *Interferometry in optical astronomy*, ed. PJ Léna, A Quirrenbach, pp. 806–14. SPIE Vol. 4006. Bellingham, WA

Frink S, Quirrenbach A, Fischer D, Röser S, Schilbach E. 2001. A strategy for identifying the grid stars for the Space Interferometry Mission (SIM). *PASP* 113:173–87

Frink S, Quirrenbach A, Röser S, Schilbach E. 2000b. Testing Hipparcos K giants as grid stars for SIM. In *Working on the fringe,* ed. S Unwin, R Stachnik, pp. 128–33. ASP Conf. Ser. Vol. 194, San Francisco, CA

Gatewood G, Eichhorn H. 1973. An unsuccessful search for a planetary companion of Barnard's Star (BD +4°3561). *AJ* 78:769–76

Gaudi BS. 1998. Distinguishing between binary-source and planetary microlensing perturbations. *ApJ* 506:533–39

Gaudi BS. 2000. Planetary transits toward the Galactic bulge. *ApJ* 539:L59–62

Gaudi BS, Albrow MD, An J, Beaulieu JP, Caldwell JAR, et al. 2002. Microlensing constraints on the frequency of Jupiter-mass companions: analysis of 5 years of PLANET photometry. *ApJ* 566:463–99

Gaudi BS, Gould, A. 1997. Planet parameters in microlensing events. *ApJ* 486:85–99

Gaudi BS, Naber RM, Sackett PD. 1998. Microlensing by multiple planets in high-magnification events. *ApJ* 502:L33–37

Gaudi BS, Sackett PD. 2000. Detection efficiencies of microlensing data sets to stellar and planetary companions. *ApJ* 528:56–73

Gay J, Rabbia Y. 1996. Principe du coronographe interférentiel achromatique. *C R Acad Sci Paris* 322:265–71

Ge J, Erskine D, Rushford M. 2002. An externally dispersed interferometer for sensitive Doppler extrasolar planet searches. *PASP* 114:1016–28

Genzel R. 1992. In *The galactic interstellar medium.* Springer-Verlag

Gilliland RL, Brown TM, Guhathakurta P, Sarajedini A, Milone EF, et al. 2000. A Lack of planets in 47 Tucanae from a Hubble Space Telescope search. *ApJ* 545:L47–51

Gilmore G, de Boer K, Favata F, Høg E, Lattanzi M, et al. 2000. GAIA: origin and evolution of the Milky Way. In *UV, optical, and IR space telescopes and instruments,* ed. JB Breckinridge, P Jakobsen, pp. 453–72. SPIE Vol. 4013, Bellingham, WA

Gladman B. 1993. Dynamics of systems of two close planets. *Icarus* 106:247–63

Goldreich P, Soter S. 1966. Q in the Solar System. *Icarus* 5:375–89

Gonzalez G, Laws C, Tyagi S, Reddy BE. 2001b. Parent stars of extrasolar planets. VI. Abundance analyses of 20 new systems. *AJ* 121:432–52

Goodwin SP. 1997. Residual gas expulsion from young globular clusters. *MNRAS* 284:785–802

Gould A, Loeb A. 1992. Discovering planetary systems through gravitational microlensing. *ApJ* 396:104–14

Gould SJ. 1989. *Wonderful life: the Burgess Shale and the nature of history.* New York: W.W. Norton. 347 pp.

Goździewski K. 2002. Stability of the 47 UMa planetary system. *A&A* 393:997–1013

Goździewski K, Maciejewski, AJ. 2001. Dynamical analysis of the orbital parameters of the HD 82943 planetary system. *ApJ* 563:L81–85

Goździewski K, Maciejewski, AJ. 2003. The Janus head of the HD 12661 planetary system. *ApJ* 586:L153–56

Grady CA, Mora A, de Winter D. 2000a. Infall, accretion, and the spectroscopic evidence for planetesimals. In *Disks, planetesimals and planets*, ed. F Garzón, C Eiroa, D de Winter, TJ Mahoney, pp. 202–14. ASP Conf. Ser. Vol. 219. San Francisco, CA

Grady CA, Pérez MR, Talavera A, Bjorkman KS, de Winter D, et al. 1996. The β Pictoris phenomenon among Herbig Ae/Be stars. UV and optical high dispersion spectra. *A&AS* 120:157–77

Grady CA, Sitko ML, Bjorkman KS, Pérez MR, Lynch DK, et al. 1997. The star-grazing extrasolar comets in the HD 100546 system. *ApJ* 483:449–56

Grady CA, Sitko ML, Russell RW, Lynch DK, Hanner MS, et al. 2000b. Infalling planetesimals in pre-main stellar systems. In *Protostars and planets IV*, ed. V Mannings, AP Boss, SS Russell, pp. 613–38. University of Arizona Press

Graff DS, Gaudi BS. 2000. Direct detection of large close-in planets around the source stars of caustic-crossing microlensing events. *ApJ* 538:L133–36

Gratton RG, D'Antona F. 1989. HD 39853: a high velocity K5 III star with an exceptionally large Li content. *A&A* 215:66–78

Gray DF. 1998. A planetary companion for 51 Pegasi implied by absence of pulsations in the stellar spectra. *Nature* 391:153–54

Greaves JS, Holland WS, Moriarty-Schieven G, Jenness T, Dent WR, et al. 1998. A dust ring around ε Eridani: analog to the young Solar System. *ApJ* 506:L133–37

Greaves JS, Mannings V, Holland WS. 2000. The dust and gas content of a disk around the young star HR 4796 A. *Icarus* 143:155–58

Griest K, Safizadeh N. 1998. The use of high-magnification microlensing events in discovering extrasolar planets. *ApJ* 500:37–50

Griffin RF, Griffin RE. 1973. On the possibility of determining stellar radial velocities to 0.01 km/s. *MNRAS* 162:243–53

Grinin VP. 1999. Infalling material on young stars. In *Planets outside the Solar System: theory and observations*, ed. JM Mariotti, D Alloin, pp. 51–63. NATO ASI Vol. 532, Dordrecht: Kluwer

Gubler J, Tytler D. 1998. Differential atmospheric refraction and limitations on the relative astrometric accuracy of large telescopes. *PASP* 110:738–46

Guilloteau S, Dutrey A. 1998. Physical parameters of the Keplerian protoplanetary disk of DM Tauri. *A&A* 339:467–76

Guirado J, Ros E, Jones D, Alef W, Marcaide J, Preston R. 2002. Searching for low mass objects around nearby dMe radio stars. In *Proceedings of the 6th European VLBI Network Symposium*, ed E Ros, RW Porcas, AP Lobanov, JA Zensus, pp. 255–58. MPIfR, Bonn, Germany

Haisch KE, Lada EA, Lada CJ. 2001a. Circumstellar disks in the IC 348 cluster. *AJ* 121:2065–74

Haisch KE, Lada EA, Lada CJ. 2001b. Disk frequencies and lifetimes in young clusters. *ApJ* 553:L153–56

Halbwachs JL, Arenou F, Mayor M, Udry S, Queloz D. 2000. Exploring the brown dwarf desert with Hipparcos. *A&A* 355:581–94

Hamilton CM, Herbst W, Shih C, Ferro AJ. 2001. Eclipses by a circumstellar dust feature in the pre-main-sequence star KH 15D. *ApJ* 554:L201–04

Han C. 2002. Astrometric method to break the photometric degeneracy between binary-source and planetary microlensing perturbations. *ApJ* 564:1015–18

Hanner MS, Lynch DK, Russell RW. 1994. The 8–13 micron spectra of comets and the composition of silicate grains. *ApJ* 425:274–85

Hardy JW. 1998. *Adaptive optics for astronomical telescopes.* New York: Oxford University Press. 438 pp.

Harrington RS, Kallarakal VV, Dahn CC. 1983. Astrometry of the low-luminosity stars VB8 and VB10. *AJ* 88:1038–39

Hatzes AP, Cochran WD. 1994. Short-period radial velocity variations of α Bootis: Evidence for radial pulsations. *ApJ* 422:366–73

Hatzes AP, Cochran WD, McArthur B, Baliunas SL, Walker GAH, et al. 2000. Evidence for a long-period planet orbiting ε Eridani. *ApJ* 544:L145–48

Heap SR, Lindler DJ, Lanz TM, Cornett RH, Hubeny I, et al. 2000. Space Telescope Imaging Spectrograph Coronagraphic observations of β Pictoris. *ApJ* 539:435–44

Heintz WD. 1971. *Doppelsterne.* München: Wilhelm Goldmann Verlag. 186 pp.

Henry GW, Donahue RA, Baliunas SL. 2002. A false planet around HD 192263. *ApJ* 577:L111–14

Henry GW, Marcy GW, Butler RP, Vogt SS. 2000. A Transiting "51 Peg-like" Planet. *ApJ* 529:L41–44

Herbst W, Hamilton CM, Vrba FJ, Ibrahimov MA, Bailer-Jones CAL, et al. 2002. Fine structure in the circumstellar environment of a young, Solar-like star: the unique eclipses of KH 15D. *PASP* 114:1167–72

Hestroffer D. 1997. Centre to limb darkening of stars: new model and application to stellar interferometry. *A&A* 327:199–206

Hill HGM, Grady CA, Nuth JA, Hallenbeck SL, Sitko ML. 2001. Constraints on nebular dynamics and chemistry based on observations of annealed magnesium silicate grains in comets and in disks surrounding Herbig Ae/Be stars. *PNAS* 98:2182–87

Hinz PM, Angel R, Hoffman W, McCarthy DW, McGuire PC, et al. 1998. Imaging circumstellar environments with a nulling interferometer. *Nature* 395:251–53

Holland WS, Greaves JS, Zuckerman B, Webb RA, McCarthy C, et al. 1998. Submillimetre images of dusty debris around nearby stars. *Nature* 392:788–91

Holland WS, Greaves JS, Dent WRF, Wyatt MC, Zuckerman B, et al. 2003. Submillimeter observations of an asymmetric dust disk around Fomalhaut. *ApJ* 582:1141–46

Hubbard WB, Fortney JJ, Lunine JI, Burrows A, Sudarsky D, Pinto P. 2001. Theory of extrasolar giant planet transits. *ApJ* 560:413–19

Hufnagel, RE. 1974. Variations of atmospheric turbulence. In *Digest of technical papers presented at the topical meeting on optical propagation through turbulence*, Optical Society of America, p. Wa1:1–4

Hui L, Seager S. 2002. Atmospheric lensing and oblateness effects during an extrasolar planetary transit. *ApJ* 572:540–55

Ida S, Makino J. 1993. Scattering of planetesimals by a protoplanet: slowing down of runaway growth. *Icarus* 106:210–27

Ignace R. 2001. Spectral energy distribution signatures of Jovian planets around white dwarf stars. *PASP* 113:1227–31

Israelian G, Santos NC, Mayor M, Rebolo R. 2001. Evidence for planet engulfment by the star HD 82943. *Nature* 411:163–66

Janes K. 1996. Star clusters: Optimal targets for a photometric planetary search program. *JGR* 101:14853–60

Jayawardhana R, Holland WS, Kalas P, Greaves JS, Dent WRF, et al. 2002. New submillimeter limits on dust in the 55 Cancri planetary system. *ApJ* 570:L93–96

Jenkins JM, Caldwell DA, Borucki WJ. 2002. Some tests to establish confidence in planets discovered by transit photometry. *ApJ* 564:495–507

Jewitt D, Luu J. 1993. Discovery of the candidate Kuiper belt object 1993 QB1. *Nature* 362:730–32

Ji J, Li G, Liu L. 2002. The dynamical simulations of the planets orbiting GJ 876

Jiang IG, Ip WH, Yeh LC. 2003. On the fate of close-in extrasolar planets. *ApJ* 582:449–54

Johnstone D, Hollenbach D, Bally J. 1998. Photoevaporation of disks and clumps by nearby massive stars: application to disk destruction in the Orion Nebula. *ApJ* 499:758–76

Jorissen A, Mayor M, Udry S. 2001. The distribution of exoplanet masses. *A&A* 379:992–98

Joshi KJ, Rasio FA. 1997. Distant companions and planets around milliseconds pulsars. *ApJ* 479:948–59

Kalas P, Graham JR, Beckwith SVW, Jewitt DC, Lloyd JP. 2002. Discovery of reflection nebulosity around five Vega-like stars. *ApJ* 567:999–1012

Kalas P, Larwood J, Smith BA, Schultz A. 2000. Rings in the planetesimal disk of β Pictoris. *ApJ* 530:L133–37

Kamp I, van Zadelhoff GJ, van Dishoeck EF, Stark R. 2003. Line emission from circumstellar disks around A stars. *A&A* 397:1129–41

Kant I. 1755. *Allgemeine Naturgeschichte und Theorie des Himmels*. Leipzig

Karlsson A, Mennesson B. 2000. The Robin Laurance nulling interferometers. In *Interferometry in optical astronomy*, ed. PJ Léna, A Quirrenbach, pp. 871–80. SPIE Vol. 4006. Bellingham, WA

Kenyon SJ, Bromley BC. 2002. Dusty rings: signposts of recent planet formation. *ApJ* 577:L35–38

Kiseleva-Eggleton L, Bois E, Rambaux N, Dvorak R. 2002. Global dynamics and stability limits for planetary systems around HD 12661, HD 38529, HD 37124, and HD 160691. *ApJ* 578:L145–48

Kley W, D'Angelo G, Henning T. 2001. Three-dimensional simulations of a planet embedded in a protoplanetary disk. *ApJ* 547:457–64

Knacke RF, Fajardo-Acosta SB, Telesco CM, Hackwell JA, Lynch DK, Russell RW. 1993. The silicates in the disk of β Pictoris. *ApJ* 418:440–50

Knittl Z. 1976. *Optics of thin films.* New York: Wiley. 548 pp.

Koch D, Borucki W, Cullers K, Dunham E, Webster L, et al. 1996. System design of a mission to detect Earth-size planets in the inner orbits of Solar-like stars. *JGR* 101:9297–302

Koch D, Borucki W, Webster L, Dunham E, Jenkins J, et al. 1998. Kepler: a space mission to detect earth-class exoplanets. In *Space telescopes and instruments V,* ed. PY Bely, JB Breckinridge, pp. 599–607. SPIE Vol. 3356. Bellingham, WA

Koerner DW. 1997. Analogs of the early Solar System. *Orig Life and Evol Biosphere* 27:157–84

Koerner DW, Sargent AI, Beckwith SVW. 1993. A rotating gaseous disk around the T Tauri star GM Aurigae. *Icarus* 106:2–10

Konacki M, Torres G, Jha S, Sasselov DD. 2003. An extrasolar planet that transits the disk of its parent star. *Nature* 421:507–09

Konacki M, Wolszczan A. 2003. Masses and orbital inclinations of planets in the PSR B1257+12 system. *ApJ* 591:L147-L150

Kornet K, Bodenheimer P, Różyczka M. 2002. Models of the formation of the planets in the 47 UMa system. *A&A* 396:977–86

Korzennik SG, Brown TM, Fischer DA, Nisenson P, Noyes RW. 2000. A high-eccentricity low-mass companion to HD 89744. *ApJ* 533:L147–50

Kovács G, Zucker S, Mazeh T. 2002. A box-fitting algorithm in the search for periodic transits. *A&A* 391:369–77

Kozai Y. 1962. Secular perturbations of asteroids with high inclination and eccentricity. *AJ* 67:591–98

Krist JE, Stapelfeldt KR, Ménard F, Padgett DL, Burrows C.J. 2000. WFPC2 images of a face-on disk surrounding TW Hydrae. *ApJ* 538:793–800

Krist JE, Stapelfeldt KR, Watson AM. 2002. Hubble Space Telescope/ WFPC2 images of the GG Tauri circumbinary disk. *ApJ* 570:785–92

Kroupa P, Aarseth S, Hurley J. 2001. The formation of a bound star cluster: from the Orion nebula cluster to the Pleiades. *MNRAS* 321: 699–712

Kürster M, Endl M, Els S, Hatzes AP, Cochran WD, et al. 2000. An extrasolar giant planet in an Earth-like orbit. Precise radial velocities of the young star ι Horologii = HR 810. *A&A* 353:L33–36

Lacy CH. 1977. Absolute dimensions and masses of the remarkable spotted dM4e eclipsing binary flare star CM Draconis. *ApJ* 218:444–60

Lagrange AM, Backman DE, Artymowicz P. 2000. Planetary material around main-sequence stars. In *Protostars and planets IV,* ed. V Mannings, AP Boss, SS Russell, pp. 639–72. University of Arizona Press

Lagrange AM, Beust H, Mouillet D, Deleuil M, Feldman PD, et al. 1998. The β Pictoris circumstellar disk. XXIV. Clues to the origin of the stable gas. *A&A* 330:1091–1108

Laplace PS. 1796. *Exposition du système du Monde.* Paris

Latham DW, Mazeh T, Stefanik RP, Mayor M, Burki G. 1989. The unseen companion of HD 114762: a probable brown dwarf. *Nature* 339:38–40

Lattanzi MG, Spagna A, Sozzetti A, Casertano S. 2000. Space-borne global astrometric surveys: the hunt for extrasolar planets. *MNRAS* 317:211–24

Laughlin G, Adams FC. 1999. Stability and chaos in the υ Andromedae planetary system. *ApJ* 526:881–89

Laughlin G, Chambers JE. 2001. Short-term dynamical interactions among extrasolar planets. *ApJ* 551:L109–13

Laughlin G, Chambers JE. 2002. Extrasolar Trojans: the viability and detectability of planets in the 1:1 resonance. *ApJ* 124:592–600

Laughlin G, Chambers JE, Fischer D. 2002. A dynamical analysis of the 47 Ursae Majoris planetary system. *ApJ* 579:455–67

Lawson PR, ed. 2000. *Principles of long baseline stellar interferometry. Course notes from the 1999 Michelson summer school.* JPL Publication 00–009, Pasadena, CA

le Poole RS, Quirrenbach A. 2002. Optimized beam-combination schemes for each channel for PRIMA. In *Interferometry for optical astronomy II,* ed. WA Traub, pp. 496–502. SPIE Vol. 4838. Bellingham, WA

Lecavelier des Etangs A. 1998. Planetary migration and sources of dust in the β Pictoris disk. *A&A* 337:501–11

Lecavelier des Etangs A, Vidal-Madjar A, Ferlet R. 1999. Photometric stellar variation due to extrasolar comets. *A&A* 343:916–22

Lecavelier des Etangs A, Vidal-Madjar A, Roberge A, Feldman PD, Deleuil, M, et al. 2001. Deficiency of molecular hydrogen in the disk of β Pictoris. *Nature* 412:706–08

Lee MH, Peale SJ. 2003. Secular evolution of hierarchical planetary systems. *ApJ* 592:1201–1216

Lestrade JF. 2000a. Stellar VLBI. In *Proceedings of the 5th European VLBI Network Symposium,* ed. JE Conway, AG Polatidis, RS Booth, Y. Pihlström, pp. 155–61. Onsala, Sweden

Lestrade JF. 2000b. Potential of ALMA for extra-solar planet search by astrometry. Unpublished.

Lewis GF, Ibata RA. 2000. Probing the atmospheres of planets orbiting microlensed stars via polarization variability. *ApJ* 539:L63–66

Lin DNC, Bodenheimer P, Richardson DC. 1996. Orbital migration of the planetary companion of 51 Pegasi to its present location. *Nature* 380:606–07

Lin DNC, Ida S. 1997. On the origin of massive eccentric planets. *ApJ* 477:781–91

Lindegren L, Perryman MAC. 1996. GAIA: global astrometric interferometer for astrophysics. *A&AS* 116:579–95

Lineweaver C, Grether D. 2002. The observational case for Jupiter being a typical massive planet. *Astrobiology* vol.2 issue 3, 325–334

Linfield RP, Colavita MM, Lane, BF. 2001. Atmospheric turbulence measurements with the Palomar Testbed Interferometer. *ApJ* 554:505–13

Liseau R, Artymowicz P. 1998. High sensitivity search for molecular gas in the β Pic disk – On the low gas-to-dust mass ratio of the circumstellar disk around β Pictoris. *A&A* 334:935–42

Liseau R, Brandeker A, Fridlund M, Olofsson G, Takeuchi T, Artymowicz P. 2003. The 1.2 mm image of the β Pictoris disk. *A&A* 402:183–187

Lissauer JJ, Rivera EJ. 2001. Stability analysis of the planetary system orbiting υ Andromedae. II. Simulations using new Lick observatory fits. *ApJ* 554:1141–50

Lloyd JP, Oppenheimer BR, Graham JR. 2002. The potential of differential astrometric interferometry from the high Antarctic plateau. *PASA* 19:318–22

Lucas PW, Roche PF, Allard F, Hauschildt PH. 2001. Infrared spectroscopy of substellar objects in Orion. *MNRAS* 326:695–721

Luhman KL, Jayawardhana R. 2002. An adaptive optics search for companions to stars with planets. *ApJ* 566:1132–46

Luu JX, Jewitt DC. 2002. Kuiper Belt objects: relics from the accretion disk of the Sun. *ARAA* 40:63–101

Lyne AG, Bailes M. 1992. No planet orbiting PSR1829−10. *Nature* 355:213

Malfait K, Waelkens C, Waters LBFM, Vandenbussche B, Huygen E, de Graauw MS. 1998. The spectrum of the young star HD 100546 observed with the Infrared Space Observatory. *A&A* 332:L25–28

Mallen-Ornélas G, Seager S, Yee HKC, Minniti D, Gladders MD et al. 2003. The EXPLORE project I: a deep search for transiting extrasolar planets. *ApJ* 582:1123–1140

Mann I. 2001. Dust in the Solar System and in other planetary systems. In *Solar and extra-solar planetary systems*, ed. IP Williams, N Thomas, pp. 218–42. Lecture Notes in Physics, Springer, Berlin

Mao S, Paczyński B. 1991. Gravitational microlensing by double stars and planetary systems. *ApJ* 374:L37–40

Marcy GW, Butler RP. 1992. Precision radial velocities with an iodine absorption cell. *PASP* 104:270–77

Marcy GW, Butler RP. 1995. The planet around 51 Pegasi. *AAS abstract* 187:70.04

Marcy GW, Butler RP. 1996. A planetary companion to 70 Virginis. *ApJ* 464:L147–51

Marcy GW, Butler RP. 1998. Doppler detection of extra-solar planets. In *Cool stars, stellar systems and the Sun, tenth Cambridge workshop*, ed.

Donahue RA, Bookbinder JA, pp. 9–24. ASP Conf. Ser. Vol. 158. San Francisco, CA

Marcy GW, Butler RP. 2000. Planets orbiting other Suns. *PASP* 112:137–40

Marcy GW, Butler RP, Fischer D, Laughlin G, Vogt SS, et al. 2002. A planet at 5 AU around 55 Cancri. *ApJ* 581:1375–88

Marcy GW, Butler RP, Fischer DA, Vogt SS. 2003a. Properties of extrasolar planets. In *Scientific frontiers in research on extrasolar planets,* ed. D Deming, S Seager. ASP Conf. Ser., 294:1–16

Marcy GW, Butler RP, Fischer D, Vogt SS, Lissauer JJ, Rivera EJ. 2001a. A pair of resonant planets orbiting GJ 876. *ApJ* 556:296–301

Marcy GW, Butler RP, Vogt SS. 2000a. Sub-Saturn planetary candidates of HD 16141 and HD 46375. *ApJ* 536:L43–46

Marcy GW, Butler RP, Vogt SS, Fischer D, Lissauer JJ. 1998. A planetary companion to a nearby M4 dwarf, Gliese 876. *ApJ* 505:L147–49

Marcy GW, Butler RP, Vogt SS, Liu MC, Laughlin G, et al. 2001b. Two substellar companions orbiting HD 168443. *ApJ* 555:418–25

Marcy GW, Butler RP, Williams E, Bildsten L, Graham JR, et al. 1997. The planet around 51 Pegasi. *ApJ* 481:926–35

Marcy GW, Fischer DA, Butler RP, Vogt SS. 2003b. Systems of multiple planets. *SSR,* in press

Mardling RA, Lin DNC. 2002. Calculating the tidal, spin, and dynamical evolution of extrasolar planetary systems. *ApJ* 573:829–44

Marley MS, Gelino C, Stephens D, Lunine JI, Freedman, R. 1999. Reflected spectra and albedos of extrasolar giant planets. I. Clear and cloudy atmospheres. *ApJ* 513:879–93

Matthews JM, Kuschnig R, Walker GAH, Pazder J, Johnson R, et al. 2000. Ultraprecise photometry from space: The MOST microsat mission. In *The Impact of Large-Scale Surveys on Pulsating Star Research,* ed. L Szabados, D Kurtz, pp. 74–75. IAU Coll. 176, ASP Conf. Ser. Vol. 203. San Francisco, CA

Mayor M, Queloz D. 1995. A Jupiter-mass companion to a Solar-type star. *Nature* 378:355–59

Mayor M., Udry S., Naef D., Pepe F., Queloz D., et al 2004, The CORALIE survey for southern extra-solar planets. XII Orbital solutions for 16 extrasolar planets detected with CORALIE *A&A* 415:391:402

Mazeh T, Naef D, Torres G, Latham DW, Mayor M, et al. 2000. The spectroscopic orbit of the planetary companion transiting HD 209458. *ApJ* 532:L55–58

Mazeh T, Zucker S, Dalla Torre A, van Leeuwen F. 1999. Analysis of the HIPPARCOS measurements of υ Andromedae: a mass estimate of its outermost known planetary companion. *ApJ* 522:L149–51

McCarthy DW, Probst RG, Low FJ. 1985. Infrared detection of a close cool companion to Van Biesbroeck 8. *ApJ* 290:L9–13

McCaughrean MJ, O'Dell CR. 1996. Direct imaging of circumstellar disks in the Orion Nebula. *AJ* 111:1977–86

McCaughrean MJ, Stapelfeldt KR, Close LM. 2000. High-resolution optical and near-infrared imaging of young circumstellar disks. In *Protostars and planets IV*, ed. V Mannings, AP Boss, SS Russell, pp. 485–507. University of Arizona Press

McLean IS. 1997. *Electronic imaging in astronomy*. Chichester: Praxis Publishing. 472 pp.

Meisner J, le Poole RS. 2002. Dispersion affecting the VLTI and 10 micron interferometry using MIDI. In *Interferometry for optical astronomy II*, ed. WA Traub, SPIE Vol. 4838, in press

Melnick GJ, Neufeld DA, Ford KES, Hollenbach DJ, Ashby MLN. 2001. Discovery of water vapour around IRC+10216 as evidence for comets orbiting another star. *Nature* 412:160–63

Men'shchikov AB, Henning T, Fischer O. 1999. Self-consistent model of the dusty torus around HL Tauri. *ApJ* 519:257–78

Mennesson B, Mariotti JM. 1997. Array configurations for a space infrared nulling interferometer dedicated to the search for earthlike extrasolar planets. *Icarus* 128:202–12

Mennesson B, Ollivier M, Ruilier C. 2002. Use of single-mode waveguides to correct the optical defects of a nulling interferometer. *JOSA A* 19:596–602

Meyer MR, Backman D, Beckwith SVW, Brooke TY, Carpenter JM, et al. 2001. The formation and evolution of planetary systems: SIRTF legacy science in the VLT era. In *The Origin of Stars and Planets: The VLT View*, ESO

Mieremet AL, Braat JJM. 2002a. Nulling interferometry without achromatic phase shifters. *Appl Opt* 41:4697–703

Mieremet AL, Braat JJM. 2002b. Towards the deepest possible null. In *Hunting for planets – GENIE VLTI instrument: a DARWIN technology demonstrator*, in press

Mieremet AL, Braat J, Bokhove H, Ravel K. 2000. Achromatic phase shifting using adjustable dispersive elements. In *Interferometry in optical astronomy*, ed. PJ Léna, A Quirrenbach, pp. 1035–41. SPIE Vol. 4006. Bellingham, WA

Miller SL. 1953. A production of amino acids under possible primitive Earth conditions. *Science* 117:528–29.

Milman MH, Turyshev SG. 2000. Observational model for the Space Interferometry Mission. In *Interferometry in optical astronomy*, ed. PJ Léna, A Quirrenbach, pp. 828–37. SPIE Vol. 4006. Bellingham, WA

Misner CW, Thorne KS, Wheeler JA. 1973. *Gravitation*. New York: W.H. Freeman and Company. 1279 pp.

Mochejska BJ, Stanek KZ, Sasselov DD, Szentgyorgyi AH. 2002. Planets In Stellar Clusters Extensive Search. I. Discovery of 47 low-amplitude variables in the metal-rich cluster NGC 6791 with millimagnitude image subtraction photometry. *AJ* 123:3460–72

Monet D, Bird A, Canzian B, Harris H, Reid N, et al. 1996. *USNO-A1.0*. U.S. Naval Observatory, Washington DC

Mora A, Eiroa C, Natta A, Grady CA, de Winter D, et al. 2003. Dynamics of the circumstellar gas in BF Orionis, SV Cephei, WW Vulpeculae and XY Persei. 2003. A&A, submitted

Mora A, Natta A, Eiroa C, Grady CA, de Winter D, et al. 2002. A dynamical study of the circumstellar gas in UX Orionis. A&A 393:259–71

Morgan RM, Burge J, Woolf N. 2000. Nulling interferometric beam combiner utilizing dielectric plates: experimental results in the visible broadband. In *Interferometry in optical astronomy,* ed. PJ Léna, A Quirrenbach, pp. 340–48. SPIE Vol. 4006. Bellingham, WA

Mouillet D, Larwood JD, Papaloizou JCB, Lagrange AM. 1997. A planet on an inclined orbit as an explanation of the warp in the β Pictoris disc. *MNRAS* 292:896–904

Moutou C, Coustenis A, Schneider J, St Gilles R, Mayor M, et al. 2001. Search for spectroscopical signatures of transiting HD 209458 B's exosphere. A&A 371:260–66

Mozurkewich D, Johnston KJ, Simon RS, Bowers PF, Gaume R, et al. 1991. Angular diameter measurements of stars. *AJ* 101:2207–19

Murray CD, Dermott SF. 1999. *Solar system dynamics.* Cambridge: Cambridge University Press. 592 pp.

Murray N, Chaboyer B. 2002. Are stars with planets polluted? *ApJ* 566:442–51

Murray N, Chaboyer B, Arras P, Hansen B, Noyes RW. 2001. Stellar pollution in the Solar neighborhood. *ApJ* 555:801–15

Naef D, Latham DW, Mayor M, Mazeh T, Beuzit JL. 2001a. HD 80606 b, a planet on an extremely elongated orbit. A&A 375:L27–30

Naef D, Mayor M, Pepe F, Queloz D, Santos NC, et al. 2001b. The CORALIE survey for southern extrasolar planets. V. 3 new extrasolar planets. A&A 375:205–18

Naef D, Mayor M, Beuzit J-L, Perrier C., Queloz et al. 2004, The ELODIE survey for northern extrasolar planets III: Three planetary candidates detected with ELODIE. it A&A 414:351–359

NASA. 1999a. *SIM Space Interferometry Mission: taking the measure of the Universe,* eds. R Danner & S Unwin. JPL 400–811. Pasadena, CA. 139 pp.

NASA. 1999b. *TPF: Terrestrial Planet Finder,* eds. CA Beichman, NJ Woolf, CA Lindensmith. JPL 99–3. Pasadena, CA. 158 pp.

Natta A, Grinin VP, Tambovtseva LV. 2000. An interesting episode of accretion activity in UX Orionis. *ApJ* 542:421–27

Nauenberg M. 2002a. Determination of masses and other properties of extrasolar planetary systems with more than one planet. *ApJ* 568:369–76

Nauenberg M. 2002b. Stability and eccentricity for two planets in 1:1 resonance, and their possible occurence in extrasolar planetary systems. *ApJ* 124:2332–38

Nelson AF, Angel JRP. 1998. The range of masses and periods explored by radial velocity searches for planetary companions. *ApJ* 500:940–57

Nelson RP, Papaloizou JCB. 2002. Possible commensurabilities among pairs of extrasolar planets. *MNRAS* 333:L26–30

Nisenson P, Contos A, Korzennik S, Noyes R, Brown T. 1999. The Advanced Fiber-optic Echelle (AFOE) and extrasolar planet searches. In *Precise stellar radial velocities,* ed. JB Hearnshaw, CD Scarfe, pp. 143–53. IAU Coll. 170, ASP Conf. Ser. Vol. 185. San Francisco, CA

Noyes RW, Hartmann LW, Baliunas SL, Duncan DK, Vaughan AH. 1984. Rotation, convection, and magnetic activity in lower main-sequence stars. *ApJ* 279:763–77

Noyes RW, Jha S, Korzennik SG, Krockenberger M, Nisenson P, et al. 1997. A planet orbiting the star ρ Coronae Borealis. *ApJ* 483:L111–14

O'Dell CR, Wen Z, Hu X. 1993. Discovery of new objects in the Orion nebula on HST images: shocks, compact sources, and protoplanetary disks. *ApJ* 410:696–700

O'Leary BT. 1966. On the occurrence and nature of planets outside the Solar System. *Icarus* 5:419–36

Olofsson G, Liseau R, Brandeker A. 2001. Widespread atomic gas emission reveals the rotation of the β Pictoris disk. *ApJ* 563:L77–80

Ozernoy LM, Gorkavyi NN, Mather JC, Taidakova TA. 2000. Signatures of exosolar planets in dust debris disks. *ApJ* 537:L147–51

Paczyński B. 1986. Gravitational microlensing by the Galactic halo. *ApJ* 304:1–5

Paczyński B. 1996. Gravitational microlensing in the Local Group. *ARAA* 34:419–59

Pancharatnam S. 1956. Generalized theory of interference, and its applications. Part I. Coherent pencils. *Proc Ind Acad Sci A* 44:247–62

Pantin E, Lagage PO, Artymowicz P. 1997. Mid-infrared images and models of the β Pictoris dust disk. *A&A* 327:1123–36

Papaloizou JCB, Terquem C. 2001. Dynamical relaxation and massive extrasolar planets. *MNRAS* 325:221–30

Patience J, White RJ, Ghez AM, McCabe C, McLean IS, et al. 2002. Stellar companions to stars with planets. *ApJ* 581:654–65

Patterson RJ, Majewski SR, Kundu A, Kunkel WE, Johnston KV, et al. 1999. The Grid Giant Star Survey for the SIM astrometric grid. *AAS abstract* 195:46.03

Pätzold M, Rauer H. 2002. Where are the massive close-in extrasolar planets? *ApJ* 568:L117–20

Paulson DB, Saar SH, Cochran WD, Hatzes AP. 2002. Searching for planets in the Hyades. II. Some implications of stellar magnetic activity. *AJ* 124:572–82

Peale SJ. 1997. Expectations from a microlensing search for planets. *Icarus* 127:269–89

Pepe F, Mayor M, Delabre B, Kohler D, Lacroix, D, et al. 2000. HARPS: a new high-resolution spectrograph for the search of extrasolar planets. In *Optical and IR telescope instrumentation and detectors,* ed. M Iye, AF Moorwood, pp. 582–92. SPIE Vol. 4008. Bellingham, WA

Perrier C, Mariotti JM. 1987. On the binary nature of Van Biesbroeck 8. *ApJ* 312:L27–30

Perryman MAC, de Boer KS, Gilmore G, Høg E, Lattanzi MG, et al. 2001. GAIA: composition, formation and evolution of the Galaxy. *A&A* 369:339–63

Perryman MAC, Peacock, A. 2000. The astronomical potential of optical STJs. In *Imaging the Universe in three dimensions,* ed. W van Breugel, J Bland-Hawthorn, pp. 487–94. ASP Conf. Ser. Vol. 195

Phinney ES, Hansen BMS. 1993. The pulsar planet production process. In *Planets around pulsars,* ed. JA Phillips, SE Thorsett, SR Kulkarni, pp. 371–90. ASP Conf. Ser. Vol. 36

Podsiadlowski P. 1993. Planet formation scenarios. In *Planets around pulsars,* ed. JA Phillips, SE Thorsett, SR Kulkarni, pp. 149–65. ASP Conf. Ser. Vol. 36

Potter DE, Close LM, Roddier F, Roddier C, Graves JE, Northcott M. 2000. A high-resolution polarimetry map of the circumbinary disk around UY Aurigae. *ApJ* 540:422–28

Pravdo SH, Shaklan SB. 1996. Astrometric detection of extrasolar planets: results of a feasibility study with the Palomar 5 Meter Telescope. *ApJ* 465:264–77

Press WH, Teukolsky SA, Vetterling WT, Flannery BP. 1992. *Numerical Recipes in C, the art of scientific computing, 2nd edition.* Cambridge: Cambridge University Press. 994 pp.

Queloz D. 1999. Indirect searches: Doppler spectroscopy and pulsar timing. In *Planets outside the Solar System: theory and observations,* ed. JM Mariotti, D Alloin, pp. 229–47. NATO ASI Vol. 532, Dordrecht: Kluwer

Queloz D, Casse M, Mayor M. 1999. The fiber-fed spectrograph, a tool to detect planets. In *Precise stellar radial velocities,* ed. JB Hearnshaw, CD Scarfe, pp. 13–21. IAU Coll. 170, ASP Conf. Ser. Vol. 185. San Francisco, CA

Queloz D, Eggenberger A, Mayor M, Perrier C, Beuzit JL, et al. 2000a. Detection of a spectroscopic transit by the planet orbiting the star HD 209458. *A&A* 359:L13–17

Queloz D, Henry GW, Sivan JP, Baliunas SL, Beuzit JL, et al. 2001. No planet for HD 166435. *A&A* 379:279–87

Quillen AC, Holman M. 2000. Production of star-grazing and star-impacting planetesimals via orbital migration of extrasolar planets *AJ* 119:397–402

Quillen AC, Thorndike S. 2002. Structure in the ε Eridani dusty disk caused by mean motion resonances with a 0.3 eccentricity planet at periastron. *ApJ* 578:L149–52

Quintana EV, Lissauer JJ, Chambers JE, Duncan MJ. 2002. Terrestrial planet formation in the α Centauri system. *ApJ* 576:982–96

Quirrenbach A. 2000a. Astrometry with the VLT Interferometer. In *From extrasolar planets to cosmology: the VLT opening symposium,* ed. J Bergeron, A Renzini, pp. 462–67. Berlin/Heidelberg: Springer-Verlag

Quirrenbach A. 2000b. Observing through the turbulent atmosphere. In *Principles of long baseline stellar interferometry. Course notes from the 1999 Michelson summer school*, ed. PR Lawson, pp. 71–85. JPL Publication 00–009, Pasadena, CA

Quirrenbach A. 2001. Optical interferometry. *ARAA* 39:353–401

Quirrenbach A. 2002a. The Space Interferometry Mission (SIM) and Terrestrial Planet Finder (TPF). In *From optical to millimetric interferometry: scientific and technological challenges*, ed. J Surdej, JP Swings, D Caro, A Detal, pp. 51–67. Université de Liège

Quirrenbach A. 2002b. Site testing and site monitoring for extremely large telescopes. In *Astronomical site evaluation in the visible and radio range*, ed. J Vernin, Z Benkhaldoun C Muñoz-Tuñón, pp. 516–22. ASP Conf. Ser. Vol. 266. San Francisco, CA

Quirrenbach A, Cooke J, Mitchell D, Safizadeh N, Deeg H, EXPORT team. 2000. EXPORT: Search for transits in open clusters with the Jakobus Kapteyn and Lick 1 m telescopes. In *Disks, planetesimals and planets*, ed. F Garzón, C Eiroa, D de Winter, TJ Mahoney, pp. 566–71. ASP Conf. Ser. Vol. 219. San Francisco, CA

Quirrenbach A, Coudé du Foresto V, Daigne G, Hofmann KH, Hofmann R, et al. 1998. PRIMA – study for a dual-beam instrument for the VLT Interferometer. In *Astronomical interferometry*, ed. RD Reasenberg, pp. 807–17. SPIE Vol. 3350. Bellingham, WA

Quirrenbach A, Mozurkewich D, Buscher DF, Hummel CA, Armstrong JT. 1994. Phase-referenced visibility averaging in optical long-baseline interferometry. *A&A* 286:1019–27

Quirrenbach A, Mozurkewich D, Buscher DF, Hummel CA, Armstrong JT. 1996. Angular diameter and limb darkening of Arcturus. *A&A* 312:160–66

Rauer H, Bockelée-Morvan D, Coustenis A, Guillot T, Schneider J. 2000. Search for an exosphere around 51 Pegasi B with ISO. *A&A* 355:573–80

Reid IN. 2002. On the nature of stars with planets. *PASP* 114:306–29

Reuyl D, Holmberg E. 1943. On the existence of a third component in the system 70 Ophiuchi. *ApJ* 97:41–45

Rhee J, Slesnick CL, Crane JD, Polak AA, Patterson RJ, et al. 2001. Preliminary results of the Grid Giant Star Survey (GGSS) for the Space Interferometry Mission (SIM) astrometric grid. *AAS abstract* 198:62.03

Rhie SH, Becker AC, Bennett DP, Fragile PC, Johnson BR, et al. 1999. Observations of the binary microlens event MACHO 98-SMC-1 by the Microlensing Planet Search collaboration. *ApJ* 522:1037–45

Rhie SH, Bennett DP, Becker AC, Peterson BA, Fragile PC, et al. 2000. On planetary companions to the MACHO 98-BLG-35 microlens star. *ApJ* 533:378–91

Ridgway ST, Roddier F. 2000. An infrared Very Large Array for the 21st century. In *Interferometry in optical astronomy*, ed. PJ Léna, A Quirrenbach, pp. 940–50. SPIE Vol. 4006. Bellingham, WA

Rivera E, Lissauer JJ. 2001. Dynamical models of the resonant pair of planets orbiting the star GJ 876. *ApJ* 558:392–402

Roberge A, Feldman PD, Lagrange AM. 2000. High-resolution Hubble Space Telescope STIS spectra of C I and CO in the Beta Pictoris circumstellar disk. *ApJ* 538:904–10

Roberge A, Feldman PD, Lecavelier des Etangs A, Vidal-Madjar A, Deleuil M, et al. 2002. Far Ultraviolet Spectroscopic Explorer observations of possible infalling planetesimals in the 51 Ophiuchi circumstellar disk. *ApJ* 568:343–51

Robichon N, Arenou F. 2000. HD 209458 planetary transits from Hipparcos photometry. *A&A* 355:295–98

Robinson LB, Wei MZ, Borucki WJ, Dunham EW, Ford CH, Granados AF. 1995. Test of CCD precision limits for differential photometry. *PASP* 107:1094–98

Roddier C, Roddier F, Northcott MJ, Graves JE, Jim K. 1996. Adaptive optics imaging of GG Tauri: optical detection of the circumbinary ring. *ApJ* 463:326–35

Roddier, F. 1981. The effects of atmospheric turbulence in optical astronomy. *Prog Opt* XIX:281–376

Roddier F. 1989. Optical propagation and image formation through the turbulent atmosphere. In *Diffraction-limited imaging with very large telescopes*, ed. DM Alloin, JM Mariotti, pp. 33–52. NATO ASI Vol. 274, Dordrecht: Kluwer

Romani RW, Miller AJ, Cabrera B, Figueroa-Feliciano E, Nam SW. 1999. First astronomical application of a cryogenic transition edge sensor spectrophotometer. *ApJ* 521:L153–56

Rosenblatt F. 1971. A two-color photometric method for detection of extra solar planetary systems. *Icarus* 14:71–93

Rouan D, Baglin A, Copet E, Schneider J, Barge P, et al. 2000. The exosolar planets program of the COROT satellite. *Earth, Moon, and Planets* 81:79–82

Rubenstein EP, Schaefer BE. 2000. Are superflares on Solar analogues caused by extrasolar planets? *ApJ* 529:1031–33

Ryan P. 2002. Scintillation correlations in the near-infrared. *PASP* 114:462–70

Saar SH, Butler RP, Marcy GW. 1998. Magnetic activity-related radial velocity variations in cool stars: First results from the Lick extrasolar planet survey. *ApJ* 498:L153–57

Saar SH, Donahue RA. 1997. Activity-related radial velocity variation in cool stars. *ApJ* 485:319–27

Saar SH, Fischer D. 2000. Correcting radial velocities for long-term magnetic activity variations. *ApJ* 534:L105–08

Sackett PD. 1999. Searching for unseen planets via occultation and microlensing. In *Planets outside the Solar System: theory and observations*, ed. JM Mariotti, D Alloin, pp. 189–227. NATO ASI Vol. 532, Dordrecht: Kluwer

Sackett PD. 2000. Results from microlensing searches for extrasolar planets. In *Planetary Systems in the UUniverse*, IAU Symp 202 in press

Safizadeh N, Dalal N, Griest K. 1999. Astrometric microlensing as a method of discovering and characterizing extrasolar planets. *ApJ* 522:512–17

Sahu KC, Anderson J, King IR. 2002. A reexamination of the "planetary" lensing events in M 22. *ApJ* 565:L21–24

Sahu KC, Casertano S, Livio M, Gilliland RL, Panagia N, et al. 2001. Gravitational microlensing by low-mass objects in the globular cluster M22. *Nature* 411:1022–24

Salaris M, Weiss A. 1998. Metal-rich globular clusters in the galactic disk: new age determinations and the relation to halo clusters *A&A* 335:943–53

Sandquist E, Taam RE, Lin DNC, Burkert A. 1998. Planet consumption and stellar metallicity enhancements. *ApJ* 506:L65–68

Santos NC, Israelian G, Mayor M. 2001a. The metal-rich nature of stars with planets. *A&A* 373:1019–31

Santos NC, Israelian G, Mayor M, Rebolo R, Udry S. 2003a. Statistical properties of exoplanets. II. Metallicity, orbital parameters, and space velocities. *A&A* 398:363–76

Santos NC, Mayor M, Naef D, Pepe F, Queloz D, et al. 2000. The CORALIE survey for southern extra-solar planets. IV. Intrinsic stellar limitations to planet searches with radial-velocity techniques. *A&A* 361:265–72

Santos NC, Mayor M, Naef D, Pepe F, Queloz D, et al. 2001b. The CORALIE survey for southern extra-solar planets. VI. New long period giant planets around HD 28185 and HD 213240. *A&A* 379:999–1004

Santos NC, Mayor M, Queloz D, Udry S. 2002. Extra-solar planets. *The Messenger*, 110:32–38

Santos NC, Udry S, Mayor M, et al. 2003b, *A&A* 406:373–381

Santos NC et al. 2004, The HARPS survey for southern extra-solar planets: II A 14 Earth-masses exoplanet around μ Arae

Sartoretti P, Schneider J. 1999. On the detection of satellites of extrasolar planets with the method of transits. *A&AS* 134:553–60

Sato B, Ando H, Kambe E, et al. 2003. A planetary Companion to the G-type Giant Star HD 104985, *ApJ* 597:L157–L160

Schneider G, Becklin EE, Smith BA, Weinberger AJ, Silverstone M, Hines DC. 2001. NICMOS coronagraphic observations of 55 Cancri. *ApJ* 121:525–37

Schneider G, Smith BA, Becklin EE, Koerner DW, Meier R, et al. 1999. NICMOS imaging of the HR 4796 A circumstellar disk. *ApJ* 513:L127–30

Schneider G, Wood K, Silverstone MD, Hines DC, Koerner DW, et al. 2003. NICMOS coronagraphic observations of the GM Aurigae circumstellar disk. *ApJ* 125:1467–79

Schneider J, Chevreton M. 1990. The photometric search for Earth-sized extrasolar planets by occultation in binary systems. *A&A* 232:251–57

Schneider P, Weiß A. 1986. The two-point-mass lens – detailed investigation of a special asymmetric gravitational lens. *A&A* 164:237–59

Schroeder DJ. 2000. *Astronomical optics, 2nd edition.* San Diego: Academic Press. 478 pp.

Schultz AB, Jordan I, Hart HM, Bruhweiler F, Fraquelli DA, et al. 2000. Imaging planets about other stars with UMBRAS II. In *Infrared spaceborne remote sensing VIII,* ed. M Strojnik, BF Andresen, pp. 132–40. SPIE Vol. 4131. Bellingham, WA

Schultz AB, Schroeder DJ, Jordan I, Bruhweiler F, DiSanti MA, et al. 1999. Imaging planets about other stars with UMBRAS. In *Infrared spaceborne remote sensing VII,* ed. M Strojnik, BF Andresen, pp. 49–58. SPIE Vol. 3759. Bellingham, WA

Seager S, Hui L. 2002. Constraining the rotation rate of transiting extrasolar planets by oblateness measurements. *ApJ* 574:1004–10

Seager S, Mallén-Ornelas G. 2002. On the unique solution of planet and star parameters from an extrasolar planet transit light curve. *ApJ* 585:1038–1055

Seager S, Sasselov DD. 1998. Extrasolar giant planets under strong stellar irradiation. *ApJ* 502:L157–61

Seager S, Sasselov DD. 2000. Theoretical transmission spectra during extrasolar giant planet transits. *ApJ* 537:916–21

Seager S, Whitney BA, Sasselov DD. 2000. Photometric light curves and polarization of close-in extrasolar giant planets. *ApJ* 540:504–20

Sekanina Z. 2002. Statistical investigation and modeling of sungrazing comets discovered with the Solar and Heliospheric Observatory. *ApJ* 566:577–98

Serabyn E. 1999. Nanometer-level path-length control scheme for nulling interferometry. *Appl Opt* 38:4213–16

Serabyn E. 2000. Nulling interferometry: symmetry requirements and experimental results. In *Interferometry in optical astronomy,* ed. PJ Léna, A Quirrenbach, pp. 328–39. SPIE Vol. 4006. Bellingham, WA

Serabyn E, Colavita MM. 2001. Fully symmetric nulling beam combiners. *Appl Opt* 40:1668–71

Serabyn E, Wallace JK, Hardy GJ, Schmidtlin EGH, Nguyen HT. 1999. Deep nulling of visible laser light. *Appl Opt* 38:7128–32

Shaklan S. 1990. Fiber optic beam combiner for multiple-telescope interferometry. *Opt Eng* 29:684–89

Shaklan SB, Roddier F. 1988. Coupling starlight into single-mode fiber optics. *Appl Opt* 27:2334–38

Shao M, Colavita MM. 1992. Potential of long-baseline infrared interferometry for narrow-angle astrometry. *A&A* 262:353–58

Shao M, Colavita MM, Hines B, Staelin D, Hutter DJ, et al. 1988. The Mark III stellar interferometer. *A&A* 193:357–71

Shu FH, Adams FC, Lizano S. 1987. Star formation in molecular clouds: observation and theory. *ARAA* 25:23–81

Shu F, Najita J, Ostriker E, Wilkin F, Ruden S, Lizano S. 1994. Magnetocentrifugally driven flows from young stars and disks. I. A generalized model. *ApJ* 429:781–96

Siess L, Livio M. 1999. The accretion of brown dwarfs and planets by giant stars – II. Solar-mass stars on the red giant branch. *MNRAS* 308:1133–49

Sigurdsson S. 1992. Planets in globular clusters? *ApJ* 399:L95–97

Sigurdsson S. 1993. Genesis of a planet in Messier 4. *ApJ* 415:L43–46

Sigurdsson S. 1995. Assessing the environmental impact on PSR B1620–26 in M4. *ApJ* 452:323–31

Skrutskie MF, Forrest WJ, Shure MA. 1987. Direct infrared imaging of VB 8. *ApJ* 312:L55–58

Smith BA, Terrile RJ. 1984. A circumstellar disk around β Pictoris. *Science* 226:1421–24

Snellgrove MD, Papaloizou JCB, Nelson RP. 2001. On disc driven inward migration of resonantly coupled planets with application to the system around GJ 876. *A&A* 374:1092–99

Sozzetti A, Casertano S, Lattanzi MG, Spagna A. 2001. Detection and measurement of planetary systems with GAIA. *A&A* 373:L21–24

Spangler C, Sargent AI, Silverstone MD, Becklin EE, Zuckerman B. 2001. Dusty debris around Solar-type stars: temporal disk evolution. *ApJ* 555:932–44

Stapelfeldt KR, Krist JE, Ménard F, Bouvier J, Padgett DL, Burrows CJ. 1998. An edge-on circumstellar disk in the young binary system HK Tauri *ApJ* 502:L65–69

Steinacker J, Henning T. 2003. Detection of gaps in circumstellar disks. *ApJ* 583:L35–38

Stern SA. 1994. The detectability of extrasolar terrestrial and giant planets during their luminous final accretion. *AJ* 108:2312–17

Stern SA, Shull JM, Brandt JC. 1990. Evolution and detectability of comet clouds during post-main-sequence stellar evolution. *Nature* 345:305–08

Stetson PB. 1987. DAOPHOT – A computer program for crowded-field stellar photometry. *PASP* 99:191–222

Strand KA. 1943. 61 Cygni as a triple system. *PASP* 55:29–32

Street RA, Horne K, Lister TA, Penny A, Tsapras Y, et al. 2002. Variable stars in the field of open cluster NGC 6819. *MNRAS* 330:737–54

Street RA, Horne K, Lister TA, Penny A, Tsapras Y, et al. 2003. Searching for planetary transits in the field of open cluster NGC 6819. *MNRAS* 340:1287–97

Street RA, Horne K, Penny A, Tsapras Y, Quirrenbach A, et al. 2000. A search for planetary transits in open clusters. In *Disks, planetesimals and planets*, ed. F Garzón, C Eiroa, D de Winter, TJ Mahoney, pp. 572–77. ASP Conf. Ser. Vol. 219. San Francisco, CA

Strom RG, Smolders B, van Ardenne A. 2001. Active adaptive arrays: the ASRTRON approach to SKA. *ApSS* 278:209–12

Struck C, Cohanim BE, Willson LA. 2002. Models of planets and brown dwarfs in Mira winds. *ApJ* 572:L83–86

Struve O. 1952. Proposal for a project of high-precision stellar radial velocity work. *The Observatory* 72:199–200

Sudarsky D, Burrows A, Pinto P. 2000. Albedo and reflection spectra of extrasolar giant planets. *ApJ* 538:885–903
Sudarsky D, Burrows A, Hubeny I. 2003. Theoretical spectra and atmospheres of extrasolar giant planets. *ApJ* 588:1121–48
Szostak JW, Bartel DP, Luisi PL. 2001. Synthesizing life. *Nature* 409:387–90
Tabachnik S, Tremaine S. 2002. Maximum likelihood method for estimating the mass and period distributions of extra-solar planets. *MNRAS* 335:151–58
Tallon M, Tallon-Bosc I. 1992. The object-image relationship in Michelson stellar interferometry. *A&A* 253:641–45
Tarter J. 2001. The search for extraterrestrial intelligence (SETI). *ARAA* 39:511–48
Tatarski, V.I. 1961. *Wave propagation in a turbulent medium.* New York: McGraw-Hill. 285 pp.
Telesco CM, Fisher RS, Piña RK, Knacke RF, Dermott SF, et al. 2000. Deep 10 and 18 micron imaging of the HR 4796 A circumstellar disk: transient dust particles and tentative evidence for a brightness asymmetry. *ApJ* 530:341
Thi WF, Blake GA, van Dishoeck EF, van Zadelhoff GJ, Horn JMM, et al. 2001. Substantial reservoir of molecular hydrogen in the debris disks around young stars. *Nature* 409:60–63
Thompson AR, Moran JM, Swenson GW. 1986. *Interferometry and synthesis in radio astronomy.* New York: Wiley. 534 pp.
Thorsett SE, Arzoumanian Z, Camilo F, Lyne AG. 1999. The triple pulsar system PSR B1620−26 in M4. *ApJ* 523:763–70
Tough A. 2000. Five strategies for detecting intelligence. In *Bioastronomy '99 – a new era in bioastronomy,* ed. GA Lemarchand, KJ Meech, pp. 445–47. ASP Conf. Ser. Vol. 213. San Francisco, CA
Trilling DE, Benz W, Guillot T, Lunine JI, Hubbard WB, Burrows A. 1998. Orbital evolution and migration of giant planets: modeling extrasolar planets. *ApJ* 500:428–39
Trilling DE, Brown RH. 1998. A circumstellar dust disk around a star with a known planetary companion. *Nature* 395:775–77
Trilling D.E., 2000, *ApJ*, 537, 61–64
Trimble V. 1999. Milky Way and galaxies. In *Allen's astrophysical quantities,* ed. AN Cox, pp. 569–83. New York: Springer
Tsapras Y, Street RA, Horne K, Penny A, Clarke F, et al. 2001. Can Jupiters be found by monitoring Galactic bulge microlensing events from northern sites? *MNRAS* 325:1205–12
Udalski A, Paczyński B, Żebruń K, Szymański M, Kubiak M, et al. 2002. The Optical Gravitational Lensing Experiment. Search for planetary and low-luminosity object transits in the Galactic disk. Results of 2001 campaign. *Acta Astron* 52:115–128
Udalski A, Szymański M, Kałużny J, Kubiak M, Krzemiński W, et al. 1993. The Optical Gravitational Lensing Experiment. Discovery of the first candi-

date microlensing event in the direction of the Galactic bulge. *Acta Astron* 43:289–94

Udalski A, Szymański M, Mao S, di Stefano R, Kałużny J, et al. 1994. The optical gravitational lensing experiment: OGLE no. 7: Binary microlens or a new unusual variable? *ApJ* 436:L103–06

Udalski A, Żebruń K, Szymański M, Kubiak M, Pietrzyński G, et al. 2000. The Optical Gravitational Lensing Experiment. Catalog of microlensing events in the Galactic bulge. *Acta Astron* 50:1–65

Udry S, Mayor M, Clausen JV, Freyhammer LM, Helt BE, et al. 2003a. The CORALIE survey for southern extra-solar planets. X. A hot Jupiter orbiting HD 73256. *A&A* 407:679–684

Udry S, Mayor M, Naef D, Pepe F, Queloz D, et al. 2000. The CORALIE survey for southern extra-solar planets. II. The short-period planetary companions to HD 75289 and HD 130322. *A&A* 356:590–98

Udry S, Mayor M, Naef D, Pepe F, Queloz D, et al. 2002. The CORALIE survey for southern extra-solar planets – VIII. The very low-mass companions of HD 141937, HD 162020, HD 168443 and HD 202206: Brown dwarfs or "superplanets"? *A&A* 390:267–79

Udry S, Mayor M, Queloz D. 2001. CORALIE-ELODIE new planets and planetary systems. Looking for fossil traces of formation and evolution. In *Planetary Systems in the Universe*, ed. A Penny, P Artymowicz, AM Lagrange, S Russel. IAU Symp. 202, ASP Conf. Ser. San Francisco, CA, in press

Udry S, Mayor M, Santos NC. 2003b. Statistical properties of exoplanets. I. The period distribution: constraints for the migration scenario. *A&A*, 407:369–376

van de Kamp P. 1963. Astrometric study of Barnard's Star from plates taken with the 24-inch Sproul refractor. *AJ* 68:515–21

van Dishoeck EJ, Blake GA. 1998. Chemical evolution of star-forming regions. *ARAA* 36:317–68

Van Hamme W. 1993. New limb-darkening coefficients for modeling binary star light curves. *AJ* 106:2096–117

Vidal-Madjar A, Lagrange-Henri AM, Feldman PD, Beust H, Lissauer JJ. 1994. HST-GHRS observations of β Pictoris: additional evidence for infalling comets. *A&A* 290:245–58

Vogt SS. 1987. The Lick Observatory Hamilton Echelle Spectrometer. *PASP* 99:1214–28

Vogt SS, Butler RP, Marcy GW, Fischer DA, Pourbaix D, et al. 2002. Ten low-mass companions from the Keck precision velocity survey. *ApJ* 568:352–62

Vogt SS, Marcy GW, Butler RP, Apps K. 2000. Six new planets from the Keck precision velocity survey. *ApJ* 536:902–14

Wahhaj Z, Koerner DW, Ressler ME, Werner MW, Backman DE, Sargent AI. 2003. The Inner Rings of β Pictoris. *ApJ* 584:L27–31

Walker GAH, Walker AR, Irwin AW, Larson AM, Yang SLS, Richardson DC. 1995. A search for Jupiter-mass companions to nearby stars. *Icarus* 116:359–75

Wallace K, Hardy G, Serabyn E. 2000. Deep and stable interferometric nulling of broadband light with implications for observing planets around nearby stars. *Nature* 406:700–02

Wallner O, Kudielka K, Leeb WR. 2001. Nulling interferometry for spectroscopic investigation of exoplanets – a statistical analysis of imperfections. In *The search for extraterrestrial intelligence (SETI) in the optical spectrum III*, ed. SA Kingsley, R Bhathal, pp. 47–55. SPIE Vol. 4273. Bellingham, WA

Wambsganss J. 1997. Discovering Galactic planets by gravitational microlensing: magnification patterns and light curves. *MNRAS* 284:172–88

Worley CE and Douglass GG, 1997, A&A 125, 523. "The Wostinjtan Double star catolog"

Weinberger AJ, Becklin EE, Schneider G, Smith BA, Lowrance PJ, et al. 1999. The circumstellar disk of HD 141569 imaged with NICMOS. *ApJ* 525:L53–56

Weinberger AJ, Becklin EE, Zuckerman B. 2003. First spatially resolved mid-infrared spectroscopy of β Pictoris. *ApJ* 584:L33–37

Weinberger AJ, Rich RM, Becklin EE, Zuckerman B, Matthews K. 2000. Stellar companions and the age of HD 141569 and its circumstellar disk. *ApJ* 544:937–43

Wetherill GW. 1990. Formation of the Earth. *Ann Rev Earth Planet Sci* 18:205–56

Wiedemann G, Deming D, Bjoraker G. 2001. A sensitive search for methane in the infrared spectrum of τ Bootis. *ApJ* 546:1068–74

Willacy K, Cherchneff I. 1998. Silicon and sulphur chemistry in the inner wind of IRC+10216. *A&A* 330:676–84

Wolf S, Gueth F, Henning T, Kley W. 2002. Detecting planets in protoplanetary disks: a prospective study. *ApJ* 566:L97–99

Wolszczan A. 1994. Confirmation of Earth mass planets orbiting the millisecond pulsar PSR B1257+12. *Science* 264:538–42

Wolszczan A. 1999. Detecting planets around pulsars. In *Pulsar timing, general relativity and the internal structure of neutron stars*, ed. Z Arzoumanian, F van der Hooft, EPJ van den Heuvel, pp. 101–11. Koninklijke Nederlandse Akademie van Wetenschappen, Amsterdam

Wolszczan A, Doroshenko O, Konacki M, Kramer M, Jessner A, et al. 2000. Timing observations of four millisecond pulsars with the Arecibo and Effelsberg radio telescopes. *ApJ* 528:907–12

Wolszczan A, Frail DA. 1992. A planetary system around the millisecond pulsar PSR 1257+12. *Nature* 355:145–47

Woolf N, Angel JRP. 1997. Planet Finder options I: new linear nulling array configurations. In *Planets beyond the Solar System and the next generation of space missions*, ed. D Soderblom, pp. 285–93. ASP Conf. Ser. Vol. 119. San Francisco, CA

Worley CE, Douglass GG. 1997. The Washington Double Star Catalog. *A & A* 125:523

Wyatt MC, Dermott SF, Telesco CM, Fisher RS, Grogan K, et al. 1999. How observations of circumstellar disk asymmetries can reveal hidden planets: pericenter glow and its application to the HR 4796 disk. *ApJ* 527:918–44

Youdin AN, Shu FH. 2002. Planetesimal formation by gravitational instability. *ApJ* 580:494–505

Yura HT, Fried DL. 1998. Variance of the Strehl ratio of an adaptive optics system. *JOSA A* 15:2107–10

Zapatero Osorio MR, Béjar VJS, Martín EL, Rebolo R, Barrado y Navascués D, et al. 2000. Discovery of young, isolated planetary mass objects in the σ Orionis star cluster. *Science* 290:103–07

Zapatero Osorio MR, Béjar VJS, Martín EL, Rebolo R, Barrado y Navascués D, et al. 2002. A methane, isolated planetary-mass object in Orion. *ApJ* 578:536–42

Zarka P, Treumann RA, Ryabov BP, Ryabov VB. 2001. Magnetically-driven planetary radio emissions and application to extrasolar planets. *ApSS* 277:293–300

Zucker S, Mazeh T. 2001a. Analysis of the Hipparcos observations of the extrasolar planets and the brown dwarf candidates. *ApJ* 562:549–57

Zucker S, Mazeh T. 2001b. Derivation of the mass distribution of extrasolar planets with MAXLIMA, a maximum likelihood algorithm. *ApJ* 562:1038–44

Zucker S, Mazeh T. 2002. On the mass-period correlation of the extrasolar planets. *ApJ* 568:L113–16

Zucker S, Naef D, Latham D, Mayor M, Mazeh T, et al. 2002. A Planet candidate in the stellar triple system HD 178911. *ApJ* 568:363–68

Zuckerman B. 2001. Dusty circumstellar disks. *ARAA* 39:549–80

Zuckerman B, Becklin EE. 1992. Companions to white dwarfs: very low-mass stars and the brown dwarf candidate GD 165 B. *ApJ* 386:260–64

Zuckerman B, Song I, Bessell MS, Webb RA. 2001. The β Pictoris moving group. *ApJ* 562:L87–90

Physics of Substellar Objects Interiors, Atmospheres, Evolution

T. Guillot

1 Introduction

All stars visible to the naked eye owe their momentary brightness to nuclear reactions occurring in their interior. While this certainly makes them jewels of the night skies, it will eventually lead them to a tragic end, in which they will explode to become either degenerate white dwarfs, neutron stars or black holes. Another, more numerous but barely visible population has chosen to lead a dull but quiet and almost eternal life: these are careful not to ever become dependent on hydrogen to shine. Some, in their youth, do burn less energetic substances as deuterium and lithium, but they rapidly get short of supply. As a consequence, they steadily cool and contract, retaining intact most of the elements that made them.

These brown dwarfs and giant planets form an entirely new class of astronomical objects. They fill a gap between stars and the planets of our Solar System. Their study informs us on our origins, the formation of stars and planets. It can also help us to understand or test theories from high pressure physics, to atmospheric dynamics, tides, condensation and cloud formation...*etc*.

The course focuses on some physical aspects related to the theoretical study of these substellar objects: I detail their hydrostatic evolution and how it is modeled, what we can learn from Jupiter, Saturn, Uranus and Neptune, how the atmospheres of brown dwarfs and giant planets are key to their appearance and cooling, what we can learn from the recent observations of brown dwarfs and extrasolar planets, and how this affects our view of planet formation.

2 "Our" Giant Planets as a Basis for the Study of Substellar Objects

2.1 Origins: Role of the Giant Planets for Planet Formation

The Solar System contains our Sun, which possesses more than 98% of the mass of the system, and eight planets orbiting around it in the same plane and

Guillot T (2006), Physics of substellar objects Interiors, atmospheres, evolution. In: Mayor M, Queloz D, Udry S and Benz W (eds) Extrasolar planets. Saas-Fee Adv Courses vol 31, pp 243–368
DOI 10.1007/3-540-29216-0_2 © Springer-Verlag Berlin Heidelberg 2006

same direction with quasi-circular orbits. The planets contain 99.5% of the angular momentum of the system. The four inner planets, Mercury, Venus, Earth and Mars have the highest densities, but more than 99.5% of the mass of the planetary system is in its four outer planets, Jupiter, Saturn, Uranus and Neptune. Most of the planets have moons, or natural satellites. Orbiting around the Sun, one also finds asteroids, Kuiper belt objects (including Pluto) and comets.

A picture emerges naturally from these observations: the formation of the planets in a circumstellar disk: the protosolar nebula. Planets formed close to the Sun naturally contain less volatiles and ices, while the outer planets were favored by the abundant presence of ices and could therefore grow fast enough to get hold of the surrounding hydrogen and helium of the nebula before its dissipation. In this picture, asteroids, Kuiper belt objects and comets all represent leftovers from an inefficient planet formation mechanism.

By their masses, the giant planets Jupiter, Saturn, Uranus and Neptune played a key role in this story. While the inner, terrestrial planets took tens of millions of years to reach their present masses, the giant planets had to form rapidly, before the gas of the protosolar nebula disappeared onto the star or was swept away from the system. They led to the ejection of numerous material, preventing the formation of a planet between Mars and Jupiter, and sending planetesimals into the Oort cloud, from where these remains of planetary formation come back once in a while as comets.

Their study therefore informs us on our origins. It also allows us to extend our knowledge beyond the frontiers of the Solar System and to model with confidence the other giant planets that have been found orbiting other stars. Before presenting the theoretical aspects of that understanding, I will detail here a few of the observations and measurements of significance for our purposes.

Most of the measurements at the basis of our understanding of the structure of our giant planets have been acquired by spacecraft missions: Pioneer 10 & 11, Voyager 1 & 2, Ulysses, Galileo, Cassini-Huygens.

2.2 Gravity Field and Global Properties

The mass of the giant planets can be obtained with great accuracy from the observation of the motions of their natural satellites: 317.834, 95.161, 14.538 and 17.148 times the mass of the Earth ($1\,\mathrm{M}_\oplus = 5.97369 \times 10^{27}$ g) for Jupiter, Saturn, Uranus and Neptune, respectively. More precise measurements of their gravity field can be obtained through the analysis of the trajectories of spacecrafts during flyby, especially when they come close to the planet and preferably in a near-polar orbit. The gravitational field thus measured departs from a purely spherical function due to the planets rapid rotation. The measurements are generally expressed by expanding the components of the gravity field on Legendre polynomials P_i of progressively higher orders:

$$V_{\text{ext}}(r,\theta) = -\frac{GM}{r}\left\{1 - \sum_{i=1}^{\infty}\left(\frac{R_{\text{eq}}}{r}\right)^i J_i P_i(\cos\theta)\right\}, \quad (1)$$

where R_{eq} is the equatorial radius, and J_i are the gravitational moments. Because the giant planets are very close to hydrostatic equilibrium the coefficients of even order are the only ones that are not negligible. We will see how these gravitational moments help us constrain the planets' interior density profiles.

Table 1 also indicates the radii obtained with the greatest accuracy by radio-occultation experiments. By convention, these radii and gravitational moments correspond to the 1 bar pressure level. The rotation periods show the relatively fast revolution of these planets: about 10 hours for Jupiter and Saturn, about 17 hours for Uranus and Neptune. The fact that this fast rotation visibly affects the figure (shape) of these planets is seen by the significant difference between the polar and equatorial radii.

A first result obtained from the masses and radii indicated in Table 1 is the fact that these planets have low densities: 1.33, 0.688, 1.27, and 1.64 g cm^{-3} for Jupiter, Saturn, Uranus and Neptune, respectively (these values are calculated using the planets' *mean* radii, as defined in Sect. 3.5). Considering the

Table 1. Characteristics of the gravity fields and radii

	Jupiter	**Saturn**	**Uranus**	**Neptune**
$M \times 10^{-29}$ [g]	18.986112(15)[a]	5.684640(30)[b]	0.8683205(34)[c]	1.0243542(31)[d]
$R_{\text{eq}} \times 10^{-9}$ [cm]	7.1492(4)[e]	6.0268(4)[f]	2.5559(4)[g]	2.4766(15)[g]
$R_{\text{pol}} \times 10^{-9}$ [cm]	6.6854(10)[e]	5.4364(10)[f]	2.4973(20)[g]	2.4342(30)[g]
$J_2 \times 10^2$	1.4697(1)[a]	1.6332(10)[b]	0.35160(32)[c]	0.3539(10)[d]
$J_4 \times 10^4$	−5.84(5)[a]	−9.19(40)[b]	−0.354(41)[c]	−0.28(22)[d]
$J_6 \times 10^4$	0.31(20)[a]	1.04(50)[b]	...	...
$P_\omega \times 10^4$ [s]	3.57297(41)[h]	3.83577(47)[h]	6.2064[i]	5.7996[j]

The numbers in parentheses are the uncertainty in the last digits of the given value. The value of the gravitational constant used to calculate the masses of Jupiter and Saturn is $G = 6.67259 \times 10^{-8}$ dyn cm^2 g^{-1} (Cohen and Taylor 1986)
[a] Campbell and Synnott (1985)
[b] Campbell and Anderson (1989)
[c] Anderson et al. (1987)
[d] Tyler et al. (1989)
[e] Lindal et al. (1981)
[f] Lindal et al. (1985)
[g] Lindal (1992)
[h] Davies et al. (1986)
[i] Warwick et al. (1986)
[j] Warwick et al. (1989)

compression that strongly increases with mass, one is led to a sub-classification between the hydrogen–helium giant planets Jupiter and Saturn, and the "ice giants" Uranus and Neptune.

2.3 Magnetic Fields

As the Earth, the Sun and Mercury, our four giant planets possess their own magnetic fields, as shown by the Voyager 2 measurements. The structures of these magnetic fields are very different from one planet to another and the dynamo mechanism that generates them is believed to be related to convection in their interior but is otherwise essentially unknown (see Stevenson 1983 for a review).

The magnetic field $\mathbf{B}$ is generally expressed in form of a development in spherical harmonics of the scalar potential W, such that $\mathbf{B} = -\boldsymbol{\nabla}W$:

$$W = a \sum_{n=1}^{\infty} \left(\frac{a}{r}\right)^{n+1} \sum_{m=0}^{n} \{g_n^m \cos(m\phi) + h_n^m \sin(m\phi)\} P_n^m(\cos\theta). \qquad (2)$$

r is the distance to the planet's center, a its radius, θ the colatitude, ϕ the longitude and P_n^m the associated Legendre polynomials. The coefficients g_n^m and h_n^m are the magnetic moments that characterize the field. They are expressed in magnetic field units (i.e. the Gauss in c.g.s. units).

One can show that the first coefficients of relation (2) (for $n = 0$ and $n = 1$) correspond to the potential of a magnetic dipole such that $W = \mathbf{M} \cdot \mathbf{r}/r^3$ of moment:

$$M = a^3 \left\{ \left(g_1^0\right)^2 + \left(g_1^1\right)^2 + \left(h_1^1\right)^2 \right\}^{1/2}. \qquad (3)$$

Jupiter and Saturn have magnetic fields of essentially dipolar nature, of axis close to the rotation axis (g_1^0 is much larger than the other harmonics); Uranus and Neptune have magnetic fields that are intrinsically much more complex. To provide an idea of the intensity of the magnetic fields, the value of the dipolar moments for the four planets are 4.27 Gauss R_J, 0.21 Gauss R_S, 0.23 Gauss R_U, 0.133 Gauss R_N, respectively (Connerney et al. 1982; Acuña et al. 1983; Ness et al. 1986, 1989).

2.4 Atmospheric Composition

The most important components of the atmospheres of our giant planets are also among the most difficult to detect: H_2 and He have a zero dipolar moment. Also their rotational lines are either weak or broad. On the other hand, lines due to electronic transitions correspond to very high altitudes in the atmosphere, and bear little information on the structure of the deeper levels. The only robust result concerning the abundance of helium in a giant planet is by *in situ* measurement by the Galileo probe in the atmosphere of Jupiter (von Zahn et al. 1998). The helium mole fraction (i.e. number of helium atoms over the total number of species in a given volume) is $q_{\rm He} = 0.1359 \pm 0.0027$. The helium mass mixing ratio Y (i.e. mass of helium atoms over total mass) is constrained by its ratio over hydrogen, X: $Y/(X+Y) = 0.238 \pm 0.05$. This ratio is by coincidence that found in the Sun's atmosphere, but because of helium sedimentation in the Sun's radiative zone, it was larger in the protosolar nebula: $Y_{\rm proto} = 0.275 \pm 0.01$ and $(X+Y)_{\rm proto} \approx 0.98$. Less helium is therefore found in the atmosphere of Jupiter than inferred to be present when the planet formed. We will discuss the consequences of this measurement later: let us mention that the explanation invokes helium settling due to a phase separation in the interiors of massive and cold giant planets.

Helium is also found to be depleted compared to the protosolar value in Saturn's atmosphere. However, in this case the analysis is complicated by the fact that Voyager radio occultations apparently led to a wrong value. The current adopted value is now $Y = 0.18 - 0.25$ (Conrath and Gautier 2000), in agreement with values predicted by interior and evolution models (Guillot 1999a,b; Hubbard et al. 1999). Finally, Uranus and Neptune are found to have near-protosolar helium mixing ratios, but with considerable uncertainty.

The abundance of other elements (that I will call hereafter "heavy elements") bears crucial information for the understanding of the processes that led to the formation of these planets. Again, the most precise measurements are for Jupiter, thanks to the Galileo probe. Most of the heavy elements are enriched by a factor 2 to 4 compared to the solar abundance (Niemann et al. 1998; Owen et al. 1999). One exception is neon, but an explanation is its capture by the falling helium droplets (Roulston and Stevenson 1995). Another exception is water, but this molecule is affected by meteorological processes, and the probe was shown to have fallen into a dry region of Jupiter's atmosphere. There are strong indications that its abundance is at least solar. Possible very high interior abundances ($\sim$10 times the solar value) have also been suggested, either to explain waves propagation after the Shoemaker-Levy 9 impacts (Ingersoll et al. 1994) or as a scenario to explain the delivery of heavy elements to the planet (Gautier et al. 2001).

Assuming that all elements are enriched by a factor $\sim$3 in Jupiter's interior, the total mass of heavy elements in the planet would be $\sim$18 $M_\oplus$. In the other planets, the case is considerably less clear as only the abundance of CH_4 can be measured with confidence. As shown in Table 2 this ratio is consistent with an increased proportion of heavy elements when moving from Jupiter to Neptune. The problem of how these elements were delivered to these planets will be discussed later.

2.5 Energy Balance and Atmospheric Temperature Profiles

Jupiter, Saturn and Neptune are observed to emit significantly more energy than they receive from the Sun (see Table 3). The case of Uranus is less clear. Its intrinsic heat flux $F_{\rm int}$ is significantly smaller than that of the other giant planets. Detailed modeling of its atmosphere however indicate

Table 2. Chemical species detected in the atmospheres of giant planets (courtesy of B. Bézard)

	Jupiter	Saturn	Uranus	Neptune
H_2	0.864	0.86 − 0.90	0.81 − 0.86	0.77 − 0.82
He	0.134	0.10 − 0.14	0.12 − 0.17	0.16 − 0.22
rare gases	Ne, Ar, Kr, Xe			
species in thermochemical equilibrium	CH_4: 2×10^{-3} NH_3: 5×10^{-3} H_2O: $> 10^{-3}$ H_2S: 8×10^{-5}	CH_4: $3-6\times 10^{-3}$ NH_3 H_2O	CH_4: $\sim 2\times 10^{-2}$ H_2S?	CH_4: $\sim 2\times 10^{-2}$ H_2S?
species in thermochemical disequilibrium	PH_3 CO GeH_4 AsH_3	PH_3 CO GeH_4 AsH_3		CO
photochemical products	C_2H_6, C_2H_2, C_2H_4, CH_3C_2H, C_6H_6	C_2H_6, C_2H_2, CH_3C_2H, C_4H_2, C_6H_6, CH_3	C_2H_2	C_2H_6, C_2H_2, C_2H_4, CH_3, HCN
meteoritic flux	H_2O, CO	H_2O	H_2O	H_2O CO_2 (from H_2O)
SL9 residuals	CO, CO_2 CS, HCN			

Table 3. Energy balance as determined from Voyager IRIS data[a]

	Jupiter	Saturn	Uranus	Neptune
absorbed power [10^{23} erg s^{-1}]	50.14±2.48	11.14±0.50	0.526±0.037	0.204±0.019
emitted power [10^{23} erg s^{-1}]	83.65±0.84	19.77±0.32	0.560±0.011	0.534±0.029
intrinsic power [10^{23} erg s^{-1}]	33.5±2.6	8.63±0.60	$0.034^{+0.038}_{-0.034}$	0.330±0.035
intrinsic flux [erg s^{-1} cm^{-2}]	5440±430	2010±140	42^{+47}_{-42}	433±46
bond albedo	0.343±0.032	0.342±0.030	0.300±0.049	0.290±0.067
effective temperature [K]	124.4±0.3	95.0±0.4	59.1±0.3	59.3±0.8
1-bar temperature[b] [K]	165±5	135±5	76±2	72±2

[a] After Pearl and Conrath (1991)
[b] Lindal (1992)

that $F_{\rm int} \gtrsim 60\,{\rm erg\,cm^{-2}\,s^{-1}}$ (Marley and McKay 1999). With this caveat, all four giant planets can be said to emit more energy than they receive from the Sun. Hubbard (1968) showed in the case of Jupiter that this can be explained simply by the progressive contraction and cooling of the planets.

A crucial consequence of the presence of an intrinsic heat flux is that it requires high internal temperatures (∼10, 000 K or more), and that consequently the giant planets are *fluid* (not solid) (Hubbard 1968; see also Hubbard et al. 1995). Another consequence is that they are essentially convective, and that their interior temperature profile are close to *adiabats*. We will come back to this in more details.

The deep atmospheres (more accurately tropospheres) of the four giant planets are indeed observed to be close to adiabats, a result first obtained by Trafton (1967), but verified by radio-occultation experiments by the Voyager spacecrafts, and by the *in situ* measurement from the Galileo probe (Fig. 1). The temperature profiles show a temperature minimum, in a region near 0.2 bar called the tropopause. At higher altitudes, in the stratosphere, the temperature gradient is negative (increasing with decreasing pressure). In the regions that we will be mostly concerned with, in the troposphere and in the deeper interior, the temperature always increases with depth. It can be noticed that the slope of the temperature profile in Fig. 1 becomes almost constant when the atmosphere becomes convective, at pressures of a few tens of bars, in the four giant planets.

It should be noted that the 1 bar temperatures listed in Table 3 and the profiles shown in Fig. 1 are retrieved from radio-occultation measurements using a helium to hydrogen ratio which, at least in the case of Jupiter and Saturn, was shown to be incorrect. The new values of Y are found to lead to increased temperatures by ∼5 K in Jupiter and ∼10 K in Saturn (see Guillot 1999a,b).

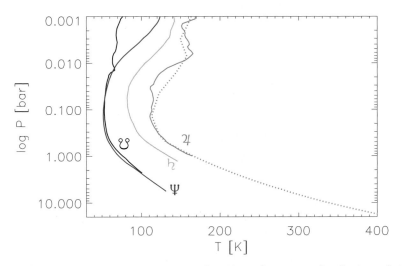

Fig. 1. Atmospheric temperatures as a function of pressure for Jupiter, Saturn, Uranus and Neptune, as obtained from Voyager radio-occultation experiments (see Lindal 1992). The *dotted line* corresponds to the temperature profile retrieved by the Galileo probe, down to 22 bar and a temperature of 428 K (Seiff et al. 1998)

However, to make things simple (!), the Galileo probe found a 1 bar temperature of 166 K (Seiff et al. 1998), and generally a good agreement with the Voyager radio-occultation profile with the wrong He/H_2 value.

When studied at low spatial resolution, it is found that all four giant planets, in spite of their inhomogeneous appearances, have a rather uniform brightness temperature, with pole-to-equator latitudinal variations limited to a few kelvins (e.g. Ingersoll et al. 1995). However, in the case of Jupiter, some small regions are known to be very different from the average of the planet. This is the case of hot spots, which cover about 1% of the surface of the planet at any given time, but contribute to most of the emitted flux at 5 microns, due to their dryness (absence of water vapor) and their temperature brightness which can, at this wavelength, peak to 260 K. This fact is to be remembered when analyzing e.g. brown dwarfs spectra.

2.6 Spectra

A spectrum of a jovian hot spot obtained from the Galileo orbiter is shown in Fig. 2. It demonstrates the complex structure of a planet, and the significant departures from a black-body radiation. At short wavelengths ($\lambda \lesssim 3\,\mu$m, the spectrum is dominated by the directly reflected solar light. At longer wavelengths, the thermal radiation dominates. The spectrum is dominated by the absorption bands of methane with some absorption by ammonia; water lines are seen around 5 µm, and a number of less abundant chemical species (e.g. phosphine) contribute to this spectrum.

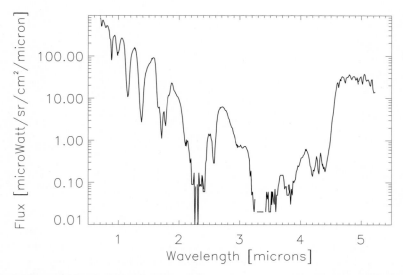

Fig. 2. Flux emitted by a Jupiter hot spot as seen by the Galileo orbiter with NIMS [From Carlson et al. 1996; Courtesy of P. Drossart]

2.7 Atmospheric Dynamics: Winds and Weather

The atmospheres of all giant planets are evidently complex and turbulent in nature. This can for example be seen from the mean zonal winds (inferred from cloud tracking), which are very rapidly varying functions of the latitude (see e.g. Ingersoll et al. 1995): while some of the regions rotate at the same speed as the interior magnetic fields ("system III"), most of the atmospheres do not. Jupiter and Saturn both have superrotating equators (+100 and +400 m s^{-1} in system III, for Jupiter and Saturn, respectively), Uranus and Neptune have subrotating equators, and superrotating high latitude jets. Neptune, which receives the smallest amount of energy from the Sun has the largest peak-to-peak latitudinal variations in wind velocity: about 600 m s^{-1}. It can be noted that, contrary to the case of the strongly irradiated planets to be discussed later, the winds of Jupiter, Saturn, Uranus and Neptune, are significantly smaller than the surface speed due to the revolution of the planet on itself (from 12.2 km s^{-1} for Jupiter to 2.6 km s^{-1} for Neptune).

The observed surface winds are believed to be related to motions in the planets' interiors, which, according to the Taylor–Proudman theorem, should be confined by the rapid rotation to the plane perpendicular to the axis of rotation (e.g. Busse 1978). Unfortunately, no convincing model is yet capable of modeling with sufficient accuracy both the interior and the surface layers.

Our giant planets also exhibit planetary-scale to small-scale storms with very different temporal variations. For example, Jupiter's great red spot is a 12000 km-diameter anticyclone found to have lasted for at least 300 years. Storms developing over the entire planet have even been observed on Saturn (Sanchez-Lavega et al. 1991). Neptune's storm system has been shown to have been significantly altered since the Voyager era. On Jupiter, small-scale storms related to cumulus-type cloud systems has been observed by Galileo, and lightning strikes can be monitored.

It is tempting to extrapolate these observations to the objects outside our Solar System as well. However, it is important to stress that an important component of the variability in the atmospheres of our giant planets is the presence of relatively abundant condensing chemical species: ammonia and water in the case of Jupiter and Saturn, and methane for Uranus and Neptune. These species can only condense (and thus provide the necessary latent heat) in very cold atmospheres. Other phenomena are however possible.

2.8 Moons and Rings

A discussion of our giant planets motivated by the opportunity to extrapolate the results to objects outside our solar system would be incomplete without mentioning the moons and rings that these planets all possess. First, the satellites/moons can be distinguished from their orbital characteristics as regular or irregular. The first ones have generally circular, prograde orbits. The latter tend to have eccentric, extended, and/or retrograde orbits.

These satellites are numerous: After the Voyager era, Jupiter was known to possess 16 satellites, Saturn to have 18, Uranus 20 and Neptune 8. Recent extensive observation programs have seen the number of satellites increase considerably. The number of satellites detected is now 60 around Jupiter, 31 around Saturn, 26 around Uranus and 11 around Neptune (see Gladman et al. 2001; Sheppard and Jewitt 2003). All of these new satellites are classified as irregular.

The presence of regular and irregular satellites is due in part to the history of planet formation. It is believed that the regular satellites have mostly been formed in the protoplanetary subnebulae that surrounded the giant planets (at least Jupiter and Saturn) at the time when they accreted their envelopes. On the other hand, the irregular satellites are thought to have been captured by the planet. This is for example believed to be the case of Neptune's largest moon, Triton, which has a retrograde orbit.

A few satellites stand out by having relatively large masses: it is the case of Jupiter's Io, Europa, Ganymede and Callisto, of Saturn's Titan, and of Neptune's Triton. Ganymede is the most massive of them, being about twice the mass of our Moon. However, compared to the mass of the central planet, these moons and satellites have very small weights: 10^{-4} and less for Jupiter, 1/4000 for Saturn, 1/25000 for Uranus and 1/4500 for Neptune. All these satellites orbit relatively closely to their giant planets. The furthest one, Callisto rotates around Jupiter in about 16 Earth days.

The four giant planets also have rings, whose material is probably constantly resupplied from their satellites. The ring of Saturn stands out as the only one directly visible with only binocular. In this particular case, its enormous area allows it to reflect a sizable fraction of the stellar flux arriving at Saturn, and makes this particular ring as bright as the planet itself. The occurrence of such rings would make the detection of extrasolar planets slightly easier, but it is yet unclear how frequent they can be, and how close to the stars rings can survive both the increased radiation and tidal forces.

2.9 Oscillations

Last but not least, the case for the existence of free oscillations of the giant planets is still unresolved. Such a discovery would lead to great leaps in our knowledge of the interior of these planets, as can be seen from the level of accuracy reached by solar interior models since the discovery of its oscillations. Observations aimed at detecting modes of Jupiter have shown promising results (Schmider et al. 1991), but have thus far been limited by instrumental and windowing effects. A recent work by Mosser et al. (2000) puts an upper limit to the amplitude of the modes at $0.6\,\mathrm{m\,s^{-1}}$, and shows an increased energy of the Fourier spectrum in the expected range of frequencies. Observations from space of through an Earth-based network should be pursued in order to verify these results.

3 Basic Equations, Gravitational Moments and Interior Structures

3.1 Hydrostatic Equilibrium

A very pleasing property of giant planets and brown dwarfs is that in spite of more than two decades of variation in mass, these objects basically obey the same physics: for most of their life, their interior is fluid and they are governed by the equilibrium between their internal pressure and their gravity. Unlike terrestrial planets, the characteristic viscosities are extremely small and can be neglected. The standard hydrostatic equation is thus:

$$\frac{\partial P}{\partial r} = -\rho g, \tag{4}$$

where P is the pressure, ρ the density, and $g = Gm/r^2$ the gravity (m is the mass, r the radius and G the gravitational constant).

Another equation is necessary to obtain the temperature as a function of pressure:

$$\frac{\partial T}{\partial r} = \frac{\partial P}{\partial r}\frac{T}{P}\nabla_T. \tag{5}$$

While the equation itself is trivial, the calculation of the temperature gradient $\nabla_T \equiv (d\ln T/d\ln P)$ is not, and depends on the process by which the internal heat is transported. This term will be analyzed in a following section.

Thirdly, a special case of the mass conservation with zero velocity is:

$$\frac{\partial M}{\partial r} = 4\pi r^2 \rho. \tag{6}$$

Again, the physics of this equation is hidden in the dependency of the density ρ with the pressure, temperature and composition, something given by the equation of state (see Sect. 3.2).

Finally, a crucial equation is derived from energy conservation considerations:

$$\frac{\partial L}{\partial r} = 4\pi r^2 \rho \left(\dot{\epsilon} - T\frac{\partial S}{\partial t} \right), \tag{7}$$

where L is the intrinsic luminosity, t the time, S the specific entropy (per unit mass), and $\dot{\epsilon}$ accounts for the sources of energy due e.g. to radioactivity or more importantly nuclear reactions. Generally it is a good approximation to assume $\dot{\epsilon} \sim 0$ for objects less massive than $\sim 13\,M_J$, i.e. too cold to even burn deuterium (but we will see that in certain conditions this term may be useful, even for low mass planets).

3.2 Boundary Conditions

At the center, $r = 0$; $m = 0$, $L = 0$. The external boundary conditions are more complex to obtain because they depend on how energy is transported in the atmosphere. One possibility is to use the Eddington approximation, and to write (e.g. Chandrasekhar 1960):

$$r = R: \quad T_0 = T_{\text{eff}},$$
$$P_0 = \frac{2}{3}\frac{g}{\kappa}, \qquad (8)$$

where κ is the opacity in $\text{cm}^2\,\text{g}^{-1}$ (see Sect. 5). Note for example that in the case of Jupiter $T_{\text{eff}} = 124\,\text{K}$, $g = 2600\,\text{cm}\,\text{s}^{-2}$ and $\kappa \approx 5\times10^{-2}(P/1\,\text{bar})\,\text{cm}^2\,\text{g}^{-1}$. This implies $P \approx 0.2\,\text{bar}$, which is actually close to Jupiter's tropopause, where $T \approx 110\,\text{K}$.

Another possibility is to use an atmospheric model and to relate the temperature and pressure at a given level to the gravity and effective temperature of the object (or equivalently luminosity and radius):

$$T_0 = T_0(T_{\text{eff}}, g); \quad P_0 = P_0(T_{\text{eff}}, g). \qquad (9)$$

In the case of Jupiter and Saturn, an approximation often used is based on old calculations by Graboske et al. (1975). It takes the form

$$T_{1\,\text{bar}} = K T_{\text{eff}}^a g^{-b}, \qquad (10)$$

where $K = 1.5$, $a = 1.243$ and $b = 0.167$, all the quantities being expressed in cgs units. As shown by Fig. 3, this approximation is relatively good for effective temperatures lower than 200 K, but it degrades substantially above that value (see also discussion in Saumon et al. 1996).

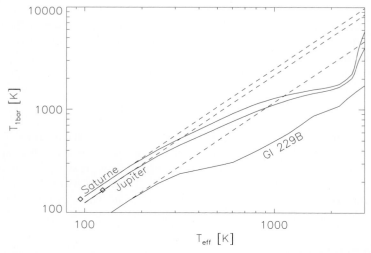

Fig. 3. Comparison of the boundary condition obtained from (10) (*dashed*) to a gray atmosphere from Saumon et al. 1996 (*plain*), in the case of Saturn ($g \approx 1100\,\text{cm}\,\text{s}^{-2}$), Jupiter ($g \approx 2600\,\text{cm}\,\text{s}^{-2}$) and Gl229B ($g \approx 10^5\,\text{cm}\,\text{s}^{-2}$)

Note that these boundary conditions assume that the object is isolated. This is not the case of the giant planets of the solar system and for extrasolar planets for which the insolation can play an important role. We leave that problem for a further discussion.

3.3 Simple Solutions

Central Pressure

In order to estimate the central pressure, it is useful to write the hydrostatic equilibrium in a form which is independent on density:

$$\frac{\partial P}{\partial m} = -\frac{Gm}{4\pi r^4}. \tag{11}$$

Approximating by $m \approx M/2$, $r \approx R/2$ (M and R being the total mass and radius, respectively) yields

$$P_c \approx \frac{2}{\pi} \frac{GM^2}{R^4}. \tag{12}$$

Another simple solution is obtained by assuming uniform density $\bar{\rho} = 3M/4\pi R^3$. Equation (4) can then be integrated to obtain

$$P_c \approx \frac{3}{8\pi} \frac{GM^2}{R^4}. \tag{13}$$

Knowing the mass and radius of a moon, planet or star, its central pressure can therefore be approximated within a factor of a few.

Using (12,13) the central pressure of the moon is found to be $17-91$ kbar, $1.7-9.1$ Mbar for the Earth, $12-64$ Mbar for Jupiter and $1.3-7.2$ Gbar for the Sun. For comparison, the corresponding values given by more elaborate models are ~ 40 kbar, 3.6 Mbar, 40 to 70 Mbar and 230 Gbar, respectively. The approximation is least successful in the case of the Sun, mostly because of the increase in density of the central regions ($\rho_c \approx 150\,\mathrm{g\,cm^{-3}}$).

When dealing with objects of small masses like planetary moons, the uniform density model is in fact a good approximation to the internal pressure, which can be shown to be:

$$P(\xi) \approx \frac{4\pi}{6} GR^2 \bar{\rho}^2 \left[1 - \left(\frac{r}{R}\right)^2\right]. \tag{14}$$

The central temperatures are more difficult to obtain *a priori* because contrary to main-sequence stars the interiors strongly depart from ideality. An *a posteriori* estimate uses the fact that these objects are mostly convective and that their temperature gradient $\nabla_T \equiv (d\ln T/d\ln P) \approx 0.3$. One then finds that $T \approx T_{\rm eff}(P/P_0)^{\nabla_T}$, with $T_{\rm eff}$ and P_0 being defined by the boundary conditions discussed in Sect. 3.2. In the case of Jupiter, starting from $T(1\,\mathrm{bar}) = 165\,\mathrm{K}$ and $P_c \approx 12\,\mathrm{Mbar}$, one gets $T_c \approx 22000\,\mathrm{K}$, a relatively accurate estimate of the temperature at the bottom of the hydrogen–helium envelope.

Polytropic Solutions

A full integration of the set of differential equations is of course necessary to obtain the necessary precision on quantities such as pressure, temperature and density. However, it is sometime useful to use approximate analytical solutions to understand the underlying physics. One of these approximations, of considerable importance for stellar physics, is to assume a polytropic relation between pressure and density:

$$P = K\rho^{1+1/n} , \tag{15}$$

where K is supposed constant, and n is the *polytropic index*. Of course, this relation implicitly assumes that either density only depends on pressure not on temperature, or that the temperature profile is well-behaved and yields K and n constants.

This property is indeed verified in the limit when the pressure is due to non-relativistic fully degenerate electrons (e.g. Chandrasekhar 1939). In that case, a pure hydrogen plasma obeys the polytropic relation (15) with $n = 3/2$ and a constant K defined by fundamental physics (i.e. independent of M, $T_{\rm eff}$...etc.).

On the other hand, a perfect gas with a constant temperature gradient can be shown to obey a polytropic relation of index $n = 1/(1+1/\nabla_T)$. In the case of a monoatomic perfect gas, $n = 3/2$. It is important to notice that in that case K is set by the atmospheric boundary condition: it depends on parameters such as the mass and effective temperature of the object considered.

A solution of the polytropic problem is obtained from the integration of the hydrostatic and Poisson equations:

$$\begin{cases} \dfrac{dP}{dr} = -\dfrac{d\Phi}{dr}\rho , \\ \dfrac{1}{r^2}\dfrac{d}{dr}\left(r^2\dfrac{d\Phi}{dr}\right) = 4\pi G\rho , \end{cases} \tag{16}$$

where Φ is the gravitational potential. The problem can be solved with some algebra. With the following change of variables,

$$z = Ar , \qquad A^2 = \frac{4\pi G}{(n+1)K}\rho_{\rm c}^{\frac{n-1}{n}} \\ w = \frac{\Phi}{\Phi_{\rm c}} = \frac{\rho}{\rho_{\rm c}} , \tag{17}$$

where $\rho_{\rm c}$ and $\Phi_{\rm c}$ are the central density and gravitational potential, respectively, one is led to the famous *Lane–Emden* equation (see Chandrasekhar 1939; Kippenhahn and Weigert 1991 for a demonstration):

$$\frac{1}{z^2}\frac{d}{dz}\left(z^2\frac{dw}{dz}\right) + w^n = 0 . \tag{18}$$

This equation possesses analytical solutions for $n = 0$, 1 and 5. For our purpose, it is sufficient to say that the solutions are characterized by the surface condition: z_n such that $w(z_n) = 0$ and by the derivative of the function w at that point: $(dw/dz)_{z_n}$. It can be shown that the total mass and surface radius of a polytrope are such that:

$$M = 4\pi \rho_c R^3 \left(-\frac{1}{z}\frac{dw}{dz} \right)_{z=z_n}, \qquad (19)$$

$$R = z_n \left[\frac{1}{4\pi G}(n+1)K \right]^{1/2} \rho_c^{\frac{1-n}{2n}}. \qquad (20)$$

If we assume that K and n are independent of the mass and surface conditions of the object considered, it is easy to show that the mass–radius relation is such that

$$R \propto M^{\frac{1-n}{3-n}}. \qquad (21)$$

First, one can notice that the exponent diverges for $n = 3$. In this case, the Lane–Emden equation has only one solution: this leads to the Chandrasekhar limit for the mass of white dwarfs. Second, for uncompressible materials, $n = 0$ and we can verify that $R \propto M^{1/3}$. Third, objects whose internal pressure is dominated by non-relativistic degenerate electrons (this is formally valid only in the white dwarfs regime) are such that $n = 3/2$ (see Sect. 4.1) and $R \propto M^{-1/3}$.

3.4 Mass–Radius Relation

The relation between mass and radius has very fundamental astrophysical applications. Most importantly is allows one to infer the gross composition of an object from a measurement of its mass and radius. This is especially relevant in the context of the discovery of extrasolar planets with both radial velocimetry and the transit method, as the two techniques yield relatively accurate determination of M and R.

Figure 4 shows as a plain line the mass–radius relation of isolated hydrogen–helium objects (of approximate solar composition) after 10 Gyr of evolution. As could have been inferred from the polytropic solutions, this curve has a local maximum: at small masses, the compression is rather small so that the radius increases with mass (corresponding to a low polytropic index). (Note for example that in the case of the Earth, the central density is $\sim 13\,\mathrm{g\,cm^{-3}}$, to be compared with a mean density of $5.52\,\mathrm{g\,cm^{-3}}$). At large masses, degeneracy sets in and the radius decreases with mass (note from Fig. 4 that it never quite reaches the white dwarf limit $R \propto M^{-1/3}$). At still larger masses (more than $70\,M_\mathrm{J}$), we get in the *stellar* regime, which is dominated by thermonuclear reactions, and thermal effects have to be taken into account.

The polytropic indexes of the isolated 0.1, 1 and 10 M_J are shown in Fig. 5. At small masses, n is effectively rather small and the tends toward a uniform

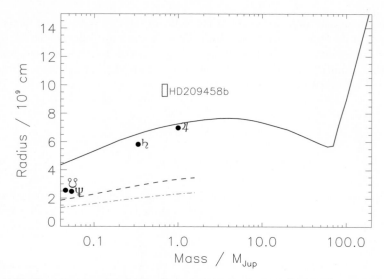

Fig. 4. Radius versus mass for hydrogen–helium planets (Y=0.25) after 10 Ga of evolution (*plain line*). An approximate mass–radius relation for zero-temperature water and olivine planets is shown as *dashed* and *dash-dotted* lines, respectively (Courtesy of W.B. Hubbard). The observed values for Uranus, Neptune, Saturn and Jupiter, as well as that for the Pegasi planet HD209458b are indicated.

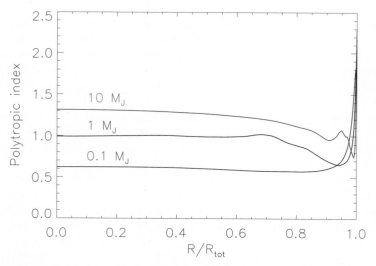

Fig. 5. Polytropic index n (such that $P \propto \rho^{1+1/n}$) as a function of internal radius, for 0.1, 1 and 10 M_J isolated planets of solar composition after 10 Ga of evolution

density solution. At around the mass of Jupiter, we get $n \sim 1$, which effectively corresponds to a maximum in the polytropic mass–radius relation (21). Above a mass of $\sim 4\,\mathrm{M_J}$, the radius starts decreasing with increasing mass, and effectively, the $10\,\mathrm{M_J}$ object has $n \approx 1.3$ in most of its interior. Equation (21) would imply $R \propto M^{-0.18}$, which is steeper than obtained on Fig. 4. This is due to the fact that even after 10^{10} years, a $10\,\mathrm{M_J}$ object still retains part of its primordial heat and that K *cannot* be considered as independent of effective temperature and mass, as assumed in (21).

Another conclusion that can be derived from Fig. 4 is that the planets in our Solar System are not of solar composition: their radii lie below that predicted for $Y = 0.25$ objects. Indeed, it can already be inferred that Jupiter, Saturn, and the two ice-giants Uranus and Neptune contain a growing proportion of heavy elements. The theoretical curves for olivine and ice planets predict even smaller radii however: even Uranus and Neptune contain 10 to 20% of their mass as hydrogen and helium.

An object is found above the hydrogen–helium curve: HD209458b. In this case, we will see that the planet has its evolution dominated by the intense stellar irradiation it receives. Thermal effects are no longer negligible: One cannot neglect the variations of the polytropic constant K with mass. Instead of (21), one is led to:

$$R \propto K^{\frac{n}{3-n}} M^{\frac{1-n}{3-n}} . \qquad (22)$$

The constant K can be estimated through the surface boundary condition, assuming that the planetary interior is tied to the surface with an approximately constant polytropic index n (a condition which is generally verified). Thus, using a perfect gas relation

$$K = P_0^{-1/n} \left(\frac{\mathcal{R} T_0}{\mu}\right)^{1+1/n} . \qquad (23)$$

Let us assume that T_0 is, in the case of irradiated planets, set by the stellar insolation (and therefore independent of M). Using the Eddington boundary condition $P_0 \propto g/\kappa$. The relation for the opacity $\kappa \propto P$ is generally valid for hot atmospheres not dominated by hydrogen–helium collision-induced absorption (see Sect. 5). Therefore, a constant insolation and constant interior n implies

$$K \propto \left(\frac{M}{R^2}\right)^{-1/2n} . \qquad (24)$$

It is then easy to show that the mass–radius relation for strongly irradiated planets becomes

$$R \propto M^{\frac{1/2-n}{2-n}} . \qquad (25)$$

Thus, for $n = 3/2$, a relation valid for an adiabatic, ideal monoatomic gas, one finds $R \propto M^{-2}$. For $n = 1$, one finds $R \propto M^{-1/2}$. Strongly irradiated hydrogen–helium planets of small masses are hence expected to have the largest radii. Note that this estimate implicitly assumes that n is constant throughout the planet. The real situation is more complex because of the growth of a deep radiative region in most irradiated planets, and because of structural changes between the degenerate interior and the perfect gas atmosphere.

3.5 Rotation and the Figures of Planets

Hydrostatic Equilibrium and Symmetry Breaking

We have thus seen that the knowledge of the mass and radius of a planet could inform us on its global composition. Fortunately, the giant planets in the Solar System are also fast rotators and their *figure* can also inform us more precisely on their internal composition. In the case of an inviscid fluid rotating with an angular velocity $\mathbf{\Omega}(\mathbf{r})$, the hydrostatic equilibrium has to be written in the frame of rest of the system (see e.g. Pedlosky 1979):

$$\frac{\nabla P}{\rho} = \nabla V - \mathbf{\Omega} \times (\mathbf{\Omega} \times \mathbf{r}) \;, \tag{26}$$

where the gravitational potential is defined as

$$V(\mathbf{r}) = G \int \frac{\rho(\mathbf{r}')}{|\mathbf{r} - \mathbf{r}'|} d^3\mathbf{r}' \;. \tag{27}$$

The resolution of (26) is generally a complex problem. It can however be somewhat simplified by assuming that $|\mathbf{\Omega}| \equiv \omega$ is such that the centrifugal force can be derived from a potential:

$$W(\mathbf{r}) = \frac{1}{2}\omega^2 r^2 \sin^2\theta \;, \tag{28}$$

where θ is the angle from the rotation axis (colatitude). This implies that ω is either constant, or a function of the distance to the axis of rotation (rotation on cylinders).

The total potential is $U = V + W$ and the hydrostatic equilibrium can be written as

$$\nabla P = \rho \nabla U \;. \tag{29}$$

The *figure* of a fluid planet in hydrostatic equilibrium is then defined by the $U = $ cte level surface. The expression of W shows that the centrifugal acceleration will be maximal at the equator. Since it tends to oppose gravity, it can be intuited that the planet's figure will depart from a sphere and become oblate, with a smaller polar radius than its equatorial radius. This was first demonstrated by Newton in 1687, but is in no way straightforward, and was contested by contemporaries, some advocating that the Earth's dimension should be larger at the poles!

Most of the problem lies in the breaking of the symmetry by rotation: the gravitational potential can no longer be integrated simply. We will summarize here one method, worked out by Lagrange, Clairaut, Darwin and Poincaré and detailed by Zharkov and Trubitsyn (1978). At its basis is a projection of the integrand of (27) onto a basis of Legendre polynomials $P_n(\cos\psi)$:

$$\frac{1}{|\mathbf{r}-\mathbf{r}'|} = \begin{cases} \dfrac{1}{r}\sum_{n=0}^{\infty}\left(\dfrac{r'}{r}\right)^{n} P_n(\cos\psi) & \text{if } r \geq r', \\ \dfrac{1}{r}\sum_{n=0}^{\infty}\left(\dfrac{r'}{r}\right)^{-n-1} P_n(\cos\psi) & \text{if } r < r', \end{cases} \quad (30)$$

where ψ is the angle between the radius vectors $\mathbf{r}$ and $\mathbf{r}'$. The Legendre polynomials are determined from the formula

$$P_n(x) = \frac{1}{2^n n!}\frac{d^n}{dx^n}\left[(x^2-1)^n\right]. \quad (31)$$

In particular, $P_0 = 1$ and $P_2(x) = (3x^2-1)/2$. These polynomials also have very important orthogonal properties that will not be detailed here.

Some geometry, the properties of Legendre polynomials and the assumption of hydrostatic equilibrium (azimuthal symmetry) allows one to write the gravitational potential in the form

$$V = \frac{G}{r}\sum_{n=0}^{\infty}\left(r^{-2n}D_{2n} + r^{2n+1}D'_{2n}\right)P_{2n}(\cos\theta),$$

$$D_{2n} = \int_{r'\leq r}\rho(r',\cos\theta')r'^{2n}P_{2n}(\cos\theta')d^3\mathbf{r}',$$

$$D'_{2n} = \int_{r'>r}\rho(r',\cos\theta')r'^{-2n-1}P_{2n}(\cos\theta')d^3\mathbf{r}'. \quad (32)$$

The potential V is thus projected on the basis of Legendre polynomials $P(\cos\theta)$. The D_{2n} and D'_{2n} coefficients are complex functions. It is to be noted that this projection, as proposed by Lagrange poses a mathematical problem of divergence of the Legendre series between the sphere and level surface. Using a method initially proposed by Lyapunov, Trubitsyn showed that this expression is however valid because of the exact cancellation of the divergent terms (see Zharkov and Trubitsyn 1978).

On the other hand, the centrifugal potential can be written on the same basis:

$$W = \frac{1}{3}\omega^2 r^2[1 - P_2(\cos\theta)]. \quad (33)$$

The total potential U thus appears as a weighted sum (however complex) of Legendre polynomials.

Equations for the Level Surfaces: Principles

The *figure* of a planet is determined by the level surfaces on which the total potential is constant. As shown by (29), in hydrostatic equilibrium ∇P and ∇U are in the same direction. Taking the curl of that equation, one finds that $\nabla \rho \times \nabla U = 0$. The surfaces of constant potential are also surfaces of constant pressure, density, and hence temperature. Hydrostatic equilibrium therefore also corresponds to barotropic equilibrium. (But remember our hypothesis that the centrifugal acceleration derives from a potential). These surfaces of constant U are sought in the form:

$$r(s, \cos\theta) = s \left[1 + \sum_{n=0}^{\infty} s_{2n}(s) P_{2n}(\cos\theta) \right] , \qquad (34)$$

where $s_{2n}(s)$ are coefficients to be determined, and s is chosen to be the radius of a sphere of equal volume (and hence, equal mass):

$$\frac{4\pi}{3} s^3 = \frac{4\pi}{3} \int_0^1 r^3(s, \cos\theta) d\cos\theta . \qquad (35)$$

This allows one to integrate the angular part entering the calculation of the coefficients D_{2n} and D'_{2n} in (32). The solution of the problem is found by noticing that the total potential can now be written

$$U(s, \cos\theta) = \frac{4\pi}{3} G \bar{\rho} s^2 \sum_{n=0}^{\infty} A_{2n}(s) P_{2n}(\cos\theta) , \qquad (36)$$

where $\bar{\rho}$ is the planet's mean density. Since by definition the gravitational potential is constant on a level surface (fixed s), all coefficients $A_{2n}(s)$ must be zero for $n \neq 0$. With (35), we thus have $n+1$ equations for the $n+1$ variables $s_0, \ldots, s_{2n}$. The problem can thus be solved for weak rotation rates ω by introducing a small parameter, q, the ratio of the centrifugal acceleration at the equator to the leading term in the gravitational acceleration:

$$q = \frac{\omega^2 R_{\text{eq}}^3}{GM} , \qquad (37)$$

R_{eq} being the equatorial radius. One can show that $s_0 \propto q$ and $s_{2n} \propto q^n$ for $n \neq 0$. This system of integro-differential equations is rather complex and will not be given here (see Zharkov and Trubitsyn 1978 for equations to third order).

With our choice of coordinates, the hydrostatic equation retains a simple form:

$$\frac{\partial P}{\partial s} = \rho \frac{\partial U}{\partial s} , \qquad (38)$$

i.e. the equation is now integrated with respect to the mean planetary radius. Furthermore, because of our assumption that the fluid remains barotropic, the other equations are unchanged. A detailed calculation of U shows that

$$\frac{1}{\rho}\frac{\partial P}{\partial s} = -\frac{Gm}{s^2} + \frac{2}{3}\omega^2 s + \frac{GM}{R^3}s\varphi_\omega , \qquad (39)$$

where φ_ω is a slowly varying function of s which, in the case of Jupiter varies from about 2×10^{-3} at the center to 4×10^{-3} at the surface.

The External Potential: Constraints from Observations

As suggested previously, the effect of rotation is not only to complexify the equation for hydrostatic equilibrium. It also provide ones with the only way (yet) to probe the interiors of the giant planets of the solar system. This was first recognized by Sir H. Jeffreys (1923), but has seen significant progresses due to the flybys of the giant planets by the Pioneer and Voyager spacecrafts that allowed for a direct measurement of the planets' gravitational potentials.

The thus measured gravitational potentials are generally written in the form

$$V_{\text{ext}}(r,\cos\theta) = \frac{GM}{r}\left[1 - \sum_{n=1}^{\infty}\left(\frac{a}{r}\right)^{2n} J_{2n} P_{2n}(\cos\theta)\right] , \qquad (40)$$

and the coefficients J_{2n} are the planet's *gravitational moments*. These are hence directly related to the coefficients D_{2n} defined by (32), from which it can be shown that

$$J_{2n} = -\frac{1}{Ma^n}D_{2n} . \qquad (41)$$

(Note that because we are always outside the planet $r > r'$ and the centrifugal potential does not appear since we are in an inertial coordinate system).

For example, the first gravitational moment can be calculated as

$$\begin{aligned}
-Ma^2 J_2 &= \int \rho(r')r'^2\left(\frac{3}{2}\cos^2\theta' - \frac{1}{2}\right)d^3\mathbf{r'} \\
&= \int \rho(r')\frac{1}{2}(2r'^2\cos^2\theta' - r'^2\sin^2\theta')d^3\mathbf{r'} \\
&= \int \rho(r')\frac{1}{2}[(y^2+z^2)+(x^2+z^2)-2(x^2+y^2)]d^3\mathbf{r'} \\
&= \frac{A+B-2C}{2} , \qquad (42)
\end{aligned}$$

where A, B and C are the principal moments of inertia of the planet with respect to axes x, y and z, respectively.

The measured gravitational moments can thus be compared to the theoretically measured ones. For a planet in hydrostatic equilibrium, the odd moments J_{2n+1} are all zero while the even moments have a magnitude $J_{2n} \propto q^n$. The high order gravitational moments also correspond to integrals with weighting functions peaking closer to the external layers of the

Table 4. Parameters constraining interior structure

	q	Λ_2	C/MR_{eq}^2
Jupiter	0.08923	0.165	0.26
Saturn	0.15491	0.105	0.22
Uranus	0.02951	0.119	0.23
Neptune	0.0261	0.136	0.24

planet. The information contained by the $\{J_{2n}\}$ is therefore limited: without other information from e.g. global oscillations of the planet, it is impossible to accurately constrain the structure of the inner regions.

Table 4 shows the values of the parameter q and of the axial moment of inertia of the giant planets calculated from J_2 using the Radau–Darwin approximation (Zharkov and Trubitsyn 1978):

$$\frac{C}{MR_{eq}^2} \approx \frac{2}{3}\left[1 - \frac{2}{5}\left(\frac{5}{3\Lambda_2+1} - 1\right)^{1/2}\right], \quad (43)$$

where we have introduced the linear response coefficient $\Lambda_2 \equiv J_2/q$, and we have neglected second order terms proportional to the planets' flattening. Our four giant planets all have an axial moment of inertia substantially lower than the value for a sphere of uniform density, i.e. $2/5\,MR^2$, indicating that they have dense central regions.

An analytical solution of the figure equation can be found for a polytropic equation of state of index $n = 1$ ($P \propto \rho^2$), which is, as we have seen relevant for most of Jupiter's interior. In that case, one finds that (see Zharkov and Trubitsyn 1978; Hubbard 1989), $\Lambda_2 = 0.173$ and thus $C/MR^2 = 0.263$, indeed very close to the value found for Jupiter. This shows already that Jupiter's core is small, relatively to the planet's total mass. It also indicates that Saturn, Uranus and Neptune have dense central regions and hence depart substantially from solar composition.

Effect of Differential Rotation

In order to be able to integrate the system of integro-differential equations, we have implicitly assumed a solid body rotation. The atmospheres of all giant planets is seen to rotate with a speed which is latitudinally dependent. These latitudinal variations amount to about 1% for Jupiter to more than 15% in the case of Neptune, from peak to peak.

A first consequence is that the gravitational calculated assuming solid body rotation will be different than if the interior rotation is, say, on cylinders. For a given structure, differential rotation such as imposed by the surface winds of Jupiter and Saturn increases the absolute values of the planets' gravitational moments. In order to account for that effect using solid body rotation, one has to use effective gravitational moments that are smaller in absolute value than those directly measured (Hubbard 1982).

Another interesting consequence concerns the high order gravitational moments, J_{10} and above. Hubbard (1999) has shown that if the observed atmospheric rotation pattern persists deep enough into the interior (say to within a few % of the total radius beneath the atmospheric layer), then the gravitational moments will stop decreasing and reach a plateau at a value $|J_n| \approx 10^{-8}$ with $n \gtrsim 10$. This lends support to space missions that would enable a detailed mapping of the gravitational fields of the giant planets. This would require both a polar-like orbit and one (or better several) very close flybys.

3.6 Equations of Evolution

We have so far expressed the differential equations in terms of the radius r. This Eulerian approach has the inconvenience that the spatial variable can be a rapidly varying function of time (when, during the evolution, the contraction is fast). It is therefore generally more convenient to use a Lagrangian approach, in which the new independent coordinates are the mass m and time t. This has the advantage that except in the case of mass loss/gain, the outer boundary condition is defined at a fixed $m = M$, the total mass of the object. Note that because of our definition of the radius as the mean radius, the effect of rotation is just to add two terms to the hydrostatic equation. Hereafter, we will use r instead of s as the *mean* radius (see e.g. Guillot and Morel 1995 for a possible method to numerically resolve the equations). The system of differential equations becomes:

$$\begin{cases} \dfrac{\partial P}{\partial m} = -\dfrac{Gm}{4\pi r^4} + \dfrac{\omega^2}{6\pi r} + \dfrac{GM}{4\pi R^3 r}\varphi_\omega \,, \\ \dfrac{\partial T}{\partial m} = \left(\dfrac{\partial P}{\partial m}\right)\dfrac{T}{P}\nabla_T \,, \\ \dfrac{\partial r}{\partial m} = \dfrac{1}{4\pi r^2 \rho} \,, \\ \dfrac{\partial L}{\partial m} = \dot{\epsilon} - T\dfrac{\partial S}{\partial t} \,. \end{cases} \qquad (44)$$

The boundary conditions are as discussed in Sect. 3.2, except that the variable is now m instead of r. Note however that when studying the present-day interiors of Jupiter, Saturn, Uranus or Neptune, the most logical surface boundary condition is at a fixed temperature $T = T_{\rm surf}$ and pressure $P_{\rm surf}$, for $m = M$. Note that in that case, there is no time dependency, and the energy conservation equation cannot be integrated. This requires a priori setting the luminosity (usually by assuming that it is uniformly equal to the measured intrinsic luminosity of the planet). In all other cases, i.e. when considering the *evolution* of substellar objects, the outer boundary condition must depend on L and R.

Most of the important physics in the system of equations (44) is hidden in several quantities: φ_ω contains the physics related to rotation discussed previously, but is generally a small perturbation. The term ∇_T depends on the process which transports the energy inside the planet and will be discussed in Sect. 5. The density ρ and specific entropy S are functions of the temperature, pressure and composition. They have to be calculated independently using an appropriate *equation of state*, the subject of Sect. 4. Finally, $\dot\epsilon$ accounts for any source of energy, e.g. thermonuclear reactions, radioactivity or heat dissipation. This term is generally neglected, but will be discussed for brown dwarfs, and also in the case of Pegasi planets.

4 Equations of State

4.1 Basic Considerations

Calculation of Equations of State

The knowledge of appropriate equations of state is at the basis of any modeling of substellar objects. Basically, for a given atomic composition, and two macroscopic thermodynamic variables, say temperature and volume, an equation of state is to provide all the other thermodynamic variables and their derivatives (pressure, internal energy, entropy, specific heat...etc.). As discussed by Fontaine, Graboske and van Horn (1977), the thermodynamic constraints that have to be satisfied for any equilibrium thermodynamic description of a single-phase material are:

I. Accuracy $\quad P^{\text{approx}}(T,V) = P^{\text{exact}}(T,V)$.
$\qquad\qquad\qquad U^{\text{approx}}(T,V) = U^{\text{exact}}(T,V)$.
II. Stability $\quad \left(\frac{\partial P}{\partial V}\right)_T < 0, \quad \left(\frac{\partial U}{\partial T}\right)_V > 0$.
III. Consistency $\quad \left(\frac{\partial P}{\partial T}\right)_V = \left(\frac{\partial S}{\partial V}\right)_T = \frac{1}{T}\left(P + \left(\frac{\partial U}{\partial V}\right)_T\right)$.
IV. "Normality" $\quad \left(\frac{\partial P}{\partial T}\right)_V > 0, \quad \left(\frac{\partial^2 P}{\partial V^2}\right)_T > 0$.

As noted by the authors, condition II is generally trivial to achieve; condition III is straightforward but often grossly violated; condition IV is not thermodynamically demanded, but holds for most ρ, T values. Indeed, we will see one possible equation of state for which condition IV is violated.

The calculation of equations of state itself can become extremely complex. For our purposes, it will suffice to say that it can be split into two main groups: the "chemical" and "physical" pictures. In the chemical picture, one assumes that bound configurations (e.g. atoms, molecules) retain a definite identity and interact through pair potentials. The system of particles of species α confined to a volume V at temperature T is conveniently described by the Helmoltz free energy F, which is itself obtained from microscopic physics through

$$F(\{N_\alpha\}, V, T) = -kT \ln Z(\{N_\alpha\}, V, T) \ , \tag{45}$$

where N_α denotes the number of particles and Z is the canonical partition function of the system. Other thermodynamical quantities are then obtained from derivatives of F. For example,

$$P = -\left(\frac{\partial F}{\partial V}\right)_{\{N_\alpha\}, T} .$$

When confronted to ionization and/or dissociation, the actual composition of the system (i.e. abundances of electrons, ions, atoms and molecules) is obtained through a minimization of the free energy of the system. As discussed by Fontaine et al. the calculation of the free energy requires several assumptions that necessarily limit its accuracy. Its main drawback is the *apriori* definition of certain classes of particles, i.e. ions, atoms and molecules which necessitates the use of effective interaction potentials. The calculation can thus fail in states where more complex systems are formed and the distinction between bound and free states is not easily made.

Another method consists in directly computing the n-body Schrödinger equation of the quantum-statistical system. This approach is generally exact in the limit set by the computationally intensive method that has to be used to solve the problem. Within this physical picture, two main approaches have been used: restricted path integral Monte Carlo simulations, and density functional theory molecular dynamics. The first approach consists in solving the full problem for a limited number of protons and electrons in a box (64 of each at the most, with today's computers). The second approach involves local solutions to the problem and fails when both short range and long range interactions have to be taken into account.

The Phase Diagram

In terms of pressures and temperatures, the interiors of giant planets and brown dwarfs lie in a region for which accurate equations of state are extremely difficult to calculate. Some of the important phenomena that occur in these objects are illustrated by the phase diagram of hydrogen (Fig. 6).

The photospheres of these objects is generally relatively cold and at low pressure, so that hydrogen is in molecular form and the perfect gas conditions apply:

$$P = \frac{\rho \mathcal{R} T}{\mu}; \quad U = C_V T \ , \tag{46}$$

with $\mu \approx 2$ (neglecting helium atoms and heavy elements) and $C_V \approx 5/2k$, due to the vibration of the hydrogen molecule.

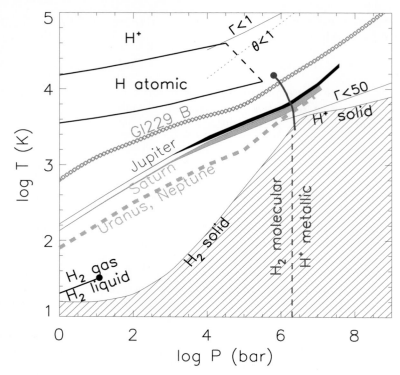

Fig. 6. Phase diagram for hydrogen with the main phase transitions occurring in the fluid or gas phase. The temperature–pressure profiles for Jupiter, Saturn, Uranus, Neptune, and Gl229B (assumed to be a 30 M$_J$ brown dwarf) are shown. The *plain*, almost vertical line near 1 Mbar represents the Plasma Phase Transition (PPT) supposed to separate molecular from metallic hydrogen as computed by Saumon et al. (1995). The region in which hydrogen is predicted to be solid is represented as a *dashed* area. Lines showing the values of the parameters θ and Γ (see text) are also shown

As one goes deeper into the interior however, the molecules become closer to one another. The system progressively becomes a liquid, in which the interactions between molecules play an important role. This occurs when the intermolecular distance becomes of the same order as the size of a hydrogen molecule. Using real equations of state, it can be estimated that the perfect gas relation tends to underestimate the pressure by 10% or more when the density becomes larger than about 0.02 g cm^{-3} (pressures above 1 kbar in the case of Jupiter).

At higher densities (or pressures) and relatively low temperatures, the electrons can become *degenerate* : in that limit, their momentum is not determined by the temperature of the mixture, but by the fact that, as fermions of spin $-1/2$ or $+1/2$, only two of them can be stacked in a cell $\Delta p \Delta V = h^3$ (Pauli's exclusion principle). The significance of this phenomenon can be measured through a *degeneracy temperature parameter*

$$\theta = \frac{T}{T_\mathrm{F}} = \frac{2m_\mathrm{e} k}{h^2} \left(\frac{8\pi}{3} \mu_\mathrm{e} m_\mathrm{u}\right)^{2/3} \frac{T}{\rho^{2/3}} , \qquad (47)$$

where T_F is the Fermi temperature, and the number density of electrons is $n_\mathrm{e} = \rho/\mu_\mathrm{e} m_\mathrm{u}$. The quantity μ_e is the mean molecular weight per electron ($\mu_\mathrm{e} \approx 2/(1+X)$). The parameter θ can be defined regardless of the presence of bound states. However, in the presence of atoms and molecules, the energy of most of the electrons is not kT nor kT_F so that its usefulness in that regime is limited. It can be seen from Fig. 6 that the interiors of substellar objects are always characterized by $\theta < 1$. It is never possible to assume that free electrons behave like a perfect gas .

Another important quantity is the coupling parameter, defined as the ratio of the Coulomb potential to the thermal energy:

$$\Gamma = \frac{e^2}{akT} = \frac{e^2}{k} \left(\frac{4\pi}{3\mu m_\mathrm{u}}\right)^{1/3} \frac{\rho^{1/3}}{T} , \qquad (48)$$

where a is the mean distance between nuclei. As Γ increases due either to an increase of the density or to a decrease of the temperature, the Coulomb forces becomes more effective. With increasing densities, the system of ions eventually favors a non-random organization and becomes bound into a lattice system. This occurs for large values of Γ (~ 100). Figure 6 shows that substellar objects always have $\Gamma > 1$: the system is dominated by the repulsive coulombian potential between nuclei. However, we will be concerned with values of $\Gamma < 50$, i.e. unlike white dwarfs, substellar objects are not expected to crystallize (this occurs for $\Gamma \gtrsim 180$). Hubbard (1968) was the first to show that Jupiter's interior should be hot enough for its interior to be *fluid*. It can also be seen in the phase diagram that it is the case of Saturn. For Uranus and Neptune, the situation is actually more complex because at large pressures they are not expected to contain hydrogen, but several studies show that ices in their interior should be fluid as well (e.g. Cavazzoni et al. 1999).

The largest fraction of the interior of brown dwarfs and giant planets is in a region in which hydrogen is *metallic*: the hydrogen molecules have been dissociated and ionized. The pressure inside this region can be expressed in the following form (e.g. Stevenson 1991):

$$P = P_\mathrm{e} + P_\mathrm{th,ion} + P_\mathrm{coul} + P_\mathrm{ex} , \qquad (49)$$

where P_e represents the contribution from the electron gas, $P_\mathrm{th,ion}$ the contribution from the thermallized ions, and P_coul and P_ex are negative terms due to the Coulombian interactions of nuclei in the sea of electrons, and the reduction in electron–electron repulsion due to the exclusion principle, respectively. P_coul is significant when Γ becomes large. The exchange pressure P_ex has to be taken into account for small values of θ. Although quantitatively, the terms due to ions are important, most of the important physics and in particular the molecular/metallic transition is due to a difference in behavior of the electrons when the density rises.

The Degenerate Electron Gas

In stars with masses larger than about $0.3\,\mathrm{M}_\odot$ the electrons always behave with a near-maxwellian distribution of the momenta. However, for objects of lower interior temperatures, the Pauli exclusion principle yields a distribution which is determined by Fermi–Dirac statistics. The number of electrons in a volume dV and with an absolute of the momentum in $[p, p+dp]$ is:

$$f(p)dpdV = \frac{8\pi p^2 dpdV}{h^3} \frac{1}{1+e^{E/kT-\psi}}, \tag{50}$$

where in the non-relativistic case $E = p^2/(2m_\mathrm{e})$ and ψ is the *degeneracy parameter*. For $\psi \to -\infty$ the distribution is identical to the Maxwell–Boltzmann one. In the limit $\psi \to +\infty$ the electrons are said to be fully degenerate.

The density of electrons, electronic pressure and internal energy can be obtained through integrations of that distribution:

$$n_\mathrm{e} = \frac{8\pi}{h^3} \int_0^\infty \frac{p^2 dp}{1+e^{E/kT-\psi}}, \tag{51}$$

$$P_\mathrm{e} = \frac{8\pi}{3h^3} \int_0^\infty \frac{vp^3 dp}{1+e^{E/kT-\psi}}, \tag{52}$$

$$U_\mathrm{e} = \frac{8\pi}{h^3} \int_0^\infty \frac{Ep^2 dp}{1+e^{E/kT-\psi}}. \tag{53}$$

The degeneracy parameters ψ obtained in the central region of substellar objects is relatively independent of the mass and age (to a factor ~ 3) and is of the order of $\psi \approx -30$ (e.g. Chabrier and Baraffe 2000). The combination of these low values of θ and ψ thus implies that a significant fraction of the electrons are indeed degenerate.

Although this is not true of regions at lower pressures, we will find it instructive to use the relations for a fully degenerate electron gas for qualitative estimates. In the limit $\psi \to \infty$, one finds that the completely degenerate non-relativistic electron gas is such that (e.g. Kippenhahn and Weigert 1991):

$$\begin{aligned} P_\mathrm{e} &= \frac{1}{20}\left(\frac{3}{\pi}\right)^{2/3} \frac{h^2}{m_\mathrm{e}} n_\mathrm{e}^{5/3} \\ &= 1.0036 \times 10^{13} \left(\frac{\rho}{\mu_\mathrm{e}}\right)^{5/3} \quad \text{(cgs)} \\ U_\mathrm{e} &= \frac{3}{2} P_\mathrm{e}. \end{aligned} \tag{54}$$

Pressure Ionization

As seen in the phase diagram, hydrogen can become ionized due to increasing pressure instead of standard ionization at increasing temperature. Basically, this occurs when the degenerate electrons get a Fermi energy which is larger than that necessary to ionize hydrogen atoms. The approximate level at which this occurs can be estimated as follows.

First, it can be noted that both free and bound electrons have to obey the Pauli principle.

The energy of each electron is hence of the order U_e/n_e. For a set value of n_e a *lower bound* on U_e can be obtained by assuming full degeneracy (54). In order to become ionized this value has to become larger than the ionization potential of hydrogen, $u_0 = 13.6\,\mathrm{eV}$. This occurs when

$$n_e \gtrsim \left(\frac{8\pi}{3}\right)^{5/2} \left(\frac{5m_e}{4\pi h^2}\right)^{3/2} u_0^{3/2}, \tag{55}$$

corresponding to an electronic pressure

$$P_e \gtrsim \frac{2}{3}\left(\frac{8\pi}{3}\right)^{5/2} \left(\frac{5m_e}{4\pi h^2}\right)^{3/2} u_0^{5/2}. \tag{56}$$

Quantitatively, hydrogen metallization is then found to occur around $n_e \sim 5 \times 10^{23}\,\mathrm{cm}^{-3}$, $\rho \sim 0.8\,\mathrm{g\,cm}^{-3}$ and $P_e \sim 7\,\mathrm{Mbar}$. Even though crude assumptions were made, this is relatively close to more elaborate calculations.

The same estimates can be used for helium ionization, assuming helium atoms are immersed in a sea of protons and electrons. Because $u_0 = 54.4\,\mathrm{eV}$, the density and electronic pressure for helium ionization rise to $6.5\,\mathrm{g\,cm}^{-3}$ and $230\,\mathrm{Mbar}$, respectively. However, at those very high densities, the distance between nuclei has become much smaller than the Bohr radius ($a_0 = 5.3 \times 10^{-9}\,\mathrm{cm}$). A very crude solution is to use an effective potential $u_{\mathrm{eff}} = u_0(1 - (a_0/d)^2)$ to account for the fact that the ionization energy is reduced due to the proximity to the other nuclei. The mean distance between hydrogen nuclei is $d \sim (3/4\pi n_e)^{1/3}$. Including that correction and solving iteratively (55), one finds that helium could ionize at a pressure as low as $P_e \sim 17\,\mathrm{Mbar}$. Applied to hydrogen, this procedure also leads to a reduced ionization pressure $P_e \sim 2\,\mathrm{Mbar}$.

The total pressure cannot be obtained through that method because one then needs to describe the system of ions. In the metallic regions of substellar objects, an order of magnitude estimate is that ions and electrons have similar contributions to the total pressure. Our assumption of full degeneracy in fact tends to overestimate the pressures at which the transition occurs. This can be understood by the fact that the Pauli distribution corresponds to the minimum energy state for a fixed density n_e. Thermal effects have the tendency to move some of the electrons to higher energies, thereby0 favoring ionization. The transition from molecular to metallic hydrogen is therefore expected to occur at lower pressures and densities when the temperature is increased. Of course, these crude estimates are given for didactic purposes, but cannot replace a full treatment of this complex problem.

4.2 Experiments and Theoretical Hydrogen EOSs

Reaching Ultrahigh Pressures: Experimental Results

The high pressures and high temperatures typical of the interiors of giant planets can be achieved in the laboratory by shock-compression of a small sample of material. The shock is typically generated by a hypervelocity impactor or by a powerful laser. Measuring the thermodynamic properties of the compressed sample is quite difficult since such dynamical experiments last only 5 – 100 ns and the sample can be very small (0.4 – 500 mm^3). For a given initial state of the sample, the family of shocked states that can be achieved follows a curve in the (P, ρ, T) phase diagram known as a Hugoniot. The Hugoniot is one of the Rankine–Hugoniot relations that result from the conservation of energy, momentum, and matter flux across the shock front. Nearly all dynamical experiments on hydrogen and deuterium performed share the same cryogenic initial state and therefore measurements from different experiments can be directly compared. By reflection of the shock wave on a back plate made of a material stiffer than the sample, a double-shocked state can be achieved that reaches even higher pressures with a modest increase in temperature. Multiple shock reflections, known as shock reverberation, lead to a succession of compressed states that approach adiabatic compression.

Since 1995, deuterium has been the subject of intense experimental study using several independent techniques.[1] Measurements of the pressure, density, temperature, reflectivity, electrical conductivity, and sound speed have been performed along the single-shock Hugoniot and, in some cases, along double-shock Hugoniots.

The most reliable experimental results come from experiments where the impactor is accelerated with a gas gun. This technique allows for larger samples ($\sim$500 mm^3) and longer lasting ($\sim$100 ns) experiments but is generally limited to pressures below 1 Mbar. Pressures and densities have been measured along the single-shock Hugoniot up to 0.2 Mbar and along the double-shock Hugoniot up to 0.8 Mbar (Nellis et al. 1983). The reshocked states reproduce the (P, T) conditions of the molecular hydrogen envelope of Jupiter and provide a direct probe of the thermodynamics of hydrogen.

Under conditions where the dissociation of molecules becomes significant, the temperature becomes a sensitive test of the EOS. Processes that can absorb substantial amounts of energy like dissociation and ionization result

---

[1] Due to its higher density, deuterium is experimentally more advantageous than hydrogen because higher shock pressures can be achieved for a given impactor speed.

in relatively cool temperatures and higher degrees of compression for a given pressure along the Hugoniot. In the absence of such processes, the energy of the shock is expended mostly in the kinetic degrees of freedom with a corresponding increase in temperature. The temperature of double-shocked deuterium (Holmes, Ross and Nellis 1995) was found to be lower than all EOS predictions by about 30–40%, indicating that dissociation plays a more important role than predicted by contemporaneous models.

Finally, the sound speed has been measured along the Hugoniot in gas gun experiments up to 0.28 Mbar (N. C. Holmes, priv. comm.). Since it is a derivative of the pressure, the sound speed is a sensitive test of EOS models with the advantage of being measurable very reliably.

With powerful lasers, deuterium can be shocked to much higher pressures than with gas guns but the small sample size and the very short duration of the experiments make accurate diagnostics very challenging. The (P, ρ, T) single shock Hugoniot has been measured recently up to 3.5 Mbar with the NOVA Laser Facility (Da Silva et al. 1997; Collins et al. 1998, 2001), reaching a maximum density of $\sim 1\,\mathrm{g\,cm^{-3}}$ at ~ 1 Mbar (Fig. 7). Such a high

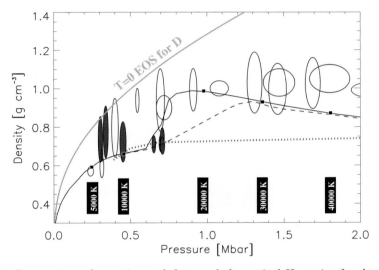

Fig. 7. Comparison of experimental data and theoretical Hugoniot for deuterium (densities are twice larger than expected for hydrogen at any given pressure). Empty ellipses correspond to data points obtained from laser compression (Collins et al. 1998). Filled ellipses were obtained by magnetic compression (Knudson et al. 2001). Theoretical calculations are represented by lines. They are respectively: the "PPT" (*solid*) and "interpolated" (*dashed*) Saumon–Chabrier equations of state (Saumon, Chabrier and Van Horn 1995), and a Path Integral Monte Carlo EOS (Militzer and Ceperley 2000). The *solid line* to the left shows the T=0 equation of state for D_2 as determined by an exp-6 potential fit to diamond-anvil cell measurements (Hemley et al. 1990). The temperatures along the Hugoniot have been calculated using the PPT-EOS. [From Guillot et al. 2004]

compressibility was not anticipated by most EOS models and this work sparked the current interest in the thermodynamics of warm dense hydrogen as well as controversy, both on the theoretical and experimental fronts. The reflectivity of shocked deuterium reaches about 60% for pressures above 0.5 Mbar along the Hugoniot (Celliers et al. 2000), a value indicative of a large density of free electrons and of a high electric conductivity characteristic of fluid metallic hydrogen. Second-shock compression up to 6 Mbar with the Nike laser give results in agreement with the NOVA (P, ρ) data (Mostovych et al. 2000). On the other hand, Knudsen et al. (2001) used a magnetic Z-accelerator to accelerate impactors to very high velocities. Their single-shock Hugoniot agrees well with the NOVA data for $P \lesssim 0.4$ Mbar but it is not as compressible at higher pressures, reaching a density of only ~ 0.7 g cm^{-3} at 0.7 Mbar (Fig. 7).

Hydrogen: EOS Calculations

While the temperatures obtained along the single-shock Hugoniot rapidly become much higher than those inside Jupiter at the same pressure (Fig. 8), these measurements provide very important, and heretofore unavailable tests

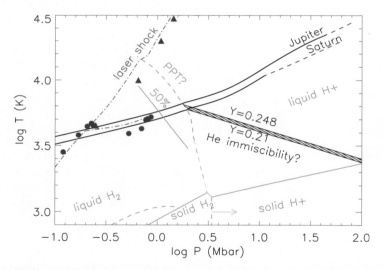

Fig. 8. Hydrogen phase diagram, with interior profiles of present-day Jupiter and Saturn overlaid, and with some experimental data shown. The boundary between liquid H$_2$ and solid H$_2$ is somewhat uncertain in the Mbar pressure range (2 estimates are shown), but is not relevant to Jupiter. The laser shock measurements of Collins et al. (2001) and the gas-gun measurements of Holmes et al. (1995) are shown as *triangles* and *filled circles* in the upper left-hand corner, respectively. Single- and double-shock hydrogen Hugoniots calculated by Saumon et al. (2000a,b) are shown as *dot-dashed lines* in the same region of the plot. The *solid line* labeled "50%" shows where 50% of molecular dissociation is obtained in the model of Ross (1998)

One approach has been to calculate the hydrogen–helium phase diagram assuming complete ionization. In that case, critical temperatures of order 8000 K at 2 Mbar can be calculated (see Stevenson 1982). Even more importantly, this leads to a critical temperature that *decreases* with increasing pressure. The consequence is that (i) this would imply that a phase separation has occurred in Jupiter, and earlier in Saturn, as suggested by the abundance of helium measured in the atmosphere (see Sect. 2); (ii) helium would be most insoluble near the molecular/metallic transition.

Other calculations have been attempted in the local density approximation (physical picture). Earlier work (Klepeis et al. 1991) suggested a unrealistically high critical temperature (40,000 K at 10.5 Mbar). However, a more careful study by Pfaffenzeller et al. (1995) with the same basic technique led to a lower critical temperature (less than 5,000 K at 4 Mbar). This value would imply no demixing of helium in Jupiter and Saturn. More importantly, the work of Pfaffenzeller et al. implies a critical temperature that *increases* with pressure. This can be explained if hydrogen is still not fully ionized at the pressures considered (4 to 24 Mbar), which seems difficult to reconcile with the more standard hydrogen EOSs. Another problem of the work of Pfaffenzeller et al. is that it does not recover the fully ionized limit. If the critical temperature increases with pressure, this would open the possibility that helium separates from hydrogen over an extended fraction of the planetary radius, with significant consequences for the interior and evolution models.

Other elements are also expected to separate from hydrogen if the temperature is low enough. However, the only estimates are for fully ionized mixtures. Table 5 shows critical temperatures and concentrations for the separation of various mixtures, as estimated by Stevenson (1976b). The low temperatures for demixing are due to the different coulombian potential for hydrogen and ions of progressively larger charges. As for helium however, these elements are not expected to be fully ionized which severely limits the applicability of these estimates to substellar objects.

5 Opacities and Heat Transport

We have seen that modeling the interiors of substellar objects requires to be able to calculate the temperature gradient ∇_T at each level. This necessitates

Table 5. Separation of fully ionized mixtures

mixture	T_c [K]	c_c
H–Li	1.4×10^4	0.18
H–C	1.1×10^5	0.086
H–O	2.6×10^5	0.064
H–Fe	5.5×10^6	0.019

to know how energy is transported. Three processes can contribute to this transport: radiation, conduction and convection.

5.1 Radiation Absorption – Basic Considerations

Let us consider a ray of radiation whose initial intensity is $I_{0\nu}$ as a function of frequency ν passing through a medium of density ρ on a distance l. The final intensity is then

$$I_\nu = I_{0\nu} e^{-\kappa_\nu \rho l},$$

where $1/\kappa_\nu \rho$ corresponds to the mean free path of photons of frequency ν, and κ_ν is the *monochromatic opacity*. As example of possible values in the interiors of giant planets and brown dwarfs are $\kappa \sim 1\,\mathrm{cm^2\,g^{-1}}$, $\rho \sim 10^{-2}$ leads to a photon mean free path of $l_\mathrm{ph} \sim 1$ meter.

As can be intuited from this very small mean free path, radiation in the interior is almost isotropic. In order to show that, let us consider the radial temperature difference between two levels separated by the photon mean free path:

$$\Delta T = l_\mathrm{ph} \frac{dT}{dr}.$$

The temperature lapse rate dT/dr cannot be calculated *a priori*. However, typical values for the Jupiter's interior are $dT/dr \approx 10^4/10^9\,\mathrm{K\,cm^{-1}}$, and $l_\mathrm{ph} \approx 10^2$ cm implying $\Delta T \approx 10^{-3}$ K. Since the energy density is proportional to T^4, the anisotropy has to be of the order $4\Delta T/T$. Using the previous estimate and $T \approx 10^4$ K, one can see that it is of the order of 4×10^{-7}, i.e. most of the interiors of giant planets and brown dwarfs can be considered as isotropic when radiation is concerned. Note that this is not the case near the photospheres of these objects, where photons can escape to space and l_ph becomes large. In that case, the full radiative transfer equation has to be solved. We refer the reader to available textbooks on the subject for further information on that problem (e.g. Goody and Yung 1989).

For modeling the interior, it is therefore justified to use the diffusion approximation, : radiation then obeys a standard diffusion equation:

$$\mathbf{j} = -D\nabla n,$$

where $\mathbf{j}$ is the radiation flux, D is the diffusion coefficient, which can be shown to be equal to $cl_\mathrm{ph}/3$ (e.g. Clayton 1968) and n represent the energy density U_ν. Because all the variables only vary radially, we can rewrite the diffusion equation as:

$$F_\nu = -\frac{c}{3\kappa_\nu \rho}\frac{\partial U_\nu}{\partial r}, \tag{61}$$

where F_ν is the net radial flux per unit wavelength[2].

---

[2] Note that when including rotation, this equation is not strictly valid any more: the surfaces of constant intrinsic flux then tend to become more spherical than those of constant pressure. In a radiative environment, this gives rise to a slow meridional circulation also known as the Eddington–Sweet circulation.

In this approximation, the energy density at each level of temperature T is proportional to the black body function $B_\nu(T)$:

$$U_\nu(T) = \frac{4\pi}{c} B_\nu(T) = \frac{8\pi h}{c^3} \frac{\nu^3}{h\nu/kT - 1}. \quad (62)$$

The total radial flux can then be obtained by integrating over all frequencies:

$$F = -\left[\frac{4\pi}{3\rho} \int_0^\infty \frac{1}{\kappa_\nu} \frac{\partial B_\nu}{\partial T} d\nu\right] \frac{\partial T}{\partial r}. \quad (63)$$

It is thus convenient to define the *Rosseland mean opacity* as

$$\kappa_R = \left[\frac{\pi}{acT^3} \int_0^\infty \frac{1}{\kappa_\nu} \frac{\partial B_\nu}{\partial T} d\nu\right]^{-1}. \quad (64)$$

Note that κ_R is a *harmonic* average of the opacity, weighted by a function which is close to a blackbody function and peaks at $\nu = 4kT/h$ (or equivalently $\sigma = 2.78T$ where σ is expressed in cm^{-1} and T in Kelvins). This has crucial consequences for its calculation, as spectral regions for which the monochromatic opacity is the smallest will tend to have the most important contribution to the mean. Physically, this can be interpreted by the fact that the cooling of any given layer in the star/planet will be governed by the photons which have the longest mean free path. Numerically, this implies that regions where the opacities are least known will have potentially very important contributions and that the final accuracy is extremely hard to estimate.

On the other hand, *in a radiative or conductive environment*, the temperature gradient will be directly given by the intrinsic luminosity, as can be seen from (63) and (64):

$$\frac{\partial T}{\partial r} = -\frac{3}{16\pi ac} \frac{\kappa_R \rho L}{r^2 T^3}. \quad (65)$$

In a radiative/conductive region, the temperature profile is hence steeper when the luminosity to be transported is larger. In the limit of a zero luminosity, it becomes isothermal as can be expected from thermodynamic principles.

5.2 Rosseland Opacities

Absorption of a Zero-Metallicity Gas

The contribution of hydrogen and helium to the overall opacities is often relatively small but fundamental, due to the nature of the Rosseland mean. At the pressures (bars or more) and temperatures (100s to 1000s K) of interest, these elements mostly have continuum opacity and therefore avoid any divergence of (64).

One of the most complete and useful work on the subject so far is certainly that of Lenzuni et al. (1991). I refer the reader to that paper for details on this problem. In this course, the materials that will be considered is relatively cool and at high density, implying that the main absorption sources are:

H_2–H_2 and H_2–He collision-induced absorption (CIA): H_2 and He in their ground state have no electric dipole and mainly absorb during collisions. The H_2 molecule has three degrees of freedom: translation, rotation[3] and vibration[4]. The largest energy transitions are between the vibrational bands, while the rotational bands imply a finer structure whose main consequence is to broaden these bands. The detailed calculation and structure is complex, especially in the case of the H_2–H_2 collision (4 other quantum numbers are then required to describe the state of the supermolecule), but to simplify it is dominated by 4 almost evenly spaced absorption bands (transitions $v:0 \to 0$ to $v:0 \to 3$) between 0 and $14000\,\mathrm{cm}^{-1}$. (See Borysow et al. 2000; Borysow 1992 and references therein).

H^- bound-free absorption: At high enough densities, the abundance of the H^- ion can become non-negligible. In this case, photons of sufficiently high energy can dissociate the ion into a hydrogen atom and a free electron. The absorption rapidly rise with increasing wavenumbers to reach a maximum at 1 micron. At higher wavenumbers (energies) it slowly decreases.

H_2^- free-free absorption: At very high densities, free electrons can "feel" the potential of the neutral H_2 molecule and therefore act as a superparticule which can absorb radiation. The cross-section for this reaction is a rapidly decreasing function of wavenumber.

Rayleigh scattering by H_2: Although this is not real absorption, Rayleigh scattering is very important for limiting the propagation of high energy radiation due to its $1/\lambda^4$ dependency.

Molecular Line Opacities

Due to the relatively low temperatures and high pressures encountered in regions where radiative heat transport matters, the opacity is dominated by molecular absorption. At low temperatures, the dominant molecules are H_2O, CH_4 and NH_3. For hotter objects CH_4 transforms into CO, and then TiO and VO, two important absorbers in the stellar regime appear (see Fegley and Lodders 1994, 1996; Lodders 1999).

Due to the complexity of the rotation and vibration modes of these molecules, one often has to rely on experimental measurements. Those can consist of measurements of mean absorptions in frequency intervals. These are however limited to a fixed number of pressures and temperature at which the measurements have been done. Another approach chosen for example for the GEISA and HITRAN data base is to measure the intensity of the largest

---

[3] Approximately, $E_{\rm rot} \sim \hbar^2/2\,Ij(j+1)$ where I is the molecule's moment of inertia and j the rotational quantum number.

[4] $E_{\rm vib} \sim \hbar\omega_{\rm osc}(v+1/2)$ where $\omega_{\rm osc}$ is the vibration frequency of the equivalent harmonic oscillator and v the vibrational quantum number.

possible number of lines. The absorption at any temperature and pressure of a given compound can then theoretically be calculated from the following relation:

$$\kappa_\nu(T,P) = \sum_i I_i(T_0,P_0) \left[\frac{1-e^{h\nu/kT}}{1-e^{h\nu/kT_0}}\right] \left[\frac{Q(T)}{Q(T_0)} e^{-\frac{E_i}{k}\left(\frac{1}{T}-\frac{1}{T_0}\right)}\right] L_\nu(T,P,\nu_i) \,, \tag{66}$$

where the monochromatic opacity κ_ν and the observed intensity of the line i are generally given in $cm^2\,molec^{-1}$ and the measured quantities have been obtained at temperature T_0 and pressure P_0. The ratio of exponential corresponds to induced emission. $Q(T)$ is the partition function at temperature T, E_i the energy of the level from which the observed line i comes from, and therefore the second term in square brackets is the ratio of the population of the initial energy level between temperature T and T_0. The line profile is L_ν and this function is such that $\int_0^\infty L_\nu d\nu = 1$.

Although theoretically reasonable, one of the main drawback of this approach is the fact that the extrapolation to high temperatures involves excited energy level transitions which are extremely difficult to detect at room temperatures. The problem of formula (66) is therefore that the population of energy levels corresponding to known lines decreases whereas the population of unknown excited levels increases. This problem, known as the "hot band" problem eventually leads to a strong (and false) decrease of the absorption with increasing temperature.

In recent years, progresses in computational power have lead to very interesting advances in *ab initio* calculations. These calculation predict the entire energy levels of a given molecule and can therefore yield the absorption spectrum at all temperatures and pressures. These kind of calculations have been successfully applied to diatomic molecules such as TiO, CO, VO...etc for quite a few years, using the principles of the harmonic oscillator. The case of the linear molecules as HCN has also been solved after that. However it is only relatively recently that convincing calculations have been performed for more complex molecules such as H_2O. Other important molecules in the context of cool objects ($T_{eff} \lesssim 1200\,K$) that still resist are CH_4 and NH_3. In the absence of crucial data for these molecules, one has to rely on hazardous extrapolation of experimental data.

Line Profiles

In the case of stellar atmospheres, the problem of the profile of absorption lines is relatively straightforward. Because the medium is at relatively high temperatures and low densities the absorption of a molecule away from the center of a line is due to the Doppler shift of the radiation as seen by the absorber. The Doppler line profile is written as a function of wavenumber ($\sigma = \nu/c$):

$$L_\sigma(T, \sigma_0) = \frac{e^{-(\sigma-\sigma_0)^2/\Delta\sigma_D^2}}{\Delta\sigma_D\sqrt{\pi}}, \qquad (67)$$

and the line halfwidth is

$$\Delta\sigma_D = \frac{\sigma}{c}\sqrt{\frac{2kT}{m}}, \qquad (68)$$

where m is the mean molecular mass.

In the case of most substellar objects, the cooler and denser conditions which prevail imply a different kind of broadening which is dominated by the effect of collisions: because the energy levels are populated only for finite periods of time (due to excitations/deexcitations caused by collisions), the transition cannot have a unique frequency. This gives rise to the so-called Lorentzian line profile, which is

$$L_\sigma(T, P, \sigma_0) = \frac{\Delta\sigma_L}{(\Delta\sigma_L)^2 + (\sigma - \sigma_0)^2}, \qquad (69)$$

and the Lorentzian half-width then depends on details of the physics of microscopic collisions. In general, it can be approximated by the following perfect gas approximation

$$\Delta\sigma_L \approx \sigma_0 \frac{P}{\sqrt{T}}, \qquad (70)$$

and σ_0 depends on the line considered. This value can be experimentally determined at room temperature, but when using *ab initio* calculation it is generally set to a fixed value corresponding to the mean of the observed ones.

The use of the Doppler line broadening can be justified when its halfwidth value is larger than that of a Lorenztian profile, i.e. when

$$T \gtrsim 7000\,\text{K}\left(\frac{\rho}{10^{-4}\,\text{g cm}^{-3}}\right).$$

At significantly lower temperature and/or larger densities, one is justified to use a pure Lorentzian profile. In between, the Voigt profile is a combination of the two:

$$L_\sigma(T, P, \sigma_0) = \int_{-\infty}^{+\infty} \frac{\Delta\sigma_L}{(\Delta\sigma_L)^2 + (\sigma - \sigma_0 - u\frac{\sigma_0}{c})^2}\left(\frac{m}{2\pi kT}\right) e^{-mu^2/2kT} du. \qquad (71)$$

The Lorentzian (or Voigt) broadening is intrinsically more complicated than the Doppler one due to the additional pressure dependency and the *a priori* unknown halfwidth. It is also more complicated due to its slow decay compared to the Doppler profile.

A cutoff to the Lorentzian profile is generally used first because it is computationally much less intensive. It has also been empirically verified that synthetic spectra of the giant planets generally fit the observations better when using a cutoff. This is for example the case of the 5 μm spectrum of Jupiter, modeled by Kunde et al. (1982) using a cutoff of $\Delta\sigma_{\text{cutoff}} \sim 120\,\text{cm}^{-1}$. Last but not least, there are theoretical grounds for which the Lorentzian profile should fail far from the line center.

The Lorentzian "core" is indeed a result of the impact approximation: it is valid when the collision time is large compared to the characteristic time of the transition:

$$\frac{r_c}{v_c} = \tau_{\text{col}} \gtrsim \tau_\omega = \frac{1}{2\pi c|\omega - \omega_0|} , \qquad (72)$$

where r_c and v_c are the mean radius and velocity at closest approach during collisions. Further from the line center, the impact approximation fails and a faster exponential decay should prevail (Birnbaum 1979). This simplification was indeed used by Guillot et al. (1994a) to predict a line cutoff proportional to $\sqrt{T}$ around $100\,\text{cm}^{-1}$ at $T = 200\,\text{K}$, consistent with spectroscopic models of Jupiter's atmosphere.

More generally, the Lorentzian profile is known to fail in a variety of conditions. Both superlorentzian and sublorentzian profiles can be observed, and spectral lines can even be shifted due to microscopic interactions and line mixing. However, a surprising result of the recent years is that, at least in the case of alkali metals, far wings can still be of significance much beyond expectation. In the case of Na lines in the visible spectral region, Nefedov et al. (1999) find that the expected exponential cutoff occurs for $\Delta\sigma \gg 1000\,\text{cm}^{-1}$. Using a Lorentzian profile convoluted with an exponentially-decaying function, Burrows et al. (2000a) find that alkali metals, and especially the potassium doublet at 0.77μm can explain the absence of flux emitted by brown dwarfs in the visible and the slope of the spectrum for wavelengths shorter than 1μ.

The consequences of these results is still to be investigated, as are many microscopic problems of line mixing and departure from ideality.

Radiative Rosseland Mean Opacities

The calculation of a Rosseland opacity table for substellar objects is a difficult task, and indeed no such table spanning the range of giant planets to M-dwarfs is yet available. An opacity table for stars have been calculated by Alexander and Ferguson (1994), but at low temperature it is dominated by the presence of interstellar-sized grains. It is then essentially designed to the study of circumstellar disks. Another effort was led by Lenzuni et al. (1991) with their zero-metallicity opacity table. As suggested by the name, the calculations includes only hydrogen and helium and its applicability to real giant planets and brown dwarfs is thus limited. The theoretical spectra of M-dwarfs and cooler objects are now relatively good but unfortunately no Rosseland opacity table has been published by modelers. Finally, a limited Rosseland table was computed for Jupiter and Saturn by Guillot et al. (1994a) and Guillot (1999a,b)

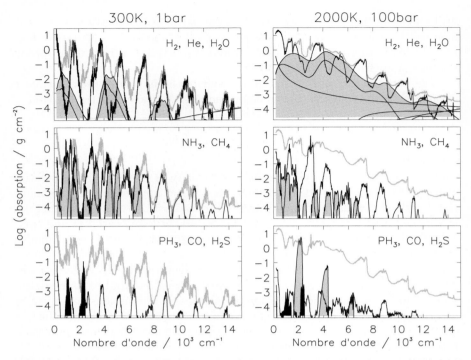

Fig. 10. Absorption of a solar-composition mixture, at 300 K and 1 bar (*left*), and at 2000 K and 100 bar. Various contributions are shown. *Top diagrams*: H_2–H_2 CIA (*shaded*) and H_2O (*gray lines*); *Middle*: NH_3 (*shaded*) and CH_4 (*gray lines*); *Bottom*: PH_3 (*shaded, left diagram*), CO (*shaded, right diagram*) and H_2S (*gray*).

but it does not include high-temperature species (such as TiO), and, as other tables calculated so far, it does not account for the absorption of alkali metals.

I won't attempt to discuss any specifics of these calculations. However the broad features can be understood by looking at the monochromatic absorptions shown in Fig. 10. At low temperatures and pressure, the spectrum is dominated by water and methane with a small contribution of H_2–H_2 collision-induced absorption, ammonia and phosphine (PH_3). The molecular bands are relatively narrow. At higher temperatures and pressures, the absorption bands become much broader. The H_2–H_2 CIA becomes more important but water still dominates the absorption spectrum. However, methane has almost disappeared in the favor of carbon monoxide, which peaks at 5 μm. The behavior of the Rosseland opacity over this range of conditions evolves mostly because of the displacement of the weighting function dB_ν/dT in (64). At 300 K, it is maximum at 830 cm^{-1}, in a spectral region where the absorption is large. At 2000 K, its peak is around 5600 cm^{-1}, and the contribution of the low-absorption region around 1 μm(10, 000 cm^{-1}) becomes important. At still higher temperatures, the increased abundance of electrons imply a very rapidly increasing H^- and H_2^- continuous absorption: the Rosseland mean opacity then rises so much that any radiative process becomes very inefficient until eventually *conduction* dominates.

Two important points are to be mentioned: first, this local minimum of the Rosseland opacity at temperatures of the order of 1500 to 2000 K which prevails for a zero-metallicity gas is conserved in the presence of water, methane, ammonia, silane and a variety of other species observed in the atmospheres of Jupiter and Saturn. This is due to the fact that these elements all have low absorptions around 1 µm. The presence of other chemical elements can alter this conclusion. First, those that have low ionization potentials can increase the number of electrons. This is the case of Al, Na...etc., but has been shown not to be sufficient to erase this minimum (Guillot et al. 1994a). Second, alkali metals have been shown to absorb precisely at these wavelengths (Burrows et al. 2000a) and can therefore greatly affect this conclusion. Finally, the presence of grains/dust/cloud particles can have a very significant effect.

Clouds and Dust

A great variety of chemical species in condensed form have been identified in the interstellar medium, and their contribution to the energy balance of interstellar clouds and circumstellar disks has been shown to be absolutely essential. Solid grains have also been shown to affect the structure and evolution of red giant stars, and in particular to be determinant for understanding the violent mass loss processes that these objects undergo. Finally, we owe our very existence to the presence of condensed species in our own atmosphere: the presence of clouds, of rain, proves that these phenomena greatly affect the energy balance in Earth atmosphere. This is also the case in the giant planets, and the presence of big particles (i.e. clouds) (generally of unknown composition) is required for a proper fit of the observed spectra.

Condensed grains have such a fundamental importance because of their ability to absorb light: Taken alone, the grains that can potentially condense out of a solar-composition mixture are capable of providing a Rosseland opacity up to $\sim 10\,\mathrm{cm}^2\,\mathrm{g}^{-1}$ (e.g. Pollack et al. 1985, 1994). This value depends on the abundance and composition of the condensed material, and hence mostly on the temperature of the mixture, but also on the size distribution.

One can define three regimes, depending on the ratio of the wavelength to the size of the grains. For grains much smaller than the wavelength of the incoming light, the opacity is essentially due to Rayleigh scattering and the Rosseland mean is independent of the size. For large grains, the cross section decreases as the grains are bigger (the total mass of condensed material being held constant), and the Rosseland opacity is consequently inversely proportional to the size of the grains. The opacity is maximal at wavelengths of the order of the size of the grain.

Astrophysical opacity tables have generally been calculated assuming a full chemical equilibrium in which the condensing species have been retained. Furthermore, their size distribution has generally been taken to be that of the interstellar medium. A good example is the widely used table provided by Alexander and Ferguson (1994), which shows several absorption jumps due to the condensation of various species, in particular silicates at temperatures lower than ~ 2000 K.

However, it is not clear that this approach is even useful in the case of giant planet and brown dwarfs atmospheres. The size distribution then obtained strongly depends on complex advective processes and has in most cases nothing to do with that of interstellar clouds. Gravity is indeed a very important factor in planetary atmospheres: it will generally lead to the removal of grains, but several complications can occur due to convection and more generally advection of material. Heterogeneity is also likely to occur, and instabilities can be generated by the presence of clouds. Finally, latent heat release can also be an important factor, as in the case of the Earth, for which cumulus clouds can penetrate the upper atmosphere because of the significant release of energy occurring during the condensation of water vapor to a liquid or solid phase.

The calculation of a mean opacity table for substellar atmospheres that would include condensed species is hence at the least impractical and limited to very special conditions. However, it may be possible to use this simplified treatment by noting that grains often dominate the absorption when they occur. A combination of two tables, one without grains, and one with grains only might be a possibility.

Conductive Opacities

We have seen that at high temperatures, the number of electrons present yields a very rapid increase of the opacity. Because we are considering environments in which the electrons become partially degenerate, conduction by these electrons can become, at high pressures, an efficient way to transport the internal heat.

In environments in which heat is entirely transported by conduction, the heat flux obeys a standard diffusion equation:

$$\mathbf{Q} = -K_c \nabla T , \quad (73)$$

where K_c is the *thermometric conductivity*, expressed in units of $\mathrm{erg\, s^{-1}\, cm^{-1}\, K^{-1}}$. An order of magnitude estimate of this quantity for the jovian interior is provided by Stevenson and Salpeter (1977a):

$$K_c \approx 10^8 \rho^{4/3} \, \mathrm{erg\, s^{-1}\, cm^{-1}\, K^{-1}} , \quad (74)$$

and ρ is expressed in $\mathrm{g\, cm^{-3}}$.

The relationship between the conductive opacity and the thermometric conductivity is

$$\kappa_c = \frac{4acT^3}{3K_c\rho}. \tag{75}$$

The diffusion equations for radiation and conduction being additive, one can define a conductive + radiative opacity κ as

$$\kappa^{-1} = \kappa_R^{-1} + \kappa_c^{-1}. \tag{76}$$

Tables of either the thermometric conductivity or the conductive opacity have been calculated by Hubbard and Lampe (1969) and more recently by Potekhin et al. (1999). The results by Potekhin et al. indicate slightly smaller opacities by $\approx 10\%$. Typically, the conductive opacity of the hydrogen gas at 10^7 K decreases from about $10^5 \, \mathrm{cm^2\, g^{-1}}$ at $\rho = 1 \, \mathrm{g \, cm^{-3}}$ to $10^3 \, \mathrm{cm^2 \, g^{-1}}$ for $\rho = 100 \, \mathrm{g \, cm^{-3}}$.

5.3 Heat Transport

We have described two ways of transporting heat: radiation and conduction. In the diffusion approximation, i.e. at levels where the medium can be considered isotropic, these fundamental physical processes can be described by relatively simple equations. However, another extremely important mechanism has been left out so far: the advection of heat by macroscopic motions. There are many ways to generate heat advection, or *convection*, and it can take many forms. We will only mention the most simple one: when convection is generated by a destabilizing temperature gradient, and the medium can be considered barotropic (surfaces of constant P and ρ coincide). The method pioneered by Prandtl, Schwarzschild and Ledoux and widely used in stellar physics is explained thoroughly in many textbooks. We will therefore only sketch it.

Convective Instability Criterion

In stellar (or in this case substellar) physics, viscosity is considered negligible and convection is predicted to occur whenever it is energetically favorable. This is unlike e.g. the Rayleigh–Bénard instability for which the system has to overcome a barrier of potential to occur. In stars and giant planets, the barrier is so small that it can be neglected (see later Sect. 5.3).

In our case, convection is supposed to occur whenever the medium is *locally* unstable to convection, i.e. when a parcel of fluid displaced upward (resp. downward) is lighter (resp. heavier) than its surrounding. Let us consider this parcel of fluid versus its environment. When it is arbitrarily displaced radially by Δr, its density changes by $\Delta \rho^\star$ and has changed by $\Delta \rho$ in the unperturbed environment. A convective instability then develops if:

$$\frac{\Delta \rho^\star}{\Delta r} < \frac{\Delta \rho}{\Delta r}. \tag{77}$$

Because pressure variations are equilibriated much faster (at the speed of sound) than temperature variations in the interior, this is equivalent to[5]:

$$\left(\frac{\partial \rho}{\partial T}\right)_{P,\mu} \left(\frac{dT}{dP}\right)^{\star} \frac{dP}{dr} < \left(\frac{\partial \rho}{\partial T}\right)_{P,\mu} \left(\frac{dT}{dP}\right) \frac{dP}{dr} + \left(\frac{\partial \rho}{\partial \mu}\right)_{T,P} \left(\frac{d\mu}{dP}\right) \frac{dP}{dr}, \quad (78)$$

where $(dT/dP)^{\star}$ corresponds to the temperature variation in the perturbed fluid parcel. We have implicitly considered that the molecular diffusivity is slower than the thermal one so that the mean molecular weight μ is held constant in the parcel and varies in the environment. Clearly this is *not* the case for fast chemical reactions as ionization. In that case, convection can be thought to occur in a homogeneous medium.

This implies that convection should develop whenever

$$\nabla_T > \nabla_T^{\star} + \frac{\varphi}{\delta} \nabla_\mu, \quad (79)$$

where φ and δ are thermodynamical derivatives of the density (equal to 1 for a perfect gas):

$$\delta = -\left(\frac{\partial \ln \rho}{\partial \ln T}\right)_{P,\mu} \quad ; \quad \varphi = -\left(\frac{\partial \ln \rho}{\partial \ln \mu}\right)_{P,T}. \quad (80)$$

This criterion is not yet in a useful form because neither $\nabla_T^{\star}$ nor ∇_T are known *a priori*. However, one should note that the following inequalities should be satisfied in a convective zone:

$$\nabla_{\rm rad} > \nabla_T > \nabla_T^{\star} > \nabla_{\rm ad}, \quad (81)$$

where $\nabla_{\rm rad}$ is the radiative (+conductive) gradient

$$\nabla_{\rm rad} = \frac{3}{4\pi\sigma G} \frac{\kappa P L}{m T^4}, \quad (82)$$

$\nabla_{\rm ad} \equiv (\partial \ln T / \partial \ln P)_S$ is the adiabatic gradient, $\nabla_T \equiv \partial \ln T / \partial \ln P$ is the real temperature gradient and $\nabla_T^{\star}$ is that gradient in the parcel of fluid. The first inequality to the left is due to the fact that given a set luminosity, the radiative gradient is a strict maximum to the temperature gradient. The second inequality is a consequence of the convection criterion. The last one is due to the fact that heat can be transported by the parcel only if its motion is slightly superadiabatic i.e. if it looses some of its heat during its ascent.

It is then easy to derive the so-called Schwarzschild–Ledoux criterion for convective instability (Note that φ and δ are positive quantities):

$$\nabla_{\rm rad} > \nabla_{\rm ad} + \frac{\varphi}{\delta} \nabla_\mu. \quad (83)$$

---

[5] For simplicity, we forget the time derivative and use dP/dr instead of $\partial P/\partial r$.

This criterion is fundamental for the evolution of planets to stars. It is important to notice that $\nabla_{\rm rad}$ is proportional to the luminosity L and to the mean opacity κ. Convection will occur if L and κ are too large so that radiation can transport the heat flux only with a steep, unstable temperature gradient.

Mixing Length Theory

While we now know when convection should occur, we haven't derived an expression for the temperature gradient ∇_T. In stellar modeling, this is generally done using the approach due to Prandtl (1925): the Mixing length theory. This is in fact a *phenomenological* approach and as such it has been widely criticized. In the case of substellar objects, a detailed treatment of convection will generally not be necessary because, as we will see, convection is almost adiabatic.

The main hypothesis of the mixing length theory is that convective elements should dissolve after a "mixing length" $l \equiv \alpha H_P$, where H_P is the pressure scale height and α is a free parameter of order unity.

We will first assume a homogeneous medium, i.e. $\nabla_\mu = 0$. By definition, the total flux F, radiative flux $F_{\rm rad}$ and convective flux $F_{\rm conv}$ obey the following relations:

$$\begin{cases} F &= F_{\rm rad} + F_{\rm conv} \\ F_{\rm rad} &= \dfrac{4acT^3}{\kappa\rho}\dfrac{T}{H_P}\nabla_T \\ F &= \dfrac{4acT^3}{\kappa\rho}\dfrac{T}{H_P}\nabla_{\rm rad}\,. \end{cases} \quad (84)$$

Prandtl's approach allows to estimate the convective flux and the convective velocity by integrating the acceleration by the buoyancy force over the mixing length. This is done in several textbooks (e.g. Kippenhahn and Weigert 1991) and I will not rederive these expressions. After some algebra, one derives the cubical equation of the mixing length:

$$\begin{cases} \dfrac{9}{4}\Gamma^3 + \Gamma^2 + \Gamma = A^2(\nabla_{\rm rad} - \nabla_{\rm ad})\,, \\ A = \dfrac{c_P\kappa\rho^2\alpha^2}{12\sqrt{2}acT^3}(g\delta)^{1/2}H_P^{3/2}\,, \end{cases} \quad (85)$$

where $0 \leq \Gamma \leq \infty$ is a parameter characterizing the efficiency of convection, and I used as second parameter of the mixing length a ratio of the volume V of convective elements over their surface S, $V/S = l/6$. The cubical equation has only one real and positive root: it defines a unique value of Γ.

The temperature gradient and convective velocity are given by

$$\nabla_T = \nabla_{\rm ad} + \frac{\Gamma(\Gamma+1)}{A^2}\,, \quad (86)$$

$$v = \frac{(g\delta H_P)^{1/2}}{2\sqrt{2}}\alpha\frac{\Gamma}{A}\,. \quad (87)$$

The cubical mixing length equation should be consistently solved when calculating the evolution of substellar objects. However, in most cases, convection is found to be very efficient and Γ is large. In that case, simpler expressions for the convective velocity and temperature gradient can be obtained:

$$\nabla_T - \nabla_{\rm ad} \sim \left[\frac{4\sqrt{2}}{\alpha^2 \delta^{1/2}} \frac{F_{\rm conv}}{c_P T (\rho P)^{1/2}} \right]^{2/3}, \tag{88}$$

$$v \sim \left[\frac{\alpha \delta}{4} \frac{P}{\rho c_P T} \frac{F_{\rm conv}}{\rho} \right]^{1/3}. \tag{89}$$

According to (84), the convective flux is $F_{\rm conv} = F(1 - \nabla_T/\nabla_{\rm rad})$. When the opacities become large, $\nabla_{\rm rad} \gg \nabla_T \approx \nabla_{\rm ad}$ and $F_{\rm conv} \approx F$. Physically, the superadiabatic gradient is an adimensional quantity involving the ratio of the energy per unit mass $F/\sqrt{\rho P}$ to be transported to that of a given layer, $c_P T$. The convective velocity is essentially proportional to $(F/\rho)^{1/3}$: since F is a slowly varying function, v should be expected to be larger near the surface, where ρ is smaller. This corresponds to the fact that in a low density material, transporting the same energy requires higher velocities.

Since we consider objects that are almost fully convective, and for which our assumption that $\Gamma \gg 1$ is verified in most of the interior, the value of α can affect the results through the change in the superadiabatic gradient. As we will see next, this value is generally extremely small. The structure of substellar objects is thus weakly dependent on the treatment of convection, except possibly for the largest masses and early in the evolution.

Properties of Convection in Substellar Objects

First, let us determine the physical reasons for which the interiors of substellar objects are found to be essentially convective (for direct applications to our giant planets, see Hubbard 1968; Stevenson and Salpeter 1977b; Guillot et al. 1994b, 1997; Guillot 1999a,b). At any given level, we define a *critical* opacity $\kappa_{\rm crit}$ as the Rosseland mean opacity for which $\nabla_{\rm rad} = \nabla_{\rm ad}$. It can be seen from the definition of $\nabla_{\rm rad}$ (82) that

$$\kappa_{\rm crit} = \frac{\nabla_{\rm ad}}{3} \frac{g}{P} \left(\frac{T}{T_{\rm eff}} \right)^4. \tag{90}$$

We now assume that ∇_T is approximately constant so that $T \propto P^{\nabla_T}$, and that the flux $\sigma T_{\rm eff}^4$ and gravity g are also constant. This yields

$$\kappa_{\rm crit} \approx \frac{\nabla_{\rm ad}}{3} \frac{g}{P_0} \left(\frac{P}{P_0} \right)^{4\nabla_T - 1}. \tag{91}$$

The critical opacity is thus only weakly dependent on the pressure level. Using the definition of the photospheric pressure, and introducing the photospheric opacity $\kappa_0 \equiv \kappa(P_0, T_{\text{eff}})$, one finds that

$$\kappa_{\text{crit}} \approx \frac{\nabla_{\text{ad}}}{2} \kappa_0 \left(\frac{P}{P_0}\right)^{4\nabla_T - 1}. \tag{92}$$

Note that this expression is relatively simple, but caution should be made regarding to its applicability. We have assumed g and T_{eff} to be constant, an assumption which generally verified, except in the central regions. The hypothesis ∇_T=cte is more *ad hoc*. However, because of the small superadiabaticities in substellar objects, $\nabla_T \leq \nabla_{\text{ad}} + \epsilon$ where ϵ is a small quantity (see (93) hereafter). With $\nabla_{\text{ad}} \sim 0.3$, regardless of variations of ∇_T, one finds that in most cases κ_{crit} is a function that weakly depends on P. We are thus led to the conclusion that *substellar objects are mostly convective because of the strong increase of their Rosseland opacities with increasing pressure and increasing temperature*.

It should be stressed however that (92) is not valid near the photosphere because it is calculated in the diffusion approximation. One can notice for example that κ_{crit} is a Rosseland mean opacity, whereas κ_0 would correspond more or less to a Planck mean opacity (straight mean weighted by the Planck function).

In the case of isolated substellar objects, we can distinguish two regimes:

1. *Cool objects* ($T_{\text{eff}} \lesssim 1500\,\text{K}$): The opacity κ is essentially due to molecular absorption which is weakly dependent on pressure and temperature. However, an important contribution is due to the H_2–H_2 collision-induced absorption which is proportional to P. The increase in this opacity guaranties convection.

2. *Hot objects* ($T_{\text{eff}} \gtrsim 1500\,\text{K}$): The increase in temperature provides a growing number of electrons which greatly contribute to the total opacity *via* H_2^- and H^- absorption. Convection is then also guaranteed in this regime.

Note that this is the case only for isolated, or weakly irradiated planets or brown dwarfs. The case of strongly irradiated objects will be discussed afterward. Furthermore, any decrease of the Rosseland opacity, such as that due to minimum 1 micron absorption in the absence of alkali metals can yield a small but important radiative region. Finally, the presence of moist convection or of gradients of composition can alter this conclusion. However, this only affects limited regions, and one can consider that substellar objects are mostly convective (however see Chabrier et al. 2000b).

Let us characterize convection in these objects. Under typical conditions, (88) can be shown to yield:

$$\nabla_T - \nabla_{\text{ad}} \approx 10^{-3} \left(\frac{T_{\text{eff}}}{1000\,\text{K}}\right)^{7/3} \left(\frac{P}{1\,\text{bar}}\right)^{-2/3}. \tag{93}$$

Convection can thus be considered adiabatic in most cases. This is due to the fact that the energy to be transported is relatively small when compared to that available from thermonuclear reactions in stars. Because of this property, the structure and evolution is found to be relatively insensitive to the treatment of convection (and e.g. to the choice of the mixing length parameter α).

The convective velocity is estimated from (89):

$$v \approx 150 \left(\frac{T_{\text{eff}}}{1000\,\text{K}}\right)^{4/3} \left(\frac{\rho}{1\,\text{g cm}^{-3}}\right)^{-1/3} \text{cm s}^{-1}. \tag{94}$$

The velocities thus derived are relatively small for giant planets such as Jupiter ($T_{\text{eff}} \sim 100\,\text{K}$) and in the interior. They can however reach $100\,\text{m s}^{-1}$ at the top of the convective region of the hottest brown dwarfs ($T_{\text{eff}} \sim 2000\,\text{K}$). Note that in the case of Jupiter, the condensation of water provides another source of energy which is not considered here. Due to that effect, updrafts can reach several 10's of m s^{-1}. (The same principle holds for the Earth's cumulus clouds).

Table 6 illustrates the properties of convection in Jupiter, based on estimates from Stevenson and Salpeter (1977a). The pressure scale height varies from a few tens of kilometers near the photosphere to a fraction of the planetary level near the center. The convective velocity is very small deep in the interior, but as discussed, it decreases significantly at smaller densities.

A few adimensional numbers characterize convection itself: the *Prandtl* number is the ratio between the opacity and the viscosity. It is small in regions where either radiation or convection are relatively efficient at transporting heat (independently of the presence of convection), i.e. near the surface and in the metallic interior where conduction becomes dominant. The *Reynolds* number compares macroscopic diffusion to the viscosity. It is very large, indicating that convection is turbulent. Finally, the *Rossby* number is a measure of the importance of rotation on convective motions. It is low, due to the rapid rotation of the planet in ~ 10 hours. This indicates that rotation will significantly affect convective motions, implying that convective motions will be mostly confined to a plane perpendicular to the axis of rotation. This gives rise to the so-called Taylor columns (e.g. Busse 1978).

Table 6. Properties of convection in Jupiter

	H_P [km]	v_{conv} [m/s]	$Pr = \nu/\kappa$	$Re = vd/\nu$	$Ro = v/\omega d$
Surface	40	1	10^{-4}	10^9	1
PPT/molecular			1		
PPT/metallic			10^{-3}		
Center	13000	0.03	10^{-3}	10^{11}	10^{-4}

Possible Inhibitions of Convection

A few phenomena are susceptible of inhibiting convection. I will enumerate a few of them:

1. *Rotation*: In the limit of very low Rossby numbers, convective motions are confined to a plane perpendicular to the axis of rotation (see e.g. Pedlosky 1979). In the case of Jupiter, it has been estimated that this yields a limited increase of the superadiabaticity and is hence negligible to first order (Stevenson 1976a).
2. *Magnetic field*: In an ionized medium, a strong magnetic field can force motions to follow the magnetic lines. All giant planets in our Solar System possess such a magnetic field. Some low-mass stars have been observed to show variability presumably related to the presence of star spots. This is a strong indication that these stars are magnetic. One can therefore surmise that magnetic dynamos occur in most or all substellar objects. The mechanisms that generate these dynamos have not been fully elucidated, and consequences for convection and heat transport remain unclear.
3. *Compositional gradients*: The presence of compositional gradients ($\nabla_\mu > 0$) can lead to the inhibition of convection. The problem becomes complex because this gradient is *a priori* unknown: on one hand, convection tends to homogenize layers, leading to $\nabla_\mu \to 0$; on the other hand, sharp interfaces can form for which $\nabla_\mu \to \infty$, yielding sharp, diffusive interfaces. This is indeed observed in the Earth oceans, where salt and heat have opposite effects.
4. *Condensation*: Phase changes of minor species, such as water can strongly modify convection. First, the latent heat released favors updrafts, as observed in Earth's cumulus clouds. In Jupiter, this leads to convective updrafts of tens of m/s. However, in hydrogen–helium atmospheres, another effect can be potentially important: in this case, because any condensing species is heavier than the surrounding air, condensation tends to yield a stable compositional gradient ($\nabla_\mu > 0$)[6]. Guillot (1995) shows that convection is locally inhibited when the abundance of the condensing species is larger than a certain critical value. This value is of the order of 5, 15 and 40 times the solar values for H_2O, CH_4 and NH_3, respectively. The temperature profiles of Uranus and Neptune retrieved from radio occultation of Voyager 2 indeed show a strong superadiabaticity in the region of methane condensation, implying that convection is probably inhibited by this mechanism (Guillot 1995).

---

[6] This is unlike the Earth's atmosphere in which the condensing molecule, water ($\mu = 18$), is lighter than air ($\mu = 29$).

6 Interior Structures of our Giant Planets: Numerical Integrations and Results

6.1 Basic Principles

Constraints on the interior structure of the giant planets of our Solar System are derived from knowledge of their mass, M, equatorial radius a, and gravitational moments J_2 and J_4. Measurements of these quantities still go back to the Pioneer and Voyager missions.

Basically, the procedure is to integrate the hydrostatic equations including rotation using appropriate equations of state, opacities and a set of observed parameters (mass, surface temperature...etc.). The *a priori* unknown composition is constrained by the gravitational moments. In the case of Uranus and Neptune, we will see that another approach has been proposed which simply relies on the computation of random density profiles. It is to be stressed that in the absence of other information such as a vibration spectrum, *only a few moments of the interior density can be constrained*. Most of the knowledge concerning the interiors of these planets is indirect: it heavily relies on the input physics.

All four giant planets appear to emit more energy than they receive from the Sun (see Table 3 and Sect. 2). As first proposed by Hubbard (1968) for Jupiter, this implies that their interior is hot, fluid and because of the large opacities (see "Possible Inhibitions of Convection"), mainly convective. This is an essential property which allow to model these objects with the same underlying physics.

I will first discuss the case of Jupiter and Saturn, which are mostly formed with hydrogen and helium. These two planets have been extensively modeled (see Hubbard and Marley 1989; Zharkov and Gudkova 1992; Chabrier et al. 1992; Guillot et al. 1994b; Gudkova and Zharkov 1999), but these works generally aimed at finding a limited sample of models matching the observational constraints. I choose to present models calculated in the purpose of extensively exploring the set of parameters (Guillot 1999a).

I will then present more briefly the cases of Uranus and Neptune. These planets are mostly made of ices, and their interior structure is consequently more difficult to grasp. They are also distantly connected to the much more massive brown dwarfs and extrasolar planets that have been detected thus far. However, they are also a crucial piece in the puzzle to understand how the Solar System was formed.

6.2 Jupiter and Saturn

Input Data

Gravitational field: The characteristics of Jupiter and Saturn's gravity fields as obtained from spacecrafts measurements are listed in Table 7. Note that

Table 7. Characteristics of the gravitational fields

	Jupiter		Saturn	
	measured[a]	adjusted[b]	measured[c]	adjusted[b]
$M/M_\oplus$	317. 83		95. 147	
$R_{eq}/10^9$ cm	7. 1492(4)		6. 0268(4)	
$\omega/10^{-4}\,\mathrm{s}^{-1}$	1. 76		1. 64	
$J_2/10^{-2}$	1. 4697(1)	1. 4682(1)	1. 6332(10)	1. 6252(10)
$J_4/10^{-4}$	−5. 84(5)	−5. 80(5)	−9. 19(40)	−8. 99(40)
$J_6/10^{-4}$	0. 31(20)	0. 30(20)	1. 04(50)	0. 94(50)

Note. The numbers in parentheses are the uncertainty in the last digits of the given value. All the quantities are relative to the 1 bar pressure level.
[a] Campbell and Synnott (1985).
[b] Campbell and Anderson (1989).
[c] Adjusted for differential rotation using Hubbard (1982).

the rotation rate ω is that of the planets' magnetic field, assumed to be tied to the rotation of the deep interior.

A complication arises from the fact that the equations derived from that theory generally assume the planet to be rotating as a solid body. Observations of the atmospheric winds show significant variations with latitude, however (e.g., Gierasch and Conrath (1993)). The question of the depth to which these differential rotation patterns extend is still open. Hubbard (1982) has proposed a solution to the planetary figure problem in the case of a deep rotation field that possesses cylindrical symmetry. It is thus possible to derive, from interior models assuming solid rotation, the value of the gravitational moments that the planet would have if its surface rotation pattern extended deep into its interior. It is *a priori* impossible to prefer one model to the other, and I will therefore present calculations assuming both solid and differential rotation. Table 7 gives both the measured gravitational moments, and those corrected for differential rotation.

Atmospheric abundances: Because Jupiter and Saturn are believed to be relatively well-mixed, precise measurements of atmospheric abundances is crucial for modeling the interior. First, helium is found in relatively small abundance: Solar evolution models indicate that the protostellar helium mass mixing ratio relative to hydrogen was $Y/(X+Y) = 0.270 \pm 0.005$ (Bahcall and Pinsonneault (1995)). In situ measurements of that quantity in Jupiter yield $Y/(X+Y) = 0.238 \pm 0.007$ (von Zahn et al. (1998)). Combined radio occultation measurements and spectra analysis from Voyager 2 indicate that, in Saturn, $Y/(X+Y) = 0.06 \pm 0.05$ (Conrath et al. (1984)). This last value has been challenged by several approaches (Guillot 1999a,b; Hubbard et al. 1999; Conrath and Gautier 2000) and could be significantly larger. However, it still appears to be smaller than the protosolar value.

The conclusion that more helium was present in the protosolar nebula gas from which Jupiter and Saturn formed than is observed today in their atmospheres seems inescapable. As discussed in Sect. 4, this implies the existence of a hydrogen/helium phase separation, in which helium droplets can grow sufficiently fast to be dragged down by gravity despite convection (Salpeter 1973; Stevenson and Salpeter 1977b). The fact that the Galileo probe measured a depleted abundance of neon is also indicative of such a phase separation, as neon tends to dissolve into the helium-rich drops (Roulston and Stevenson 1995).

The measured abundances of other elements also provide important clues to the composition of the planets. Both Jupiter and Saturn are globally enriched in heavy elements compared to the Sun. In Jupiter, the *in situ* measurements of the Galileo probe are compatible with a $\sim$3 times solar enrichment of carbon, sulfur, argon, xenon and krypton (Niemann et al. 1998; Owen et al. 1999). It is still unclear as to whether nitrogen is close to solar (by a factor 1 to 1.5; de Pater and Massie 1985), moderately (2.2 to 2.4; Carlson et al. 1992) or strongly enriched (3.5 to 4.5 times solar; Folkner et al. 1998). Water is still a problem because of its condensation at deep levels, and only a lower limit of $\sim$0.1 times solar can be inferred from the measurements. The enrichment in noble gases (except neon) is problematic and bears directly on formation issues. It has been proposed that these elements are brought to the planet in the form of clathrates (Gautier et al. 2001).

Unfortunately, the uncertainties for Saturn are still relatively large. Its atmosphere is enhanced in carbon by a factor of 2 to 7, and in nitrogen by a factor 2 or more (Gautier and Owen 1989). Observationally, it could therefore be more rich in heavy elements than the jovian atmosphere. This will be tested by the Cassini–Huygens mission.

Atmospheric temperatures: The temperatures at the tropopause (at pressures of about 0.3 bar) are relatively well constrained by direct inversions of infrared spectra. These predict relatively large latitudinal temperature changes of the order of 10 K (Conrath et al. 1989). The temperature gradients decrease with tropospheric depth, as interior convection presumably becomes more efficient in redistributing the heat. However, the accuracy of this method drops rapidly with increasing pressure and does not reach levels deep enough to be used as surface condition for interior models. So far, the only reliable measurement of the deep tropospheric temperature of a giant planet is that from the Galileo probe in Jupiter: 166 K at 1 bar (Seiff et al. 1998). It is not clear however how representative of the whole planet this measurement is. Previous analyses have relied upon (local) radio occultation data acquired with the Pioneer and Voyager spacecrafts (Lindal et al. 1981, 1985) that predicted 1 bar temperatures of 165 ± 5 K in Jupiter and 134.8 ± 5 K in Saturn. The temperatures inferred from these data are however dependent on the assumed mean molecular weight $\overline{m}$. The Galileo helium mixing ratio, applied to the Voyager data would yield a temperature of 170.4 K at 1 bar in Jupiter. It is therefore reasonable to assume that the uncertainty on these temperatures if of the order of ~5 K.

Equations of state: Ideally, one should use an equation of state valid for any chemical composition. This is of course unrealistic. The most recent astrophysical equation of state for hydrogen and helium provided by Saumon, Chabrier and Van Horn (1995) does not account for interactions between the two species. The presence of other species can be added using generally less reliable equations of state.

Figure 11 compares different pressure–temperature profiles for Jupiter and Saturn, using the various equations of state described here. The figure is intended to provide an estimate of the uncertainties on the various equations of state (hydrogen–helium, heavy elements). It is important to notice at this point that Saturn's interior lies mostly in a relatively well-known region of the hydrogen-helium EOS, i.e. in which hydrogen is molecular, whereas a

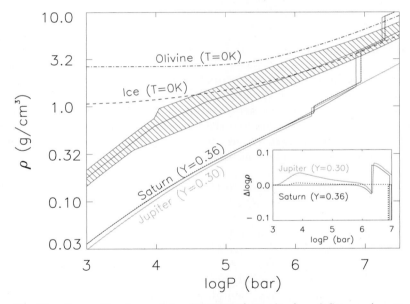

Fig. 11. Density profiles in models of Jupiter (*gray line*) and Saturn (*continuous lines*: adiabatic i-EOS and PPT-EOS models; *dashed*: non-adiabatic i-EOS model). Upper curves (*dashed* and *dot-dashed*) are $T = 0$ K density profiles for water ice and olivine (from Thompson 1990). The dashed region represents the assumed uncertainty on the EOS for heavy elements ($\rho_Z(P,T)$). Within this region, the continuous line corresponds to our "preferred" profile for ρ_Z. *Inset*: Differences of the decimal logarithm of the Saturn density profiles with the same profile using the *i*-EOS and an adiabatic structure (*plain and dotted lines*). The gray line corresponds to the same difference but for a PPT-EOS non-adiabatic Jupiter model (From Guillot 1999a)

significant fraction of Jupiter's interior is at intermediate pressures (one to a few Mbar) for which the EOS is most uncertain.

Opacities: As discussed in Sect. 5 the Rosseland opacities available for models of Jupiter and Saturn are still uncertain. A small table is provided by Guillot (1999a) but does not include the absorption due to alkali metals. It therefore predicts the existence of a radiative region in both Jupiter and Saturn at temperatures around 1500 K and pressures of 1–10 kbar. However, including the absorption of sodium and potassium, as observed in brown dwarfs (Burrows et al. 2000a) provides the required opacity source and the radiative regions then disappear for both planets (Freedman et al. in preparation). The models presented hereafter include uncertainties on the opacities as follow: a minimum value is set by the calculation of a Rosseland opacity table assuming no alkali metals, and thus include the presence of a radiative zone. Other models simply assume that the planets are fully convective. The differences between these models are in fact, in term of interior structure, relatively limited. However, we will see that the presence of a radiative zone affects the evolution more significantly.

Construction of Models

Most models of Jupiter and Saturn assume a three-layer structure: a helium-poor molecular region, a helium-rich metallic region and a central dense core. The fact that the molecular/metallic transition coincides with a jump in the abundance of helium is related to the idea that helium is most insoluble in low-pressure metallic hydrogen, as obtained from calculations assuming full ionization (see Sect. 4). The consequences of a different phase diagram have not been calculated so far.

The three regions are linked to three parameters: $M_{\rm core}$ the mass of the core, $Z_{\rm mol}$ and $Z_{\rm met}$ the mass mixing ratio of heavy elements in the molecular and metallic envelopes, respectively. (The helium mixing ratio in the molecular envelope is set equal to the atmospheric value; That in the metallic region is constrained by the fact that the total helium/hydrogen ratio should be equal to the protosolar value).

The total mass of the planet being fixed, the observational constraints are the equatorial radius $R_{\rm eq}$ and gravitational moments J_2, J_4 and J_6, measured with respective observational uncertainties $\sigma_{R_{\rm eq}}$, σ_{J_2}, σ_{J_4} and σ_{J_6}. In the framework of the three-layer models, the adjustable parameters are $Z_{\rm mol}$, $Z_{\rm met}$ and $M_{\rm core}$. A way of finding models matching the observational constraints is therefore to minimize the following function:

$$\chi^2(Z_{\rm mol}, Z_{\rm met}, M_{\rm core}) = \frac{1}{4}\left[\left(\frac{\Delta R_{\rm eq}}{\sigma_{R_{\rm eq}}}\right)^2 + \left(\frac{\Delta J_2}{\sigma_{J_2}}\right)^2 + \left(\frac{\Delta J_4}{\sigma_{J_4}}\right)^2 + \left(\frac{\Delta J_6}{\sigma_{J_6}}\right)^2\right], \tag{95}$$

where $\Delta R_{\rm eq}$, ΔJ_2, ΔJ_4, ΔJ_6 are the differences between observed and theoretical $R_{\rm eq}$, J_2, J_4 and J_6. The non-uniqueness of solutions matching the observed gravitational fields is found to be mostly due the uncertainty on J_4. So far, no useful constraint can be derived from the values of J_6, owing to their large observational uncertainties (see Table 7).

Results

The resulting interior models of Jupiter matching all available observations are shown in Fig. 12. Hundreds of models have been calculated, but the solution is represented as a filled area instead as dots for an easier interpretation of the figure. A striking result obtained from Jupiter's modeling is the large uncertainty due to our relatively poor knowledge of the behavior of hydrogen at Mbar pressures. As a consequence, two kind of solutions are found depending on one using the Saumon–Chabrier PPT EOS, or the one that is smoothly interpolated between the molecular and the metallic fluids. The uncertainties in the solutions are *not* due to the qualitative difference at the molecular/metallic transition but instead by the quantitatively different density profiles, as seen in Fig. 11. Any solution between the two regions in Fig. 12 would be valid, provided the "true" EOS for hydrogen lies between the PPT and interpolated EOSs.

More quantitatively, Fig. 12 shows that an upper limit to Jupiter's core mass is rather small, i.e. about $10\,{\rm M}_\oplus$ only. This is significantly smaller than found $\sim$20 years ago, the main difference being due to the improved EOSs. The lower limit on the core mass is found to be zero: in this case, Jupiter could have no core, or a very small one. This corresponds however to rather extreme models, assuming a hydrogen EOS close to the interpolated one, and a large J_4 value. The lower panel of Fig. 12 also indicates that this corresponds to a planet that is enhanced in heavy elements by 4 to 6 times over the solar value (assuming $Z_{\rm mol} = Z_{\rm met}$, a consequence of the presence of no physical discontinuity of the EOS). Generally, it is found that Jupiter's molecular region is enriched in heavy elements by 1.5 to 6.5 times the solar value, in agreement with the observations that indicate a $\sim$3 times solar enrichment for C, N, S.

In the case of Saturn (Fig. 13), the solutions depend less on the hydrogen EOS because the Mbar pressure region is comparatively smaller. The total amount of heavy elements present in the planet can therefore be estimated with a better accuracy than for Jupiter. It is interesting to see that presently, we do not know which of Jupiter and Saturn contain more heavy elements in absolute value! However, because Saturn's metallic region is deeper into the planet, it mimics the effect that a central core would have on J_2. The uncertainty on $M_{\rm core}$ is therefore large. In Fig. 13 constraints obtained from the evolutionary models have been used to eliminate models that otherwise satisfied the static constraints (see Guillot 1999a for details). Saturn's core is therefore found to be between 6 and $17\,{\rm M}_\oplus$. Saturn's enrichment in heavy

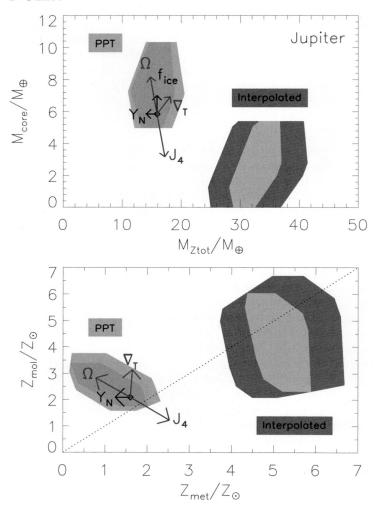

Fig. 12. Constraints on Jupiter's interior structure. The upper panel shows values of the core mass (M_{core}) and total mass of heavy elements (M_{Ztot}) of models matching all available observational constraints. The lower panel shows the mass mixing ratio of heavy elements of the molecular (Z_{mol}) and metallic (Z_{met}) regions, in solar units ($Z_\odot = 0.0192$). The two different regions correspond to different EOSs for hydrogen (see text). Arrows indicate the direction and magnitude of the assumed uncertainties, if J_4 or Y_{proto} are increased by 1σ, rotation is assumed to be solid ("Ω"), the core is assumed to be composed of ices only ("f_{ice}") and if Jupiter's interior becomes fully adiabatic ("∇_T"). *The dashed line* in the lower panel indicates a homogeneous abundance of heavy elements ($Z_{mol} = Z_{met}$) [Adapted from Guillot 1999a]

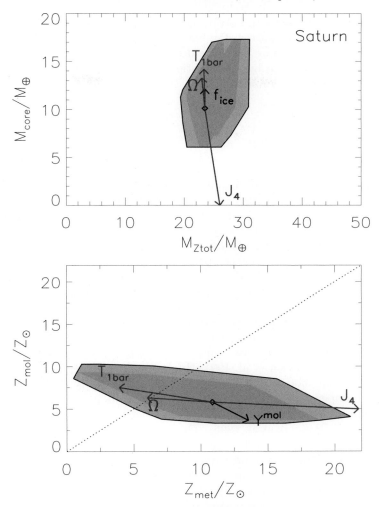

Fig. 13. Same as Fig. 12 in the case of Saturn. The solutions for the PPT and interpolated hydrogen EOSs are very similar and are thus not highlighted. The arrow labeled $T_{1\,\mathrm{bar}}$ corresponds to an increase of Saturn's 1 bar temperature from 135 to 145 K. The arrow labeled Y^{mol} corresponds to an increase of the helium mass mixing ratio from 0.16 to 0.21 [Adapted from Guillot 1999a]

elements is found to be generally larger than in the case of Jupiter, but with a considerable uncertainty in the metallic region.

Figures 12 and 13 also show as arrows the significance of various sources of uncertainties for estimating precisely the parameters of the interior structure. The uncertainty on the measured value of J_4 is shown to significantly affect the results: in the case of Saturn in particular, a more accurate measurement (that the Cassini–Huygens mission will probably provide) is expected to substantially narrow the ensemble of viable models. Note that in the case of Saturn, a more accurate determination of the surface temperature and of the helium to hydrogen ratio would also be invaluable for better constraining the interior models.

6.3 Uranus and Neptune

Spectroscopic measurements indicate that their hydrogen–helium atmospheres contain a large proportion of heavy elements, mainly CH_4, which is enriched by a factor ~ 30 compared to solar composition (see Table 2). The two planets have similar masses (14.53 $M_\oplus$ for Uranus, 17.14 $M_\oplus$ for Neptune) and radii. Neptune's larger mean density is partly due to greater compression, but could also be the result of a slightly different composition. The gravitational moments impose that the density profiles lie close to that of "ices" (a mixture initially composed of H_2O, CH_4 and NH_3, but whose composition most probably does not consist of intact molecules in the planetary interior), except in the outermost layers, which have a density closer to that of hydrogen and helium (Marley et al. 1995; Podolak et al. 2000). Three-layer models of Uranus and Neptune consisting of a central "rocks" core (magnesium-silicate and iron material), an ice layer and a hydrogen–helium gas envelope have been calculated (Podolak et al. 1991; Hubbard et al. 1995).

The fact that models of Uranus assuming homogeneity of each layer and adiabatic temperature profiles fail in reproducing its gravitational moments seem to imply that substantial parts of the planetary interior are not homogeneously mixed (Podolak et al. 1995). This could explain the fact that Uranus' heat flux is so small: its heat would not be allowed to escape to space by convection, but through a much slower diffusive process in the regions of high molecular weight gradient. Such regions would also be present in Neptune, but much deeper, thus allowing more heat to be transported outward. The existence of these non-homogeneous, partially mixed regions are further confirmed by the fact that if hydrogen is supposed to be confined solely to the hydrogen–helium envelope, models predict ice/rock ratios of the order of 10 or more, much larger than the protosolar value of ~ 2.5. On the other hand, if we impose the constraint that the ice/rock ratio is protosolar, the overall composition of both Uranus and Neptune is, by mass, about 25% rocks, 60–70% ices, and 5–15% hydrogen and helium (Podolak et al. 1991, 1995; Hubbard et al. 1995). An upper limit to the total amount of hydrogen and helium present in these planets is 3 $M_\oplus$ for Uranus and 5 $M_\oplus$ for Neptune (Podolak et al. 2000).

The characteristics of typical models of the four giant planets are summarized in Fig. 14, including corresponding uncertainties in the temperature profiles. The distinction between the "gas giants" Jupiter and Saturn and the smaller "ice giants" Uranus and Neptune is evident.

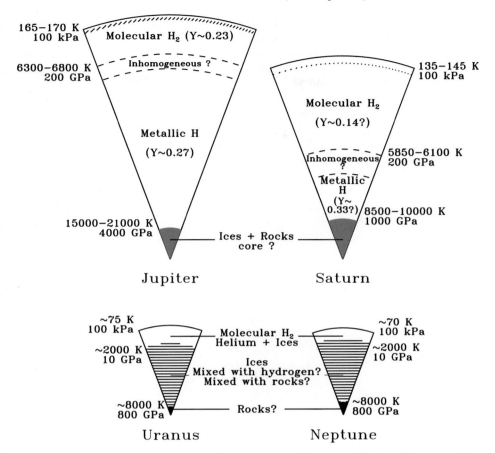

Fig. 14. Schematic representation of the interiors of Jupiter, Saturn, Uranus and Neptune. The hashed region indicate a possible radiative zone (in Jupiter, it corresponds to $P \sim 0.15$ to $0.6\,\text{GPa}$, $T \sim 1450$ to $1900\,\text{K}$, and $R \sim 0.990$ to $0.984\,R_J$; in Saturn, it is located around $P \sim 0.5\,\text{GPa}$, $T \sim 1700\,\text{K}$, $R \sim 0.965\,R_S$). The radiative zone are expected to disappear in the presence of alkali metals. The range of temperatures for Jupiter and Saturn is for models neglecting the presence of the inhomogeneous region. Helium mass mixing ratios Y are indicated. In the case of Saturn, it is assumed that $Y/(X+Y) = 0.16$ in the molecular region. The size of the central rock and ice cores of Jupiter and Saturn is very uncertain. Two representative models of Uranus and Neptune are shown, but their actual interior structure may be significantly different (see text) [From Guillot 1999b]

6.4 Consequences for Formation Models

The Minimum Mass Solar Nebula

The composition of the giant planets provides crucial information to understand the formation of planets in general. A useful first indication of the structure of the early protosolar nebula comes from the estimation of the *minimum* amount of gas that initially had to be present in the disk in order to form the planets that we see today. The result is commonly called the *minimum mass nebula* (see Weidenschilling 1977; Hayashi 1981). Figure 15 shows the minimum surface density (g/cm^2 projected onto the plane of the Solar System) of hydrogen and helium, as a function of the distance to the Sun, *assuming that the planets were formed at their present locations*, and using the most recent interior models for Jupiter, Saturn, Uranus and Neptune. The gas to solids ratio was assumed to lie between 55 and 90; these two extremes correspond to (1) the condensation of all species except H, He, and noble gases, and (2) the condensation of only water and metals and only small amounts of condensed C (which is then assumed to mostly remains in the form the gaseous CO) and N (remaining mostly bound up as N_2), respectively.

In Fig. 15, the surface density required by two models of formation of the giant planets are indicated. The most "standard" model of formation of the giant planets is based on the formation of a solid protoplanetary embryo followed by the capture of the surrounding hydrogen and helium, on a few million year time scale (Pollack et al. 1996). The density required to form

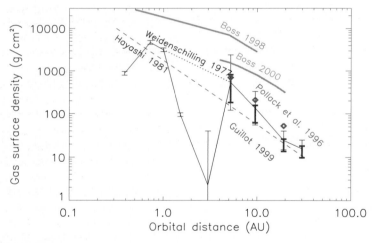

Fig. 15. Surface density of hydrogen and helium as a function of distance to the Sun, as estimated by various workers. The *dashed* and *plain black lines* correspond to the minimum mass protosolar nebula as derived by Hayashi (1981) and Weidenschilling (1977) respectively. The *thicker vertical error bars* outside 5 AU are updates of the Weidenshilling values using interior models of the outer planets from Guillot (1999a). The *diamonds* are the optimal surface densities for giant planet formation in a coreaccretion scenario, assuming a gas to solids ratio of 70 (Pollack et al. 1996). The two *upper lines* (Boss 1998, 2000) correspond to a scenario of formation of Jupiter and possibly Saturn by direct gravitational instability in the gas

Jupiter in less than 10 million years is just slightly over the the new median estimate for the minimum mass nebula. Neglecting any migration process, this implies that between 20 and 75% of the solids in that region have been lost, probably due to dynamical evolution. (These numbers account for the fact that the gas to solids ratio has to be the same when comparing different results). The density increases over the minimum one required to form Saturn, Uranus and Neptune is higher, implying an even larger ejection efficiency (or other loss mechanism) for solids in those regions.

An alternative model that might explain the formation of Jupiter is by direct gravitational instability of the gas itself, on much shorter time scales (Boss 1998, 2000). This requires even larger densities than in the core-accretion scenario, but it could be advocated that part of the gas present in the disk at that early times has been accreted onto the star. The subsequent formation of central cores in these models would require an early settling of heavy elements.

Delivering Planetesimals to the Giant Planets

To study the delivery of heavy-element rich planetesimals to the forming giant planets, Guillot and Gladman (2000) performed extensive numerical dynamical simulations of the fate of 10,000 massless particles distributed between 4 and 35 AU. In a baseline model, the masses and radii of the giant planets were set to their present-day values, exploring a scenario in which the planets "suddenly" reach nearly their current masses by a rapid gas accretion onto a much smaller core. After 100 Myr, 61% of the initial particles had been ejected out of the system, 23% had been sent to the Oort cloud (aphelia larger than 10,000 AU), with only 13% remaining in the system. Only 4% of the particles impacted one of the four giant planets. In this physical scenario, the probability of impact is low compared to that of ejection, mainly due to the presence of Jupiter, to which the other giant planets efficiently 'pass' their planetesimals. However, this inefficiency of planetesimal accretion poses grave problems when we consider the known mass of heavy elements in the giant planets.

Focusing on the core-accretion scenario, Fig. 16 shows the *accretion efficiency*, defined as the ratio between the inferred amount of heavy elements in the giant planets (Guillot 1999a) and the amount of solids required for their formation (Pollack et al. 1996). Planets with the present characteristics are found to be too efficient at ejecting material from the system compared to accreting it, as indicated by the diamonds on Fig. 16.

Guillot and Gladman (2000) therefore propose that the heavy elements present in the giant planets today were captured first during a runaway growth phase, probably yielding the cores that are observed today and second during an extended phase during which the planets had large effective capture radii ($\sim 3\,\overline{\rm R}_{\rm J}$) but relatively small masses ($\sim 20\,{\rm M}_\oplus$). This case corresponds to triangles in Fig. 16 and agrees with the accretion efficiencies needed for the giant planets' envelopes (lower thick error bars).

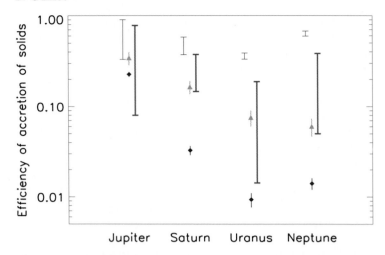

Fig. 16. Accretion efficiencies required to form the giant planets in the core accretion scenario. The accretion efficiency for a given planet is defined as the quantity of heavy elements that had to impact the planet divided by the total amount of heavy elements within the annulus that extends halfway to the next planet. The *upper left error bar* for each planets corresponds to the ratio between the total mass of heavy elements and the quantity of solids required by Pollack et al. (1996). The *lower thick error bars* account for core formation by runaway growth, and include only heavy elements in the envelopes. Diamonds are accretion probabilities in the Jupiter, Saturn, Uranus and Neptune regions, respectively, as calculated in the standard case. *Triangles* correspond to the reduced-mass case (see text) [From Guillot and Gladman 2000]

Possible Formation Scenarii

Three possible scenarii may explain Jupiter and Saturn's core mass *and* total mass of heavy elements:

1. *Rapid formation with a small core or no core* (Gas instability scenario). The formation of giant planets by gas instability is very fast, i.e. 10^4 to 10^5 years at the most (e.g. Boss 2000). Because the final mass of the planet is rapidly reached, the hydrostatic equilibrium imposes a fast contraction of the planet. Although this would have to be quantified, it appears that this planet would have a very low accretion efficiency. In order to explain the structure of Jupiter and Saturn, one needs to invoke a very large mass of solids. This scenario also would predict that Saturn has a smaller core than Jupiter, which isn't implied by the interior models. A possibility would be the capture of very small particles by the planet during their fast migration toward the Sun. All in all, this scenario cannot be ruled out but seems to be unlikely.
2. *Slow formation with extended-phase.* In the model of Pollack et al. (1996), the growth of a giant planet by capture of a hydrogen–helium envelope

onto a core cannot proceed extremely fast due to the feedback mechanism produced by the release of gravitational energy heating the envelope. A phase in which the planet has 20 to 30 $M_\oplus$ and its effective capture radius can be of the order of 2 to 3 times the present radius of Jupiter can persist for millions of years. This scenario is found to be consistent with the dynamical calculations. Most of the heavy elements are hence captured during the first Myr. An important consequence is that in order to explain the enrichment observed in the atmosphere, an efficient upward mixing is required.

3. *Formation from an initially massive core.* Another alternative could be to form a giant planet from an initially very massive core (this happens e.g. if the surface density of solids is higher than the values indicated in Fig. 15). Part of the core would then have to be mixed upward. The advantage is that the phase of extended capture radius is not required anymore and that the formation can be considerably faster.

As discussed in Sect. 5, the problem of mixing is a difficult one. It has been advocated by Stevenson (1982) that the energy available in Jupiter is barely sufficient to mix the any significant fraction of the core, owing to the strongly stabilizing compositional gradient. This would imply that the cores of the giant planets are primordial. Due to the properties of runaway growth that I will not discuss here (see e.g. Wuchterl et al. 2000), one would then expect the giant planets to have relatively similar cores.

However, an analysis of the accretion efficiency has shown us that the heavy elements observed in the atmosphere must have been mixed upward. If we report the 3 times solar value measured in Jupiter's atmosphere in the entire planet, this means that $18 M_\oplus$ had to be transported upward. Efficient mixing mechanisms had therefore to exist to be capable of overcoming the gradient of molecular weight. Two possibilities exist: one is the adiabatic compressional heating during the rapid contraction phase. This phenomenon could heat water more than hydrogen and therefore ease the mixing process. Another possibility consist in advocating Kelvin–Helmoltz instability (e.g. Chandrasekhar 1961) during the accretion of the envelope: their is indeed no reason that the primordial core and the gaseous envelope would have the same angular momentum. The powerful shear would be progressively erased by mixing the different layers and transporting part of the heavy elements upward. This scenario would also explain why Jupiter would have a smaller core than Saturn: its envelope being more massive, it would be capable of mixing more of the central core.

7 Evolution of Giant Planets and Brown Dwarfs

The problem of the formation of giant planets and brown dwarfs is still obscure. That of their evolution, which, as we will see, is only weakly dependent on the initial conditions, is relatively better understood. The accuracy of these calculations in fact mostly reflects our limited knowledge of their atmospheres as we will see in the next sections. They however allow us to predict probable characteristics of extrasolar giant planets, and appear to describe reasonably well brown dwarfs, based on their measured spectra.

7.1 The Virial Theorem

Integrating the hydrostatic equation leads to a well known relation linking the internal energy to the potential gravitational energy. It is very basic to stellar evolution, but I choose to rederive it here, following the formulation of Kippenhahn and Weigert (1991), because it is fundamental to understand the evolution of substellar objects. For simplicity, rotation or the presence of magnetic fields is neglected. Starting from (11), we write

$$\int_0^M 4\pi r^3 \frac{\partial P}{\partial m} dm = \int_0^M 4\pi r^3 \frac{Gm}{4\pi r^4} dm. \tag{96}$$

The left hand side can be integrated by part to yield:

$$3 \int_0^M \frac{P}{\rho} dm = \int_0^M \frac{Gm}{r}. \tag{97}$$

The right hand of (97) corresponds to the gravitational energy with a minus sign:

$$E_{\rm g} \equiv -\int_0^M \frac{Gm}{r} dm \tag{98}$$

and $-E_{\rm g}$ is the energy required to bring all the mass to infinity. The left hand side of (97) is related to the internal energy

$$E_{\rm i} \equiv \int_0^M u\, dm = \int_0^M \frac{3}{\xi} \frac{P}{\rho} dm, \tag{99}$$

where $\xi \equiv 3P/u\rho$ and u is the internal specific energy. In the case of a perfect gas, $\xi = 3(\gamma - 1)$ where $\gamma = c_P/c_V$. In the case of a monoatomic perfect gas, $\xi = 2$. If we furthermore assume that ξ is uniformly constant throughout the star/planet considered, (97) takes the following form:

$$\xi E_{\rm i} + E_{\rm g} = 0, \tag{100}$$

known as the *virial theorem*.

If we furthermore consider the total energy of the system $W = E_{\rm i} + E_{\rm g}$ ($W < 0$ for a gravitationally bound system), and assume that the luminosity is entirely due to the loss of energy (i.e. we neglect thermonuclear reactions, radioactivity...etc.),

$$\frac{dW}{dt} + L = 0 \tag{101}$$

and hence
$$L = (\xi - 1)\frac{dE_i}{dt} = -\frac{\xi - 1}{\xi}\frac{dE_g}{dt}. \tag{102}$$

This relation is valid in a variety of cases, including giant planets for their entire life and brown dwarfs for the parts of their life when thermonuclear reactions represent a small fraction of the total luminosity.

Let us consider a contracting brown dwarf or giant planet beginning its life mostly as a perfect H_2 gas. In this case $\gamma \approx 7/5$, hence $\xi = 16/5 = 3.2$. Two third of the energy gained by contraction is therefore radiated away, one third being used to increase the internal energy. This being proportional to the temperature, the effect is to heat the object. This represents the slightly counter-intuitive but well known effect that a star or giant planet initially increases its luminosity while heating up.

Let us now move further in the evolution, when the contraction has proceeded to a point where the electrons have become degenerate. The problem then becomes relatively complex because of the interplay between ions and electrons. It is instructive however to consider the ideal case, formerly valid only in the white dwarf regime, in which most of the pressure is provided by non-relativistic degenerate electrons. In that case, $P/\rho \approx (2/3)u$ and therefore $\xi \approx 2$: Half of the gravitational potential energy is radiated away and half of it goes into internal energy. The problem is to decide how this energy is split into an electronic and an ionic part. The gravitational energy changes with some average value of the interior density as $E_g \propto 1/R \propto \rho^{1/3}$. The energy of the degenerate electrons is essentially the Fermi energy: $E_i \approx E_F \propto \rho^{2/3}$. One is therefore led to a simple relation between E_g and E_e:

$$\dot{E}_e \approx 2\frac{E_e}{E_g}\dot{E}_g = -\frac{E_e}{E_i}\dot{E}_g, \tag{103}$$

where E_i is introduced via the virial theorem ($E_g = -2E_i$). In the case of white dwarfs, $E_{\text{ion}} \ll E_e$ and therefore $E_i = E_{\text{ion}} + E_e \approx E_e$. This means that $\dot{E}_e \approx -\dot{E}_g \approx 2L$. The energy balance $L = -\dot{E}_{\text{ion}} - \dot{E}_e - \dot{E}_g$ becomes

$$L \approx -\dot{E}_{\text{ion}} \propto -\dot{T}. \tag{104}$$

In this case, the gravitational energy lost is entirely absorbed by the degenerate electrons, and the observed luminosity is due to the thermal cooling of the ions.

For brown dwarfs and giant planets, the problem is more complex because the electrons are only partially degenerate, and the contribution of the ions to the pressure and internal energy cannot be neglected. However this only affects the solution through numerical factors: qualitatively, most of the gravitational energy lost is used up to increase the energy of the degenerate electron gas, while the luminosity is essentially provided by the cooling of the ions.

7.2 A Semi-Analytical Model

Solution for Isolated Objects

It is possible given certain assumptions to solve analytically the evolution problem. A more detailed numerical solution is of course eventually required, but the analytical solution is a tool to comprehend the physical problem. The solution that is presented here is due to Hubbard (1977).

We consider an already dense giant planet or brown dwarf without thermonuclear reactions, and assume that its metallic region provides the essential contribution to its cooling. Second, we will assume that it is adiabatic. One can then show that the internal temperature profile obeys a relation of the form

$$T \approx C T_{1\,\mathrm{bar}} \rho^\gamma . \qquad (105)$$

In the case of Jupiter, $C \approx 42.8$ when ρ is expressed in $\mathrm{g\,cm^{-3}}$, and $\gamma \approx 0.64$ is the Grüneisen parameter.

In the set of equations (44) governing the evolution of substellar objects, only the energy conservation equation involves time, through the $-T\partial S/\partial t$ term. This equation can be rewritten in the form

$$\frac{\partial L}{\partial m} = -c_V \frac{\partial T}{\partial t} + c_V \left(\frac{\partial T}{\partial \rho}\right)_S \frac{\partial \rho}{\partial t} . \qquad (106)$$

The term $(\partial T/\partial \rho)_S$ being positive, the luminosity is provided both by the contraction and cooling of the planet.

Let us first neglect insolation. Integrating (106), we obtain

$$L = 4\pi R^2 \sigma T_{\mathrm{eff}}^4 = -\int C_V \left(\frac{\partial T}{\partial t} - \gamma \frac{T}{\rho}\frac{\partial \rho}{\partial t}\right) dm . \qquad (107)$$

Furthermore, (10) and (105) imply that

$$\frac{\partial T}{\partial t} = T \left(-b\frac{\partial \ln g}{\partial t} + a\frac{\partial \ln T_{\mathrm{eff}}}{\partial t} + \gamma \frac{\partial \ln \rho}{\partial t}\right) . \qquad (108)$$

The gravity dependence is weak. The term proportional to $\partial \ln g/\partial t$ can hence be neglected. Reporting (108) into (105), one finds

$$dt = -\alpha(T_{\mathrm{eff}}) T_{\mathrm{eff}}^{a-5} dT_{\mathrm{eff}} , \qquad (109)$$

$$\alpha(T_{\mathrm{eff}}) = \frac{aCK}{4\pi R^2 \sigma g^b} \int C_V \rho^\gamma dm . \qquad (110)$$

In the case of Jupiter, $C_V \approx 1.66 k_\mathrm{B}/m_\mathrm{H}$ yielding $\alpha(T_{\mathrm{eff}} = 124.4\,\mathrm{K}) \approx 2.8 \times 10^{23}$ cgs.

Let us assume α constant (i.e. we neglect the evolution of the planet's structure during the contraction). The time necessary to cool from an effective temperature $T_{\text{eff},0}$ to $T_{\text{eff},1}$ is therefore

$$\Delta t = \frac{\alpha}{4-a}(T_{\text{eff},1}^{a-4} - T_{\text{eff},0}^{a-4}). \tag{111}$$

Using $a \approx 1.24$, one can see that the time for the planet to cool from an infinite temperature to $T_{\text{eff},1}$ is approximately 50 times smaller than that required for cooling from $T_{\text{eff},1}$ to $T_{\text{eff},1}/4$. The evolution problem is very weakly dependent on initial conditions.

Jupiter's cooling time from an initially infinite effective temperature to its present value $T_{\text{eff}} = 124.4\,K$ is found, using (111) to take about $5.4\,\text{Gyr}$. Saturn's cooling time is much shorter, i.e. about $2\,\text{Gyr}$.

Correction due to Irradiation

Let us now include the absorbed stellar luminosity. The total luminosity of the planet (or irradiated brown dwarf) has then three components: a directly reflected stellar part which does not contribute to the heating of the planet and is hence often not mentioned when studying the evolution; a part corresponding to the absorbed stellar luminosity, that I choose to note $L_\odot$; the intrinsic luminosity L_{int}. The effective temperature now has to be redefined. The definition tying most closely the effective temperature to the temperature at the photosphere is

$$4\pi R^2 \sigma T_{\text{eff}}^4 \equiv L_\odot + L_{\text{int}}. \tag{112}$$

In the interior of the planet, the only relevant quantity is the intrinsic luminosity (and it thus convenient to forget the *int* suffix when considering the internal structure). The stellar flux is generally very rapidly absorbed and contributes in fact only to heating the outer boundary. The problem is therefore to derive the new boundary temperature. A simple approach is to use the same boundary condition [(9,10)] but with the new definition of T_{eff}. We will come back on that assumption when discussing the case of Pegasi planets (Sect. 9).

We therefore rewrite (107) by taking account of the absorbed stellar luminosity:

$$4\pi R^2 \sigma (T_{\text{eff}}^4 - T_\odot^4) = -\int C_V \left(\frac{\partial T}{\partial t} - \gamma \frac{T}{\rho}\frac{\partial \rho}{\partial t}\right) dm, \tag{113}$$

where $T_\odot$ is the effective temperature that the planet would have if its intrinsic luminosity would drop to zero while conserving the same atmosphere and overall structure. It is defined by $L_\odot = 4\pi R^2 \sigma T_\odot^4$.

It is easy to show that (109) is now replaced by

$$dt = -\alpha(T_{\text{eff}}) \frac{T_{\text{eff}}^{a-5}}{1 - (T_\odot/T_{\text{eff}})^4} dT_{\text{eff}}. \tag{114}$$

An expansion in powers of $(T_\odot/T_{\rm eff})^4$ leads to

$$t = \frac{\alpha}{4-a} T_{\rm eff}^{a-4} \left[1 + \frac{4-a}{8-a}\left(\frac{T_\odot}{T_{\rm eff}}\right)^4 + \frac{4-a}{12-a}\left(\frac{T_\odot}{T_{\rm eff}}\right)^8 + ...\right]. \qquad (115)$$

The value of $(T_\odot/T_{\rm eff})^4$ is 0.60 for Jupiter and 0.56 for Saturn. Using $a = 1.243$, the term between the square brackets is therefore of the order of 1.3 for both Jupiter and Saturn. The contribution to the evolution of Solar radiation is hence far from negligible. Equation (115) furthermore demonstrates that the evolution is slowed by the stellar radiation.

Numerically, these equations predict that Jupiter should take about 7 Gyr to cool from an infinite effective temperature to today's value. This value is however overestimated due to the fact that α was held constant, i.e. variations in the structure of the planet itself were neglected. Obviously, more sophisticated models have to be developed anyway to go beyond the approximations made in this semi-analytical model.

Influence of a Radiative Zone

The possible presence of a radiative zone can have deep, structural changes on the evolution of a planet. These modifications are not always easy to intuit. Of course, the most fundamental question is to know whether a radiative zone leads to a quicker cooling, or if on the contrary it slows the evolution of the planet. In fact, we shall see that the answer depends on the evolution of the radiative zone itself.

Let us consider two adiabatic models separated in time by an unknown interval $\Delta t_{\rm ad}$. Let us also consider two non-adiabatic models (possessing an internal radiative zone), which have the same external conditions as the adiabatic ones (i.e. same surface temperature, intrinsic luminosity...etc.), but separated by in time by an unknown amount $\Delta t_{\rm nad}$. The energy conservation equation tells us that

$$\Delta t_{\rm ad} \approx -\frac{M}{L} T_{\rm ad} \Delta S_{\rm ad}, \qquad (116)$$

$$\Delta t_{\rm nad} \approx -\frac{M}{L} T_{\rm nad} \Delta S_{\rm nad}, \qquad (117)$$

where $T_{\rm ad}$, $T_{\rm nad}$, $S_{\rm ad}$ and $S_{\rm nad}$ are characteristic values of temperature and specific entropy of adiabatic and non-adiabatic models, respectively.

The external boundary conditions being identical, the condition of convective instability necessarily implies (neglecting small compositional differences between the two models) $T_{\rm ad} > T_{\rm nad}$: the radiative model is always cooler than the fully convective model. However the difference in entropy variation between the two kind of models will depend on the evolution of the characteristics of the radiative zone during the planet's cooling.

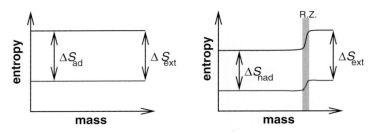

Fig. 17. Example of evolution of the specific entropy profile in the purely adiabatic case (*left*) and in the presence of a radiative zone (*gray area in the right hand side figure*). In each case, the entropy is represented in function of the mass variable at two different times. Here the entropy jump is chosen to decrease with time. In spite of the fact that the initial surface conditions are the same in the two cases for the initial and final models, the mean entropy difference is smaller in the non-adiabatic case than in the adiabatic one ($\Delta S_{\rm nad} < \Delta S_{\rm ad}$). In this example, the cooling of the non-adiabatic model is faster (see text)

In the entirely adiabatic case, Fig. 17 shows that the entropy variation $\Delta S_{\rm ad}$ between the models at ages t_0 and $t_0 + \Delta t_{\rm ad}$ is equal to the entropy variation imposed by the external conditions $\Delta S_{\rm ext}$. In the non-adiabatic case, the presence of a radiative zone induces a decrease of the entropy in the planet's interior. The evolution of this entropy decrease (shaded area in Fig. 17) is crucial. In the case of the giant planets and opacities with no alkali metals, the entropy variation in the radiative zone is greater when the planet is hotter. This implies that $\Delta S_{\rm nad} < \Delta S_{\rm ext}$. Consequently, for the case illustrated by Fig. 17, one can see that

$$\Delta t_{\rm ad} > \Delta t_{\rm nad}. \tag{118}$$

In other terms, the presence of a radiative zone tends, in this case, to accelerate the evolution.

7.3 Evolution of Jupiter and Saturn

Results of Numerical Simulations

The evolution of Jupiter and Saturn to their present state is represented on Figs. 18 and 19. As indicated by the analytic calculation, the initial contraction is very fast, and the initial conditions are forgotten after a few million years or less. The ages of the models which reproduce the observed radii, and effective temperature are for Jupiter 3.7 to 4.5 Gyr for models with a radiative zone, and 4.5 to 5.2 Gyr for fully-convective homogeneous models. In the case of Saturn, these values are 2.0 to 2.4 and 2.2 to 2.6 Gyr, respectively (Guillot et al. 1995; Guillot 1999a,b). Because, as discussed in Sect. 5, the opacities do not account for the presence of alkali metals, one would expect values for fully-convective models to be closer to reality. These corresponds to the largest

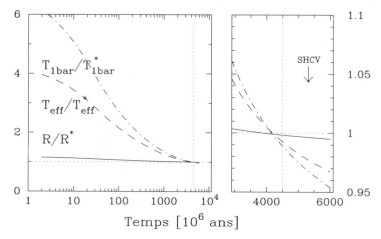

Fig. 18. Contraction and cooling of a non-adiabatic model of Jupiter (opacities not including alkali metals). The 1 bar temperature, effective temperature and mean radius are represented as a function of time. All these quantities are normalized to their present value, $T^*_{1\mathrm{bar}}$, T^*_{eff}, and R^*. The right figure is an enlargement of the left one. Note that time is then represented linearly. The *vertical dotted line* indicates the age of the solar system. The *arrow* labeled SHCV corresponds to the age obtained for a fully convective model by Saumon et al. (1992) [From Guillot et al. 1995]

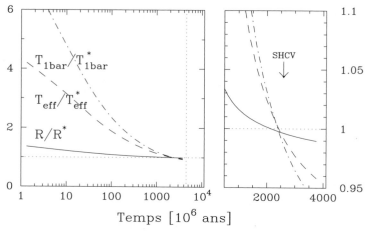

Fig. 19. Same as Fig. 18 for a model of Saturn

ages (and indeed, one can verify that the alkali-free opacities lead to a faster cooling, as discussed previously).

The "real" age of Jupiter and Saturn should be relatively well constrained, unless our understanding of planet formation is utterly wrong. Isotopic dating of meteorites shows that the first condensates appeared in the Solar System

4.56 Gyr ago. Jupiter and Saturn are mostly made with hydrogen and helium, and had to be formed when these elements were still present in the nebula. Observation of forming stars as well as models of circumstellar disks (see Chapter by Pat Cassen) show that hydrogen and helium should have been present for at most ~10 Myr. The age of both Jupiter and Saturn must be about 4.55 Gyr.

There is therefore a problem both for Jupiter and Saturn. In the case of Jupiter, the more-realistic adiabatic models seem too old. It can be argued however that the atmospheric model is crude and introduces an uncertainty on the final age of at least 10%. Imprecisions in the equations of state also introduce a probably significant uncertainty. In the case of Saturn, the discrepancy with the age of the Solar System is large and cannot be explained by inaccuracies in the calculation. Another source of energy has to be invoked: that due to the slow release of gravitational energy thus appears as the most likely one (Salpeter 1973; Stevenson and Salpeter 1977b).

It has been suggested that deuterium-deuterium reaction in a deuterium-enhanced shell around the central core may explain "Jupiter's excess heat" (Ouyed et al. 1998). The models that I have presented explain Jupiter's luminosity naturally by the slow release of gravitational energy through the planet's contraction and cooling. The corollary is that any additional source of energy, and in particular the putative D-D thermonuclear reactions would pose a very difficult problem to relate Jupiter's evolution to that of the Solar System. Furthermore, the model of Ouyed et al. requires the presence of an ad hoc deuterium shell and is therefore extremely unlikely.

Including a Hydrogen/Helium Phase Separation

The energy released by helium sedimentation per unit time can be very significant. It is of the order of

$$L_{\rm grav} \sim \left(\frac{dM}{dt}\right)_{\rm He} gH, \qquad (119)$$

where H is the mean distance over which the helium has fallen. In the case of Saturn, if 10% of the helium atoms were to be transported on a distance equal to half the planet radius in 1 Gyr then $L_{\rm grav} \approx 4 \times 10^{24}\,{\rm erg\,s^{-1}}$, to be compared with Saturn's present intrinsic luminosity $8.6 \times 10^{23}\,{\rm erg\,s^{-1}}$. This process indeed provides the right amount of energy. Note that the energy is essentially proportional to the distance H. If the hydrogen/helium phase separation is tied to the molecular/metallic transition (see discussion in Sect. 4), this could explain relatively naturally why this effect is more important in Saturn than in Jupiter since Saturn's metallic zone is much deeper than that of Jupiter ($R/R_{\rm tot} \sim 0.43$ for Saturn compared to 0.80 for Jupiter).

However, the fact that a phase separation occurs is not sufficient in itself to explain helium sedimentation. Since the planets are also convecting, the helium drops also have to grow fast enough before they are carried away by convective cells. The problem has been estimated by Stevenson and Salpeter (1977b). First, one has to estimate the size above which the free-fall velocity of drops is larger than the convective speed. In the Stokes limit, they show that this speed is, as a function of the drop size b:

$$v_b \approx \frac{b^2 g}{24\nu}. \tag{120}$$

The convective velocity being of order $10\,\mathrm{cm\,s^{-1}}$, the free-fall time becomes larger for sizes $b \gtrsim 1\,\mathrm{cm}$.

Furthermore, the molecular diffusivity of helium in hydrogen being estimated to be $D \sim 10^{-3}\,\mathrm{cm^2\,s^{-1}}$, the time scale for the drops to grow to $\sim 1\,\mathrm{cm}$ is found to be of order of $10^3\,\mathrm{s}$, i.e. much faster than convective time scales.

Let us now estimate analytically how helium sedimentation affects the evolution, using an analytic model inspired by the one of Sect. 7.2. We assume that at a mass m_t, which can vary in time, an evolving jump of the helium mass mixing ratio ΔY occurs. ΔY is chosen to be positive when more helium is present at deeper levels (small values of m).

An integration the energy equation (107), but splitting the entropy derivative in a homogeneous and an inhomogeneous part yields

$$L = \int -T\left[\left(\frac{\partial S}{\partial t}\right)_Y + \frac{dY}{dt}\left(\frac{\partial S}{\partial Y}\right)_t\right] dm. \tag{121}$$

Neglecting the entropy of mixing and the presence of species other than hydrogen and helium implies that

$$\left(\frac{\partial S}{\partial Y}\right)_t = S_{\mathrm{He}} - S_{\mathrm{H}} \equiv -\delta_Y S. \tag{122}$$

Note that the larger mass of the helium atom implies that $\delta_Y S$ is positive.

Let us now assume that both ΔY and m_t vary with time. Mass conservation implies that

$$\frac{dX}{dt} = \begin{cases} \Delta Y \dfrac{dm_t}{dt} + m_t \dfrac{d\Delta Y}{dt} & \text{if } m > m_t(t), \\ \Delta Y \dfrac{dm_t}{dt} - (1-m_t)\dfrac{d\Delta Y}{dt} & \text{if } m < m_t(t+dt). \end{cases} \tag{123}$$

dX/dt is infinite between $m_t(t)$ and $m_t(t+dt)$ but its integral is finite:

$$\int_{m_t(t)}^{m_t(t+dt)} -\frac{dY}{dt} T \delta_Y S dm = T(m_t)\delta_Y S(m_t) \Delta Y \frac{dm_t}{dt}. \tag{124}$$

We therefore obtain that the luminosity of the model at a time t is the sum of three contributions:

$$L = \int_0^M -T\left(\frac{\partial S}{\partial t}\right)_Y dm$$
$$+ \frac{d\Delta Y}{dt}\left\{\int_0^{m_t} T\delta_Y S dm - \frac{m_t}{M}\int_0^M T\delta_Y S dm\right\}$$
$$- \Delta Y \frac{dm_t}{dt}\left\{\int_0^M T\delta_Y S dm - T(m_t)\delta_Y S(m_t)\right\}. \quad (125)$$

The first part is the contribution from the homogeneous evolution assumed not to change as a result of the helium sedimentation (obviously a zero-order assumption). The second part is proportional to $d\Delta Y/dt$ and is always positive if ΔY increases with time (more helium is bought to deeper levels). The third part can be either positive or negative depending on the displacement of the transition region. The term between curved brackets is usually positive for large enough values of m_t ($m_t/M \gtrsim 0.45$ for models of Jupiter and Saturn). In this case, the contribution of this third part is thus to add to the luminosity if the transition region moves to deeper levels (small m_t).

Numerically one finds that

$$\frac{1}{M}\left\{\int_0^{m_t} T\delta_Y S dm - \frac{m_t}{M}\int_0^M T\delta_Y S dm\right\} \approx 5 \times 10^{11} \text{ erg g}^{-1} \quad (126)$$

for both Jupiter and Saturn and

$$\frac{1}{M}\left\{\int_0^M T\delta_Y S dm - T(m_t)\delta_Y S(m_t)\right\} \approx 2.5 \times 10^{12} \text{ erg g}^{-1} \quad \text{Jupiter}$$
$$\approx 10^{11} \text{ erg g}^{-1} \quad \text{Saturn} \quad (127)$$

If the transition follows the PPT, then

$$\frac{1}{M}\left|\frac{dm_t}{dt}\right| \lesssim 2 \times 10^{-2} \text{ Gyr}^{-1}, \quad (128)$$

and the contribution of the displacement of the transition region is negligible.

One therefore finds that the lifetime added to the planet through a phase transition from an initially homogeneous planet to one that has a helium jump ΔY is approximately

$$\Delta t \approx \frac{\Delta Y}{L}\left\{\int_0^{m_t} T\delta_Y S dm - \frac{m_t}{M}\int_0^M T\delta_Y S dm\right\}, \quad (129)$$

where L is the planet's present intrinsic luminosity. Numerical applications indicate that

$$\Delta t \approx 9 \Delta Y \,\mathrm{Gyr}, \tag{130}$$

for both Jupiter and Saturn. This is consistent with more detailed numerical calculations (Hubbard et al. 1999). From our evolution models, we can infer that $\Delta Y \lesssim 0.01$ for Jupiter, and that $0.2 \lesssim \Delta Y \lesssim 0.3$ in Saturn. Interior models of Jupiter are in contradiction with that upper limit since they lead to $\Delta Y \sim 0.04$ (Guillot 1999b). This problem is still not resolved. In the case of Saturn, this implies that the Voyager value for the atmospheric helium mixing ratio is too low, something recognized independently from interior models (Guillot 1999a), evolutionary models (Hubbard et al. 1999) and a reanalysis of Voyager IRIS data (Conrath and Gautier 2000).

7.4 From Giant Planets to Brown Dwarfs

Giant planets and brown dwarfs share the same physics. It is so much the case that it is difficult to make the distinction between the two classes of objects. Presumably they should be formed by different mechanisms (see e.g. Wuchterl, Guillot and Lissauer 2000). However, I will here arbitrarily define giant planets as substellar objects in which thermonuclear reactions do not occur, and brown dwarfs as objects which burn some deuterium during their life, but which never attain the equilibrium phase (main sequence) in which most of their energy is provided by hydrogen burning.

Nuclear Reactions

For brown dwarfs, the occurrence of thermonuclear reactions is almost entirely due to a truncated PPI cycle (e.g. Burrows and Liebert 1993):

$$\mathrm{p} + \mathrm{p} \to d + e^+ + \nu_e$$
$$\mathrm{p} + \mathrm{d} \to {}^3He + \gamma.$$

Note that because of the relatively low central temperatures, only ^{3}He, *not* ^{4}He is formed through these reactions. The pp and pd reactions release 1.442 and 5.494 MeV, respectively.

The energy released through these reactions is, assuming no screening (Fowler et al. 1975):

$$\dot{\epsilon}_{pp} = 2.5 \times 10^6 \frac{\rho X^2}{T_6^{2/3}} e^{-33.8/T_6^{1/3}} \,\mathrm{erg\,g^{-1}\,s^{-1}}, \tag{131}$$

and

$$\dot{\epsilon}_{pd} = 1.4 \times 10^{24} \frac{\rho X Y_d}{T_6^{2/3}} e^{-37.2/T_6^{1/3}} \,\mathrm{erg\,g^{-1}\,s^{-1}}, \tag{132}$$

where X and Y_d are the mass mixing ratios of hydrogen and deuterium, respectively, and $T_6 = T/10^6$ K. Note that the primordial deuterium abundance is of the order of $Y_d = 2 \times 10^{-5}$.

However, (131, 132) are underestimates because the fact that the plasma is strongly coupled significantly softens the repulsive potential of the nuclei. A detailed analysis of this is discussed by Saumon et al. (1996) and Chabrier and Baraffe (1997). In the case of brown dwarfs, the enhancement factor is of the order of $\sim$2 for both reactions. An estimate of the final sensitivity of the reactions to temperature and density variations is provided by Burrows and Liebert (1993):

$$\dot{\epsilon}_n \approx 5.9 \times 10^{10} \left(\frac{T}{3 \times 10^6 \text{ K}}\right)^{6.3} \left(\frac{\rho}{10^3 \text{ g cm}^{-3}}\right)^{1.28} \text{erg g}^{-1} \text{s}^{-1}. \quad (133)$$

This expansion around $T = 3 \times 10^6$ K and $\rho = 10^3$ g cm^{-3} shows the strong sensitivity of the energy production to the temperature. Note however that other thermonuclear reactions can have much steeper temperature dependences in the case of more massive stars (e.g. Clayton 1963).

Brown Dwarf Models and Results

The two most cited evolution models of giant planets and brown dwarfs to date are those of the "Tucson group" (e.g. Burrows et al. 1997) and of the "Lyon group" (e.g. Chabrier and Baraffe 1997). These models share many similarities, and in particular have the same equation of state and same nuclear burning rates. They however differ on a few points:

- On the atmospheric model: The 'Tucson' model uses the k-coefficient approach from Marley et al. (1996) and an approximate treatment of clouds using Lunine et al. (1989). The 'Lyon' model is based on a detailed line by line approach (e.g. Allard and Hauschildt 1995).
- On the treatment of convection: The 'Tucson' model is essentially adiabatic, whereas the 'Lyon' one uses the mixing length theory.
- On conduction: implemented in the 'Lyon' model only, using conductive opacities from Potekhin et al. (1999). Figure 20 shows the evolution of the conductive core of a 0.06 $M_\odot$ brown dwarf (from Chabrier et al. 2000b).

Other models by D'Antona and Mazitelli (see Montalbán et al. 2000) are available but will not be discussed since they have generally concerned objects of larger masses.

Figures 21 and 22 show the evolution of isolated giant planets, brown dwarfs and stars. The distinction between stars and brown dwarfs can be readily seen from the fact that stars reach a long equilibrium period during which the tendency of the star to contract under the action of gravity is balanced by thermonuclear hydrogen fusion. For brown dwarfs, even if thermonuclear reactions are indeed possible (even of hydrogen for the most massive ones), they are never energetic enough to reach this balance: brown dwarfs and giant

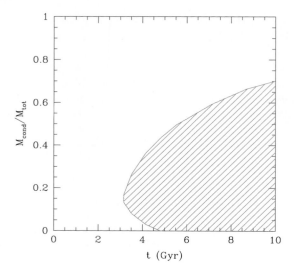

Fig. 20. Evolution of the conductive core Mcond/Mtot (*shaded area*) as a function of time for a $0.06\,M_\odot$ brown dwarf [From Chabrier et al. (2000b)]

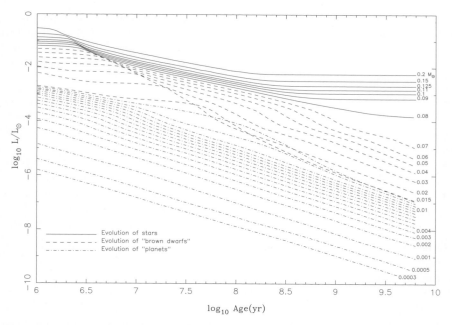

Fig. 21. Evolution of the luminosity (in $L_\odot$) of solar-metallicity M dwarfs and substellar objects vs. time (in yr) after formation. The stars, "brown dwarfs" and "planets" are shown as *solid*, *dashed*, and *dot-dashed curves*, respectively. In this figure, we arbitrarily designate as "brown dwarfs" those objects that burn deuterium, while we designate those that do not as "planets." The masses (in $M_\odot$) label most of the curves, with the lowest three corresponding to the mass of Saturn, half the mass of Jupiter, and the mass of Jupiter [From Burrows et al. (1997)]

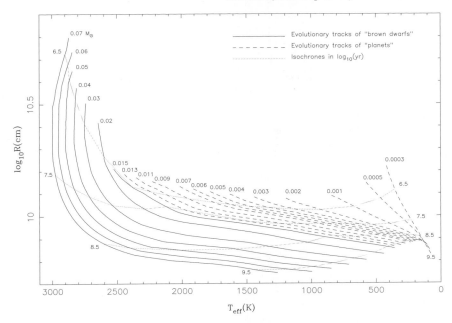

Fig. 22. log10 radius (in cm) vs. effective temperature (Teff, in K), with Teff decreasing to the right. The isochrones are the almost horizontal lines and are labeled in log10 yr. In all cases, the radius decreases with time. Initially, for the more massive brown dwarfs, the effective temperature is roughly constant, or slightly increasing, before decreasing inexorably at later times [From Burrows et al. (1997)]

planets contract inexorably. Note that the crossing of evolution lines in Fig. 21 is due to deuterium burning. It occurs later for brown dwarfs of small masses ($\sim 15\,M_J$), and those can hence be, for a small period of time, more luminous than slightly more massive brown dwarfs that have already consumed all their deuterium.

Figure 22 shows the relative constancy of the radius both as a function of time and mass, as well as the range of effective temperatures spanned by brown dwarfs and isolated giant planets. After 0.1 Gyr of evolution, it is found that all isolated brown dwarfs and giant planets should have a radius ranging between 10^{10} and 5×10^9 cm. For comparison, Jupiter's mean radius is 7×10^9 cm. The effective temperatures can range from about 3000 K for a young (~ 10 Myr) massive brown dwarf to only ~ 100 K for a 5 Gyr isolated Jupiter-mass planet.

Figure 23 is a theorist's H–R diagram for the "brown dwarfs" and giant "planets." The inset is a continuation of the figure down to low luminosities and T_{eff}s. The current Jupiter and Saturn are superposed for comparison (Pearl and Conrath 1991). Importantly, constant mass trajectories never

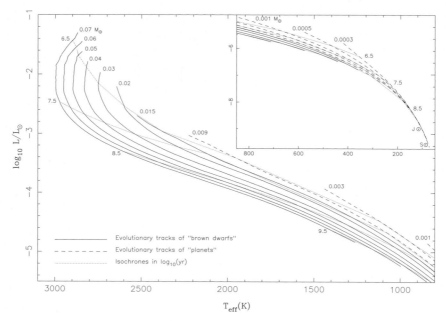

Fig. 23. H–R diagram: luminosity (in $L_\odot$) versus $T_{\rm eff}$ (in K) for various masses labeled on the figure in $M_\odot$. Due to the large range in luminosity and the near degeneracy of the tracks of substellar objects at late stages of evolution, it is not possible to represent with adequate detail the whole H–R diagram as one figure. Accordingly, the low-temperature and low-luminosity tail of the H–R diagram is shown in the inset. The observed positions of Jupiter and Saturn are labeled as points "J" and "S," respectively [From Burrows et al. (1997)]

cross and it is only for objects below $25\,M_{\rm J}$ that temperatures below 400 K are reached within 10^{10} years. All substellar objects decrease in luminosity monotonically, though during the early phases deuterium burning slows the evolution. As the "brown dwarfs" and "planets" cool to their cold radii, their tracks in the lower right of the H–R diagram correspond closely to curves of constant radius.

The consequences of a different helium to hydrogen ratio, of rotation or of the presence of a central dense core on the final luminosity and radius of 1 and $5\,M_{\rm J}$ planets are indicated by Saumon et al. (1996). They show that a 10% variation of Y generally translate into a 5% variation of R and L (Y and R being always anti-correlated, while Y and L are generally correlated). The presence of rapid rotation (with rotational speeds similar to that of Jupiter) can also significantly affect both the radius, increasing it by up to 20%, and the luminosity, which can be decreased by the same amount (but complex behavior can be found).

Deuterium, Lithium and Hydrogen Burning

The calculation of central temperatures shows at what masses and when various key-species are burned in brown dwarfs and stars. Because thermonuclear reactions are mostly temperature dependent, Fig. 24 shows as horizontal lines the burning temperatures of deuterium, lithium and hydrogen. It can be seen that objects of $0.012\,M_\odot$ ($\sim 12\,M_J$) fail to reach the deuterium-burning limit. Using this property to distinguish brown dwarfs and planets sets the realm of brown dwarfs beyond $13\,M_J$ (see Burrows et al. 1997; Chabrier et al. 2000a).

In objects of about $0.06\,M_\odot$, lithium starts burning. The signature of lithium in the spectrum of an object is thus an important sign to prove its substellar nature. There are various caveats however: at low temperatures, lithium pairs with hydrogen to form the mostly undetectable LiH (Lodders 1999). Furthermore, stars retain some of their primordial lithium for a few million years for the less massive of them. Therefore, lithium can be observed in young objects without these being brown dwarfs.

Hydrogen starts burning for masses higher than $0.7\,M_\odot$. This however depends on the metallicity of the object: this limit is valid for solar-metallicity,

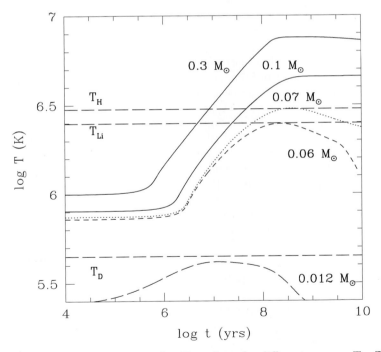

Fig. 24. Central temperature as a function of age for different masses. T_H, T_{Li} and T_D indicate the hydrogen, lithium and deuterium burning temperatures, respectively [From Chabrier and Baraffe 2000]

but goes up to $0.8\,M_\odot$ for objects with [Fe/H]= -2 (Chabrier and Baraffe 2000). This can be understood as follows. We have seen that the photospheric pressure is proportional to g/κ. Objects with a smaller metallicity, hence smaller atmospheric κ have a larger photospheric pressure, for a given gravity. They hence have a colder interior at a given pressure. One therefore needs to go to higher gravities, hence higher masses to reach a central temperature above the deuterium fusion point.

7.5 Extrasolar Giant Planets

The extrasolar planets that have been discovered so far by the radial velocimetry technique orbit all relatively close to a star; the amount of radiation that they receive has to be taken into account. I present here results obtained in the weak-irradiation approximation discussed in Sect. 7.2: these results are valid when most of the stellar flux is absorbed in a convective, adiabatic zone. We will see however that this is *not* true of strongly irradiated planets: in this case specific calculations have to be performed (Sect. 9)

Figure 25 gives examples of effective temperatures and radii predicted for some of the recently found extrasolar giant planets and brown dwarfs, assuming solar composition, a factor 2 uncertainty on the mass (due to the fact that radial velocity measurements only yield $M \sin i$, where i is the inclination of the orbital plane), and including uncertainties on the ages and albedo (between 0.1 and 0.5). It illustrates the diversity of planets detected so far. Because of the range of temperatures, many different condensates (from ammonia to silicates) are expected in planetary atmospheres. However, the calculated radii are always close to that of Jupiter, until the mass is large enough to sustain hydrogen fusion, at about $75\,M_J$. A local maximum of the radius at a mass of $\sim 4 M_J$ for isolated planets is due to the competition between additional volume and increased gravity. (This is because, when considering planets of larger masses, the degenerate metallic hydrogen region grows at the expense of the molecular region.) Planets that are significantly heated by their star have larger radii for smaller masses because their cooling is strongly suppressed. This case will be discussed in more detailed in Sect. 9, in connection with the constraints obtained for HD209458b.

8 Spectra and Atmospheres

8.1 Direct Observations of Substellar Objects

Gliese 229 B

Numerous observations of brown dwarfs are now available, but the first object whose substellar nature has been recognized as such beyond any doubt is Gliese 229 B. That object was discovered in 1995 (Nakajima et al. 1995;

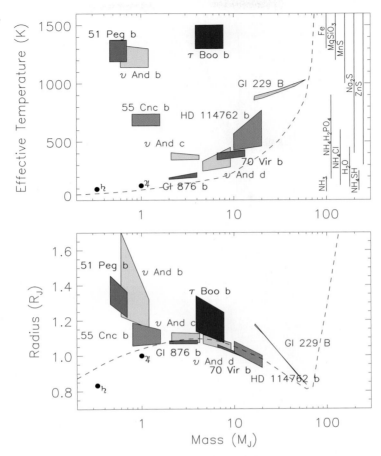

Fig. 25. Predicted effective temperatures and radii (in units of Jupiter radii, $R_J \sim 70,000$ km) of some extrasolar planets and brown dwarfs, including reasonable uncertainties for their mass, albedo and age (see text). The *dashed line* is for isolated H–He ($Y = 0.25$) objects after 10 Ga of evolution. The upper panel also shows potentially important chemical species expected to condense near the photosphere in the indicated range of effective temperatures [From Guillot (1999b)]

Oppenheimer et al. 1995), but it is still one of the most studied object of the field. It has the particularity of being companion to a cool M1-dwarf star, only 5.77 parsecs away. Its projected separation is only 45 AU, i.e. about the distance between Pluto and the Sun. (Note that its *real* mean orbital distance is still unknown.) One of the key features of the spectrum of Gl229B was the presence of methane absorption (Fig. 26). Because this molecule turns into CO at temperatures above 1000 to 1500 K for realistic photospheric pressures (see Sect. 8.2 hereafter), this implied that Gl229B was a genuine brown dwarf.

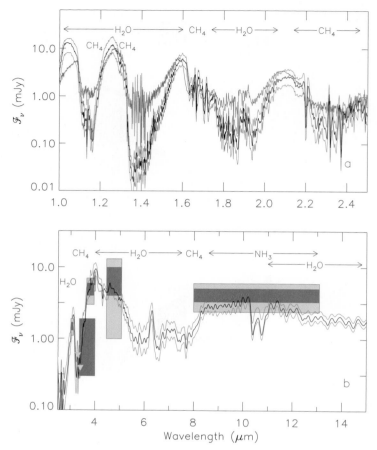

Fig. 26. Synthetic spectra for (*bottom to top*) $T_{\rm eff} = 890\,{\rm K}$, $960\,{\rm K}$, $1030\,{\rm K}$ and $g = 10^5\,{\rm cm\,s^{-2}}$ together with observed data from Geballe et al. (1996) (*top panel*) and photometric broadband measurements also from Geballe et al. (*bottom panel*). In both panels, spectral intervals are labeled with the molecules primarily responsible for the opacity in that interval [From Marley et al. 1996]

Two theoretical attempts to model the brown dwarf's spectrum were performed independently by Allard et al. (1996) and Marley et al. (1996). These works concluded to an effective temperature $T_{\rm eff} \sim 900 \pm 100\,{\rm K}$ and a gravity $\log g \sim 5 \pm 0.5$ with g in $\rm cm\,s^{-2}$. They identified several water and methane bands, and were able to correctly reproduce most of the spectrum. Another work from Tsuji et al. (1996), also coincided with the discovery. The comparison of this work with the observations showed that clouds were not present, in Gl229B, or that they were patchy. A few problems were however found at the time:

- The fit to the observed spectrum were relatively poor in the region of methane absorption. This problem is still present, in the lack of reliable opacity data for that molecule at temperatures of ~ 1000 K and more.
- The predicted fluxes were much too large, by several orders of magnitude, for wavelength $\lambda \lesssim 1\,\mu\text{m}$. The works of Tsuji et al. and Allard et al. did not include the condensation and settling of the very efficient absorber TiO: their fluxes were coincidentally similar to the observed ones, but for the wrong reasons. None of the distinctive features of TiO absorption have indeed been found in Gl229B's spectrum (Oppenheimer et al. 1998). Marley et al. used an arbitrary cutoff of the absorption for $\lambda < 1\,\mu\text{m}$. It was later advocated that photochemical processes due to the irradiation of the weak Gl229A could create large-enough particles to explain that absorption but no component could be found to provide the right slope (Griffith et al. 1998). The problem was later shown quite convincingly to be due to the absorption of alkali metals, and in particular to the potassium doublet (Burrows et al. 2000a). Although this theory depends on an *ad hoc* parameter, the slope of the non-lorentzian line profile (see Sect. 5.2), it has been shown to properly reproduce the spectrum of another cool brown dwarf (Liebert et al. 2000).
- The theoretical spectra predict absorption minima that are much deeper (by one to two orders of magnitude) than the observed one. Although one possible explanation could be the contamination of the observations by light scattered from Gl229A, a careful analysis of the measured spectrum (e.g. Oppenheimer et al. 1998) shows that this is unlikely. No convincing explanation has yet been proposed to explain the discrepancy.

The New Spectral Classes

As discussed in the introduction, the technological progresses made possible the detection of hundreds of brown dwarfs in only a few years. This led to the definition of a new spectral type, the first one in more than a hundred years. This spectral type, "L" was first proposed by Martín et al. (1997), and later worked out by Kirkpatrick et al. (1999) and Martín et al. (1999). I will not discuss in detail the classification itself (or rather the classifications, there being some divergences between the two groups). In a nutshell, M-type stars are identified by distinctive signatures of molecules, especially TiO and VO. The spectra of L-type objects see the progressive disappearing of TiO and VO lines and the advent of K, H_2O, Cs...etc. Cooler than the L-dwarfs, one then finds the T-dwarfs, whose detailed classification scheme has obviously to be worked out, but which are characterized by the presence of CH_4 absorption. Finally, it has been proposed that even cooler objects similar to Jupiter and Saturn (who mainly show features of CH_4 and NH_3) be termed "Y-dwarfs" (see Basri 2000 for a review).

The correspondence between hydrogen-burning and spectral-type is not a simple one because it depends on factors such as the gravity and metallicity of the object. For solar composition, the limit between stars and brown dwarfs is at M10 around 1 Gyr, M7 around 100 Myr, and M6 around 10 Myr.

The Colors of Brown Dwarfs

The abundance of brown dwarfs now discovered makes us almost forget how faint these objects are, and how difficult it is to find them. They are indeed 10,000 to 100,000 times fainter than our Sun. Since high-resolution spectra imply long exposures, programs aiming at the detection of brown dwarfs have relied on color information, i.e. images taken with several broadband filters that combined together provide information on the effective temperature of the targets. Because of the coolness of the sources, brown-dwarf surveys (e.g. DENIS, 2MASS) have generally been done using infrared I, J and K bands (see Basri 2000).

It is a well known property of blackbodies that they get redder and redder as one looks at objects that are cooler and cooler: the peak of the Planck function is then displaced to longer wavelengths. This implies that, e.g. relative to the K band, less flux is emitted in the J band. Because astronomers are nostalgic of outdated conventions, this implies an increasing J–K value (J and K being the magnitudes in the J and K bands, respectively).

Figure 27 shows a color-magnitude diagram in which theoretical calculations are compared to observations. Main sequence stars are easily spotted by their low, relatively uniform J–K value. This is due to the fact that at high temperatures, the J and K bands are in the Wien tail of the Planck function and their difference is independent of temperature. Objects of smaller absolute magnitude M_K then progressively move to the red-part (right) of the diagram. This tendency is well reproduced by a "dusty" atmospheric model, i.e. one in which all the condensed particles are assumed to remain in the atmosphere. The effect of the presence of dust is effectively to reduce spectral variations so that the spectrum is more similar to that of a black body (Allard et al. 2001).

However, at still lower temperatures, Gl229B sticks clearly out of this tendency, and is in fact significantly bluer in J and K than main sequence stars! Furthermore, it is not an isolated case: several other cool brown dwarfs have now been detected to have the same characteristics (e.g. Burgasser et al. 1999; Strauss et al. 1999). As we will see, this rapid variation in color is indicative of a transition from "dusty" to "clear" atmospheres, probably sharpened by the additional cooling provided by the apparition of methane at low temperatures.

Detection of Very Young Substellar Objects

Brown dwarfs can be discovered in the field, as for the DENIS and 2MASS surveys. They can also be discovered in known star-forming regions: because they are young, they can be considerably hotter than the average field objects.

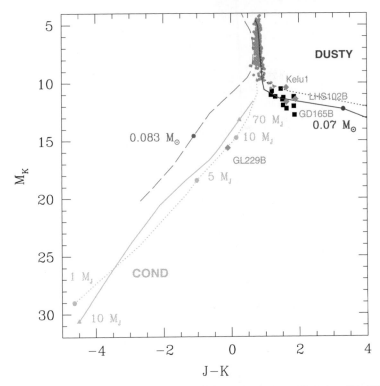

Fig. 27. M_K vs (J–K) diagram for different ages and metallicities: [Fe/H]=0 for 10^8 yr (*dotted lines*) and 5×10^9 yr (*solid lines*); [Fe/H]= −2, t=10 Gyr (*dashed line*). The curves on the upper right correspond to the DUSTY models for [Fe/H]=0. The curves on the down left correspond to the COND models for [Fe/H]=0. Filled circles and triangles on the isochrones indicate masses either in $M_\odot$ or M_J (1 $M_J \approx 10^{-3}$ $M_\odot$) [From Chabrier and Baraffe 2000]

However, most of these regions are relatively far away (100's of parsecs), and extinction then becomes a problem.

A very interesting region is the σ−Orionis cluster, which is only a few Myr old. Very faint, low-mass objects have been successfully discovered (see Martín et al. 2001 and references therein). These authors have shown that a continuous sequence of brown dwarfs is present, down to very low masses (perhaps ∼8 M_J). Similar results have been obtained for other clusters, were numerous brown dwarfs were detected down to and beyond the deuterium burning limit. Mass functions down to these low masses have been derived for the IC348 cluster (Najita, Tiede, and Carr 2000) and the Trapezium (Hillenbrand and Carpenter 2000; Luhman et al. 2000). The mass functions thus obtained are extremely interesting as they bear on formation theories, but will not be discussed in this course.

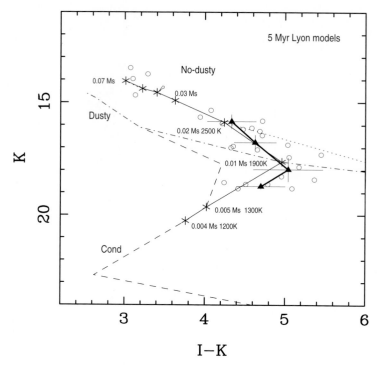

Fig. 28. K vs $I-K$ color-magnitude diagram. *Empty circles* are substellar members in σ Orionis. The *thick solid line* and *filled triangles* represent the mean locus of the σ Orionis objects. The 5 Myr isochrones from Chabrier et al. (2000b) are displayed (Nextgen no-dusty models—*dotted line*, Dusty models—*dot-dashed line*, and Cond models—*dashed line*). The *thin solid line* with asterisks is a best-guess isochrone that combines the models [From Martín et al. 2001]

Figure 28 shows the observations of Martín et al. (2001) in σ–Orionis as a color-magnitude diagram. The theoretical interpretation using the Lyon model is shown by the dotted, dash-dotted and dashed lines. (The models computed by the Lyon and Tucson groups yield very similar results). As previously, a tendency arise with hotter, more massive objects being closer to a theoretical sequence that accounts for the presence of dust in the atmosphere, and one assuming a clear atmosphere for cooler objects. We will come back to that transition in the following sections. The very low masses obtained for the fainter objects are interesting and one has to consider the uncertainties that are associated with them.

Because the age of the cluster is not precisely known (note that all stars do not form *exactly* at the same time), the intrinsic uncertainty of 1–5 Myr leads to an uncertainty on the predicted masses. It is interesting to see however that the uncertainty of the models themselves, as computed by two groups and with several assumptions relative to atmospheric composition give strikingly similar results.

First, the age of the cluster is not known. In the case of σ Orionis, it is 1–5 Myr. This translates into a factor ~ 2 uncertainty on the mass, the young ages corresponding to the lowest values of the masses. Another important source of uncertainty is due to the relative closeness to unknown initial conditions. If in the process of formation the proto-brown dwarfs are able to loose some of their entropy, their interior entropy will be lower than the external one, until cooling (possibly over millions of years) readjusts the profile to a uniform one. For some time the brown dwarf or planet will therefore have a cooler interior, for the same external boundary conditions. (See the entropy profiles and discussion in Sect. 7.2). The objects could then be found for some time to the left of the Hayashi track (see Fig. 23), implying for already compact objects that they could be interpreted to be less massive than they really are. However, the magnitude of this effect has to be investigated.

Rotation, Magnetic Activity and Variability

Rotation is a particularly important astrophysical parameter to understand formation processes. In the case of stars, most of the angular momentum present in the molecular cloud had to be lost. Two physical processes can be advocated: the formation of a circumstellar disk and angular momentum transport within that disk, and magnetic braking. The situation is similar for substellar objects.

The rotation rates of brown dwarfs can be inferred through radial-velocity measurements. Contrary to the case of the detection of companions however, one then seeks the intrinsic line broadening due to the rotation. The measured parameter is $v \sin i$, the rotation speed at the equator multiplied by the sine of the inclination. Observations of type M (see e.g. Basri 2000 for a review) indicates that objects all down to $\sim$M10 have very widespread values of $v \sin i$, and show intense chromospheric heating, as characterized by their Hα emission. On the other hands, the observed L-type objects are all very fast rotators ($v \sin i = 20$ to $80 \,\mathrm{km\,s^{-1}}$), and have a weak Hα emission. This probably indicates that low-mass objects have a weaker magnetic braking. Several interpretations are possible (see Basri 2000).

Photometric variations of up to 5% in fluxes have also been detected in M and L dwarfs (e.g. Bailer-Jones and Mundt 2001). In a few cases, these variations are periodic with a period comparable to that of the dwarfs' rotation. They can then be attributed to surface features. Non-periodic variability is also observed, indicating a more complex, time-variable activity. Interestingly, a greater occurrence of variability is found in objects later than M9 indicating that it is not correlated with chromospheric activity. We will see hereafter that in atmospheres at low temperatures (corresponding to dwarfs of later types), condensation sets in. The observed variability could thus well be due to the presence of non-homogeneously distributed, time-variable, photospheric dust clouds.

8.2 Atmospheric Models: Importance of Condensation

Modeling the atmospheres of substellar objects is a complex subject that would require several courses. I will focus on a problem which is particular to cool atmospheres: that of condensation. Although condensation has long ago been recognized as an important astrophysical phenomenon, it has been mostly developed in a low- or no-gravity framework. In planetary and substellar atmospheres, condensates are expected to be formed, transported, and vaporized continuously, as is the case on Earth. Their study consequently requires new tools, with a particular emphasis on problems related to the transport of material in these atmospheres.

Basics of Condensation

Let us consider the equilibrium between a condensed phase and a vapor phase of a given chemical species. The thermodynamical condition for equilibrium is that the pressures, temperatures and Gibbs free energies of the two phases should be equal. The last condition implies that:

$$v^{(v)} dP_s - S^{(v)} dT = v^{(c)} dP_s - S^{(c)} dT , \qquad (134)$$

where the (v) and (c) superscripts indicate the vapor and condensed phases, respectively. v and S are the volume and entropy per unit mass. Equation (134) implicitly neglects any surface tension that would appear on a finite size drop formed of condensed material. It is hence valid for equilibrium of the vapor over an infinitely long surface of condensed material. Strictly, the formation of droplets will involve a slightly larger saturation pressure, but this effect will be neglected.

The latent heat is defined as the difference in enthalpies of the two species, hence

$$L = (S^{(v)} - S^{(c)})T . \qquad (135)$$

One therefore obtains from (134) the *Clausius–Clapeyron* equation:

$$\frac{dP_s}{dT} = \frac{L}{T(v^{(v)} - v^{(c)})} . \qquad (136)$$

Using the perfect gas equation and neglecting the specific volume of the condensed phase over that of the vapor leads to the following equation:

$$\frac{d \ln P_s}{d \ln T} = \frac{L}{kT} . \qquad (137)$$

This equation allows one, from a known condensation temperature at a given pressure to derive the condensation temperatures at any other pressure. Note that P_s is the *saturation pressure*, i.e. the pressure corresponding to an equilibrium between the condensate and vapor phases.

The saturation pressure obtained from (137) will be useful in the following for a derivation of the composition gradient, but is only correct to first order. In the case of giant planets and brown dwarfs, the following relations have an accuracy of order 10%:

$$H_2O: \quad \log_{10} p_s = 5.0587 - \frac{1630.91}{T - 50.396}, \quad (138)$$

$$CH_4: \quad \log_{10} p_s = 4.3180 - \frac{451.64}{T - 4.66} \quad \text{if } T \leq 91\,\text{K}, \quad (139)$$

$$\log_{10} p_s = 3.8205 - \frac{405.42}{T - 5.37} \quad \text{if } T \geq 91\,\text{K}, \quad (140)$$

$$NH_3: \quad \log_{10} p_s = 7.0887 - \frac{1617.91}{T - 0.60}, \quad (141)$$

$$MgSiO_3: \quad p_s = \exp\left(-\frac{58663}{T} + 25.37\right). \quad (142)$$

The pressures are expressed in bars and the temperatures in kelvins. In the case of water, the following approximation for the equilibrium of the vapor with liquid water and ice, respectively, give still better accuracies (better than 0.3%) in the temperature interval $-30°C \leq T \leq 40°C$ for liquid water, and $-80°C \leq T \leq 0°C$ for ice:

$$\text{liquid } H_2O: \quad \ln p_s = 46.77181 - \frac{6743.769}{T} - 4.8451 \ln T, \quad (143)$$

$$\text{ice } H_2O: \quad \ln p_s = 16.42311 - \frac{6111.72784}{T} + 0.15215 \ln T, \quad (144)$$

where pressures are still in bars and temperatures in kelvin.

Abundance of Condensing Species in an Atmosphere

Let us consider an atmosphere in which the condensing species is not the dominant one. The saturation abundance of the condensing species is determined by the ratio of the saturation partial pressure to the total pressure: $x_s = P_s/P$. We introduce the following adimensional quantity:

$$\beta = \frac{L}{kT}. \quad (145)$$

For most species of interest in substellar atmospheres, $\beta \approx 10 - 20$. Assuming that β is constant, one can derive the compositional gradient in the atmosphere:

$$\frac{d \ln x_s}{d \ln P} = \beta \nabla_T - 1. \quad (146)$$

One can thus see that in most cases (e.g. convective tropospheres), $\nabla_T \approx 0.3$ and $d\ln x_{\rm s}/d\ln P \approx 2-5$. The abundance at saturation is decreasing with altitude faster than the pressure itself. To the contrary, in nearly isothermal regions, and in stratospheres ($\nabla_T < 0$), the abundance at saturation increases with altitude.

Possible abundance profiles are depicted in Fig. 29. In all cases the abundance x has its maximal value (the bulk abundance) and is constant at large depths. However, the composition at upper levels strongly depends on physical mechanisms. In one extreme case (labeled (a) in Fig. 29), solid species are immediately removed by gravity and atmospheric circulation is not fast enough to oppose the effect of upward diffusion of the condensing species. The other extreme (b) corresponds to a situation in which solids are transported by convection so rapidly throughout the atmosphere that they can never grow to a size at which they would fall. The total (vapor + solid) abundance is then constant.

In reality, a third situation (c) is more likely. In the presence of advection and sedimentation of part of the condensed material, downward motions will tend to produce an undersaturated mixture, while upward motions will lead to the formation of clouds. This necessarily leads to a non-homogeneous

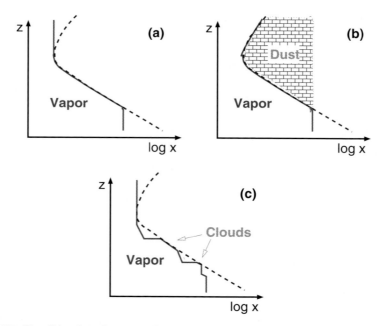

Fig. 29. Possible abundance profiles x of a condensing species in a substellar or planetary atmosphere (*plain lines*). The *dashed lines* corresponds to the saturation profiles. Three cases are: (**a**) clear atmosphere (immediate "rainout" of condensates); (**b**) dusty atmosphere (no "rainout"); (**c**) a possibly more realistic situation (see text)

atmosphere, where clouds of the same condensing species appear at various altitudes and do not cover the entire atmosphere. This patchiness is observed in the four giant planets, but is maximal when clouds occur in a convective and not radiative region.

Note that downward motions can lead to a lower-than expected abundance at great depths. This has been observed in Jupiter by the Galileo probe for NH_3, H_2S and H_2O (Niemann et al. 1998), but is also evident in Voyager 2 radio-occultation data for Neptune, in the case of CH_4 (Lindal 1992; see also Guillot 1995).

It is important to stress that depending on the condensation model (a, b or c), the chemistry of the atmosphere will be very different. In model (b), condensed particles are still present in the atmosphere and can react with other species. In model (a) they are completely removed. Therefore, model (a) cannot be consistently calculated by simply using model (b) for the chemistry and removing the opacity arising from condensed particles. In both models (a) and (c), the chemical equilibrium must be consistently calculated, accounting for the full or partial removal of the elements that have condensed.

An important example is titanium: this atom is expected to form $CaTiO_3$ and thus become solid at temperatures $\sim$2000 K. However, chemical equilibrium calculations predict that the abundance of the solid is very small, and that most of the titanium is in form of the strong absorber TiO . If solid $CaTiO_3$ particles are kept in the atmosphere and allowed to react with the environment, TiO will remain as the most important absorber. It is however not observed in Jupiter and Saturn, a sign that condensation, grain growth and subsequent sedimentation have occurred and removed Ti from the upper levels (see e.g. Fegley and Lodders 1994, 1996). This is also observed in the case of Gl229B, which shows no sign of TiO absorption (Oppenheimer et al. 1998; Marley et al. 1996).

This problem would therefore require to consistently calculate atmospheric models using a microphysical description of the clouds and including a description of vertical mixing and a fully self-consistent chemical equilibrium model.

Temperature Profiles

An important consequence of condensation is to modify heat transport by providing latent heat. The adiabatic temperature gradient is thus modified. Neglecting the heat capacity of condensed species (or equivalently, assuming any condensed material to be left behind during an ascending motion), one can derive the *moist pseudo-adiabatic* temperature gradient. Using the formulation of Emanuel (1994), but simplified notations:

$$\nabla_{\text{pseudo}} = \nabla_{\text{ad}} \left[\frac{1 + \beta \frac{f}{\epsilon}}{1 + \frac{\beta^2}{\bar{c}_P} \frac{f}{\epsilon}} \right], \qquad (147)$$

where ∇_{ad} is the adiabatic gradient when neglecting latent heat effects (*dry adiabatic gradient*), $\epsilon = m_v/m_d$ is the ratio between the molecular mass of the condensable species over that of dry air, $f = \rho_v/\rho_d = \epsilon x/(1-x)$ is the mass mixing ratio of the vapor over dry air, and $\tilde{c}_P$ is the adimensional mean specific heat per molecule (including dry air and vapor).

In the cases that are of interest to us, $f/\epsilon \ll 1$ so that

$$\nabla_{\text{pseudo}} \approx \nabla_{\text{ad}} \left[1 - \left(\frac{\beta}{\tilde{c}_P} - 1\right)\beta\frac{f}{\epsilon}\right]. \qquad (148)$$

Because $\beta \approx 10-20$ and $\tilde{c}_P \approx 3.5$, we obtain that $f/\epsilon \approx x \gtrsim 2\times 10^{-4}$ in order to change the adiabatic gradient by $\sim 1\%$ or more. In objects of approximately solar composition, the only potentially condensable species that are abundant enough are H_2O and CH_4. Note that NH_3 can induce a change of $\sim 0.6\%$ and compounds formed from Mg, Si and Fe a change of $\sim 0.2\%$.

The phenomenon of *moist convection*, i.e. of convection powered by latent heat release, such as that observed in cumulus clouds on Earth is therefore likely to be limited to atmospheres in which water and methane can condense, i.e. to relatively cold atmospheres. In the case of condensation of more refractory species, the limited effect of latent heat release on the temperature gradient is likely to be outweighted by the strongly inhibiting condensate loading.

Another consequence of condensation in substellar hydrogen atmospheres is that it yields a *stable* molecular weight gradient. Assuming that the atmosphere is saturated and using (146),

$$\nabla_\mu = \varpi f(\beta \nabla_T - 1), \qquad (149)$$

where $\varpi = (1 - 1/\epsilon)/(1+f)$. In the case of hydrogen atmospheres, $\varpi \sim 1$, $\beta \sim 20$, $\nabla_T \sim 0.3$ so that $\nabla_\mu > 0$. In the case of the Earth, $\varpi \sim -0.5$ due to the smaller weight of the water molecule than of N_2: in that case, moist air tends to rise, thereby favoring the occurrence of moist convection.

In the ideal case of a saturated atmosphere in which condensed species are removed instantaneously by gravity, the criterion for convective is slightly modified compared to (79). Because of condensation occurring both in the environment and the upwelling parcel, the local criterion becomes (Guillot 1995):

$$(1 - \varpi\beta f)(\nabla_T - \nabla_{\text{pseudo}}) > 0. \qquad (150)$$

Convection is thus inhibited when the abundance of the condensable is such that

$$f > (\varpi\beta)^{-1}. \qquad (151)$$

Physically, this condition results from the fact that the abundance of condensable species drops faster in the environment than in the rising parcel. In spite of its higher temperature, the parcel thus becomes negatively buoyant. This occurs however only for condensing species whose mass mixing ratio can raise above ~ 0.03, i.e. enrichment over the solar value of ~ 5 for H_2O, ~ 15 for CH_4 and $\sim 10-20$ in silicates and iron. This is potentially interesting in the case of water condensation in Jupiter and Saturn and of methane condensation in Uranus and Neptune. In the case of the two latter planets, it could explain the superadiabatic gradients obtained from the Voyager radio-occultation in the region of methane condensation (Guillot 1995).

For objects of solar composition however, the molecular weight gradient effect is limited, i.e. $\nabla_\mu \lesssim 10^{-2}$ as long as water condensation is not involved. To first order, the effect of condensation on the temperature gradient can be neglected.

Figure 30 shows several atmospheric temperature profiles calculated by Burrows et al. (1997) for isolated substellar objects. The figure shows as dashed lines the limits for condensation of water, ammonia, MgSiO$_3$ and iron, assuming solar composition. The lines where CO and CH$_4$ and where N$_2$ and NH$_3$ have the same abundances are also indicated. As can be intuited, depending on the effective temperatures and gravity, various elements are expected to condense near the photospheres of substellar objects having great effects on their spectra: Particularly important are the condensation of methane, ammonia and water for low effective temperatures, and iron and silicates in relatively warm atmospheres. Note that these are only the most abundant species to be formed: other potentially condensing species include numerous sulfides and chlorides (e.g. K$_2$S, Na$_2$S...etc.).

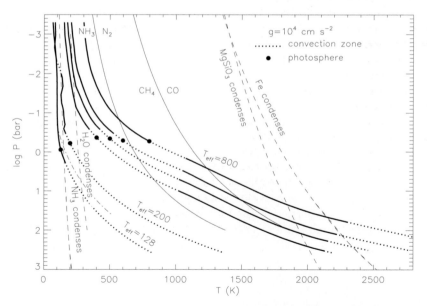

Fig. 30. Atmospheric pressure–temperature profiles for non-irradiated substellar atmospheres with surface gravity fixed at 10^4 cm s^{-2} and $T_{\rm eff} = 800$, 600, 500, 400, 200, and 128 K. Note that the inner radiative zones disappear in the presence of alkali metals (not accounted for in the calculation) [From Burrows et al. (1997)]

The temperatures profiles of Fig. 30 do not account for the presence of any clouds. When present, these would lead to a significant warming of the atmosphere (see Allard et al. 2001). Note that the inner radiative zones shown in Fig. 30 disappear when alkali metals are included in the calculation.

Dust and Clouds: Relevant Physical Processes

The formation of dust and clouds in substellar atmospheres is governed by several processes:

1. *Chemistry*: given the composition, temperatures and pressures of a substellar atmosphere, a set of chemical species is predicted to undergo vapor/liquid or vapor/solid phase changes. The deepest level at which that occurs is the condensation level, sometime improperly called cloud base (as we can see from model (c) in Fig. 29, clouds do not necessarily originate from that level). Among the different kinds of condensation, one might distinguish the condensation of a minor species, as it involves a chemical reaction (e.g. $CaTiO_3$ with TiO remaining in gaseous form), and the condensation of a major species (e.g. H_2O).
2. *Grain growth*: this groups all the mechanisms that affect the size of the condensed grains or droplets. Those include condensation (vapor molecules/atoms sticking upon an already condensed site), coagulation (due to Brownian motion) and coalescence (merging of big droplets with slightly different vertical velocities). It also includes evaporation which occurs in an undersaturated environment (i.e. when the partial pressure of the condensing species is smaller than its saturation pressure).
3. *Sedimentation*: condensed particles are affected by gravity forces, the more massive ones falling more rapidly than the lighter ones.
4. *Mixing*: the advection of saturated/undersaturated gas and small particles due to various effects (convective instability, meridional circulation, waves...etc.) inevitably influences grain growth and sedimentation. As we have seen, in the case of water and methane, this is complicated by the significant latent heat effect that tends to favor updrafts in which condensation occurs.
5. *Radiative heating/cooling*: the presence of solid/liquid particles modifies the radiative properties of the medium, which can in turn affect mixing (by creating small-scale or large-scale instabilities) and condensation (by modifying the temperature profiles).

The processes of grain growth and sedimentation can be approximated using the timescales provided by Rossow (1978). These estimates are applied to the case of an iron cloud in a typical 2000 K brown dwarf and shown in Fig. 31 (see also Lunine et al. 1989). Grain growth is dominated by condensation for sizes larger than several microns. However, before they reach those sizes, they are expected to be removed efficiently by sedimentation. Let us define a time

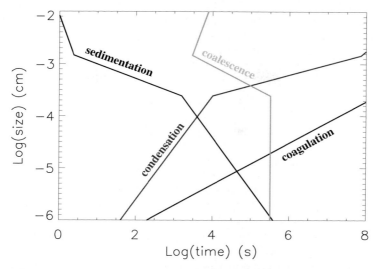

Fig. 31. Time scales for grain growth and sedimentation at the basis of an "iron-cloud" in a brown dwarf of $T_{\rm eff} = 2000\,\rm K$, $g = 3 \times 10^5\,\rm cm\,s^{-2}$ [Brown dwarf model: courtesy of F. Allard; timescales from Rossow 1978]

$\tau^\star$ and size $a^\star$ as satisfying the following condition:

$$\tau_{\rm sedimentation}(a^\star) = \tau_{\rm growth}(a^\star) \equiv \tau^\star. \qquad (152)$$

These two quantities are defined by the point at which the sedimentation and condensation lines cross in Fig. 31.

The amount of condensed particles and their sizes will then depend on the mixing of vapor and small particles from levels above and below the one considered. In terrestrial clouds, this is modeled through the solution of a complex set of differential equations (see e.g. Cotton and Anthes 1989). In giant planets and substellar objects this problem is far from being well understood, and a much simpler approach is generally sufficient. One thus generally defines a turbulent diffusivity as relevant of mixing processes in the atmosphere. In our very simple case, we will define $\tau_{\rm mix}$ as the characteristic time scale for mixing over one pressure scale height H_P.

Two cases occur:

1. $\tau_{\rm mix} \lesssim \tau^\star$: This is the case of efficient mixing (i.e. case (b) in Fig. 29). Fresh particles and vapor are constantly supplied by the mixing before they can fall under the action of gravity. In the limit when mixing does not affect the largest grains, the final mean size is expected to be of the order of $a^\star$. If the largest grains are also transported, then the final mean radius will depend mostly on the time during which they remain above the condensation level. The situation can become complex: in the Earth's

clouds, a bi-modal distribution of water droplets is generally observed (e.g. Cotton and Anthes 1989).

2. $\tau_{\rm mix} \gtrsim \tau^\star$: Here, the relatively sluggish mixing prevents the formation of particles as large as $a^\star$ because they are removed by sedimentation. The mean size is expected to be largely set by the equilibrium between mixing and sedimentation, i.e. $\tau_{\rm mix} = \tau_{\rm sedimentation}$. This yields of course particles of sizes smaller than $a^\star$. It also yields a lower abundance of particles (i.e. case (a) or (c) in Fig. 29).

A Transition from Dusty to Clear Atmospheres

As shown in Fig. 27, there is a transition from high effective temperature, red, and dusty brown dwarfs, to brown dwarfs of lower effective temperatures that are much bluer and appear to have clear atmospheres (or more accurately, to deviate from case (b) of Fig. 29). This transition is thought to be due to the sedimentation of dust in brown dwarfs of low temperatures and also to the additional cooling due to the CO to CH_4 transition.

Let us focus first on the high temperature, dusty brown dwarfs. In the case of our $T_{\rm eff} = 2000\,\rm K$ brown dwarf, one finds that $a^\star \approx 1\,\mu\rm m$, and $\tau^\star \approx 3 \times 10^3\,\rm s$. The time scale for mixing in the convective zone can be estimated from $\tau_{\rm mix} \approx H_P/v$ and v is the convective velocity from (94). At the basis of the iron cloud, we have $T \approx 1700\,\rm K$ and $P \approx 0.4\,\rm bar$. The convective velocity would then be $v \approx 2 \times 10^4\,\rm cm\,s^{-1}$, $H_P \approx 2 \times 10^5\,\rm cm$ and hence $\tau_{\rm mix} \approx 10\,\rm s$. However, convection is found to start at much deeper levels, i.e. around 10 bar (F. Allard, pers. communication). The good results obtained from stellar models using an *ad hoc* interstellar grain size distribution (Allard et al. 2001) indicates that the mean particle size is indeed probably between 0.1 and $1\,\mu\rm m$. Therefore, a mixing process is needed to explain the presence of these particles in the brown dwarfs upper photosphere. Several possibilities exist, and the fact that these objects are generally fast rotators (e.g. Basri 2000) is interesting because it could yield enough meridional circulation to provide the right amount of mixing.

The transition to brown dwarfs of lower temperatures ("T-dwarfs") is still unclear. A possible model based on the timescales discussed here is provided by Ackerman and Marley (2001), and with a free parameter reproduces relatively well the observations. However, the model assumes an eddy mixing time scale that even in radiative regions is arbitrarily large. One may therefore wonder whether the real problem to solve may instead be "why are grains present in brown dwarfs of high effective temperatures?"

Observational constraints on the amount of atmospheric mixing exist at least for one well-studied brown dwarf: Gl229B. The detection of chemical species that are out of thermochemical equilibrium informs us on how fast these species are transported throughout the atmosphere. This is in particular the case of CO which partially escapes a transformation into CH_4 as it is transported upward in Gl229B's atmosphere (roughly from levels of $\sim 10\,\rm bar$

to ~1 bar where it is detected). Griffith and Yelle (1999) estimate that the diffusion coefficient of mixing is $K_{\rm mix} \sim 3 \times 10^2$ to $10^4\,{\rm cm}^2\,{\rm s}{-}1$. The mixing time scale is (very inaccurate!) $\tau_{\rm mix} \approx H_P^2/K_{\rm mix} \approx 1.5 \times 10^6$ to 5×10^9 s. This is to be compared to the mixing times of 10–10^3 s required to keep grains up in the atmosphere.

Dust and Variability

The presence of dust opens new possibilities for atmospheric variability. As discussed previously, moist convection is not a likely possibility in the case of brown dwarfs and hot giant planets. The situation is therefore different than for our giant planets. However, variability is linked to the spatial heterogeneities. The fact that in Jupiter, small regions of the planet can emit much more than others because of a lack of clouds there has to be kept in mind.

One possibility for the presence of Jupiter's hot spots is the presence of a planetary wave (Showman and Dowling 2000). In the same frame of mind, waves could well affect the distribution of dust in the atmosphere of brown dwarfs. A potential interesting source of waves is again in the rapid rotation of these objects and the possibility of Rossby and Kelvin waves. Baroclinic instabilities linked to the rotation and the presence of meridional circulation are also a possibility. Finally, a coupling between dust formation and heat transfer may be envisioned: we have seen that the presence of dust indeed greatly increases the opacity.

9 Pegasi Planets ("51 Peg b-like" Planets)

9.1 Introduction

The detection of planetary-mass companions in small orbits around solar-type stars has been a major discovery of the past decade. More than 100 extrasolar giant planets (with masses $M \sin i < 13\,{\rm M_J}$, i being the inclination of the system) have been detected by radial velocimetry. A significant fraction have distances less than 0.1 AU. This is for example the case with the first extrasolar giant planet to have been discovered, 51 Peg b (Mayor and Queloz 1995). These close-in planets form a statistically distinct population: all planets with semi-major axis smaller than 0.06 AU have near-circular orbits while the mean eccentricity of the global population is $<e> \approx 0.27$. This is explained by the circularization by tides raised on the star by the planet (Marcy et al. 1997). One exception to this rule, HD83443b ($e = 0.079 \pm 0.033$), can be attributed to the presence of another eccentric planet in the system (Mayor et al. 2001). As we shall see, the planets inside ~0.1 AU also have very specific properties due to the closeness to their star and the intense radiation they receive. For this reason, following astronomical conventions, I choose to name them after the first object of this class to have been discovered: "51 Peg b-like" planets, or in short "Pegasi planets".

Such planets provide an unprecedented opportunity to study how intense stellar irradiation affects the evolution and atmospheric circulation of a giant planet. Roughly 1% of stars surveyed so far bear Pegasi planets in orbit, suggesting that they are not a rare phenomenon. Their proximity to their stars increases the likelihood that they will transit their stars as viewed from Earth, allowing a precise determination of their radii. (The probability varies inversely with the planet's orbital radius, reaching ~10% for a planet at 0.05 AU around a solar-type star.) One planet, HD209458b, has already been observed to transit its star every 3.524 days (Charbonneau et al. 2000; Henry et al. 2000). The object's mass is $0.69\pm0.05\,\mathrm{M_J}$. Hubble Space Telescope measurements of the transit (Brown et al. 2001) imply that the planet's radius is $96300\pm4000\,\mathrm{km}$ (see light curve on Fig. 32). An analysis of the lightcurve combined with atmospheric models shows that this should correspond to a radius of 94430 km at the 1 bar level (Hubbard et al. 2001). This last estimate corresponds to $1.349\,\overline{\mathrm{R}}_\mathrm{J}$, where $\overline{\mathrm{R}}_\mathrm{J} \equiv 70,000\,\mathrm{km}$ is a characteristic radius of Jupiter. This large radius, in fair agreement with theoretical predictions (Guillot et al. 1996), shows unambiguously that HD209458b is a gas giant.

One expects that the evolution of Pegasi planets depends more on the stellar irradiation than is the case with Jupiter. HD209458b and other Pegasi planets differ qualitatively from Jupiter because the globally-averaged stellar flux they absorb is $\sim 10^8\,\mathrm{erg\,cm^{-2}}$ $(10^5\,\mathrm{W\,m^{-2}})$, which is $\sim 10^4$ times greater than the predicted intrinsic flux of $\sim 10^4\,\mathrm{erg\,cm^{-2}}$. (In contrast, Jupiter's absorbed and intrinsic fluxes are the same within a factor of two.)

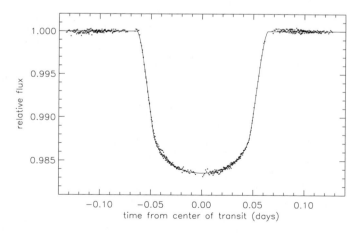

Fig. 32. Phased light curve of four planetary transits across the star HD209458 observed with the HST. The orbital period of the planet is 3.52474 days. [Figure from Brown et al. (2001)]

9.2 Evolution of Strongly Irradiated Giant Planets Including Stellar Heating: Definitions

As discussed in Sect. 7.2, the total luminosity of an irradiated planet or brown dwarf consists of three parts: (i) the part of the stellar flux which is directly reflected and does not contribute to the atmosphere's energy budget; (ii) $L_\circleddash$, the part which is absorbed, thermalized and reemitted (we assume that no seasonal effects take place and that the system is in equilibrium); (iii) L, the *intrinsic* luminosity due to the object's contraction and cooling.

First, the effective temperature of any irradiated planet is defined by:

$$4\pi R^2 \sigma T_{\text{eff}}^4 = L + L_\circleddash \tag{153}$$

and the equilibrium effective temperature $T_\circleddash$ by

$$4\pi R^2 \sigma T_\circleddash^4 = L_\circleddash . \tag{154}$$

$T_\circleddash$ is the effective temperature toward which the planet tends as it cools and $L \to 0$. It is a function of the Bond albedo A (i.e. the ratio of the luminosity directly reflected to the total luminosity intercepted by the planet):

$$T_\circleddash = T_\star \left(\frac{R_\star}{2D} \right)^{1/2} (1-A)^{1/4} , \tag{155}$$

where $T_\star$ and $R_\star$ are respectively the star's effective temperature and radius, and D is the star–planet distance.

It is important to stress that these definitions are valid independently of heat absorption and heat transport in the atmospheres of these objects. Most of the physics is hidden in the Bond albedo A. The values for our giant planets are listed in Table 3. They all lie between 0.29 (Neptune) and 0.35 (Jupiter). In the case of extrasolar giant planets, simulations indicate similar values of the albedo when alkali metals are not present (Marley et al. 1999), but very low values $A \sim 0.05$ when alkali metals contribute to the absorption in the optical (Sudarsky et al. 2000). This would imply that extrasolar planets are very difficult to detect in the optical, since they reflect little of the incoming flux. However, the albedo can be significantly modified by the presence of grains in the atmosphere (see Marley et al. 1999). Our understanding of grain and cloud formation being far from complete, these estimates of A have to be taken with caution.

Finally, in the case of Pegasi planets we will see that the stellar heat is absorbed very inhomogeneously and is not necessarily well redistributed over the entire atmosphere. This does not affect the above definitions, but it strongly modifies any calculation of the atmospheric structure. The bond albedo, and surface boundary conditions then have to account for this. I will focus on cases in which this effect is neglected. However, the effect of day/night temperatures variations both on the evolution and on chemistry in Pegasi-planet atmospheres will be discussed.

Temperature of Irradiated Atmospheres

Most of the course has been concerned with mostly-convective objects. We have seen in Sect. 7.2 that in this case, which corresponds to weak irradiation, a relatively trivial modification of the external boundary condition was sufficient to obtain a relatively good estimate of the evolution. We derived (115) an evolution time scale for weakly irradiated object that was equal to the time scale in isolation plus an expansion in powers of $(T_\mathrm{\odot}/T_\mathrm{eff})^4$. In the case of Pegasi planets however, the strong stellar irradiation leads to $T_\mathrm{eff} \approx T_\mathrm{\odot}$ (the absorbed stellar flux is typically about 10^4 times stronger than the intrinsic heat flux). With the assumptions of Sect. 7.2, one would find a cooling time scale tending to infinity. This is because when $T_\mathrm{eff} \to T_\mathrm{\odot}$, $L \to 0$, and the planetary interior necessarily becomes partly radiative.

The strong irradiation thus not only significantly slows the cooling of the planet, it also profoundly modifies its very structure. The growth of a radiative zone located just below the "atmosphere" (defined as the region which is penetrated by the stellar photons) implies that standard boundary conditions cannot be used. The problem hence becomes relatively complex, and requires a detailed treatment of the radiative transfer equations in the atmosphere.

In the absence of adequate atmospheric models, Guillot et al. (1996) however derived evolution models for Pegasi planets using the approximation of Sect. 7.2. This was also later used by Burrows et al. (2000b) for the evolution of HD209458b. In these papers, the atmospheric boundary condition is at the same pressure and temperature than that of an isolated object of the same effective temperature:

$$T(P = 10 \text{ bars}) = T_\mathrm{isolated}(T_\mathrm{eff}, g). \tag{156}$$

This approximation is exact in the limit when the stellar luminosity is *entirely* absorbed at the 10 bar level, or if the region of absorption is connected to the 10 bar level by an isentrope (i.e. the 10 bar level is in a nearly-adiabatic convective zone).

Unfortunately, the approximation becomes incorrect in the case of strongly irradiated planets because of the growth of a thick external radiative zone. Another boundary condition has therefore to be sought: either part of the stellar flux is able to penetrate to deeper levels ($P_0 > 10$ bar) and lead to a boundary condition defined by $T(P_0) > T_\mathrm{isolated}$, or most of the stellar flux is absorbed at $P_0 < 10$ bar, yielding $T(P_0) < T_\mathrm{isolated}$. (This is due to the fact that in the radiative zone $dT/dP \propto F$, where F is the flux to be transported). Indeed, more detailed models of the atmospheres of Pegasi planets have shown that most of the starlight is absorbed at pressures less than 10 bar, and that (156) overestimates the atmospheric temperatures by as much as 300 to 1000 K (Seager and Sasselov 1998, 2000; Goukenleuque et al. 2000; Barman et al. 2001).

A similarly incorrect approach has been used by Lin, Bodenheimer and Richardson (1996) and Bodenheimer et al. (2001): they also use the same approximation as described in Sect. 7.2, but instead of (156), they use the

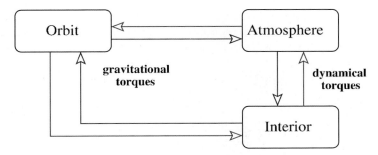

Fig. 36. Angular momentum flow between orbit, interior, and atmosphere for a Pegasi planet in steady state. *Arrows* indicate flow of prograde angular momentum (i.e., that with the same sign as the orbital angular momentum) for two cases: *Anticlockwise:* Gravitational torque on atmosphere is retrograde (i.e., adds westward angular momentum to atmosphere). For torque balance, the gravitational torque on the interior must be prograde (i.e., eastward). These gravitational torques must be balanced by fluid-dynamical torques that transport retrograde angular momentum from atmosphere to interior. *Clockwise:* Gravitational torque on atmosphere is prograde, implying a retrograde torque on the interior and downward transport of prograde angular momentum from atmosphere to interior. Atmosphere will superrotate if gravitational torques push atmosphere away from synchronous (as on Venus). It will subrotate if gravitational torques synchronize the atmosphere (e.g., gravity-wave resonance)

from the atmosphere to the interior is $\eta F_\odot$, where η is small and dimensionless. If this kinetic energy flux is balanced by dissipation in the interior with a spindown timescale of $\tau_{\rm syn}$, then the deviation of the rotation frequency from synchronous is

$$\omega - \omega_s = \left(\frac{4\pi \eta F_\odot \tau_{\rm syn}}{k^2 M} \right)^{1/2}. \tag{162}$$

Experience with planets in our solar system suggests that atmospheric kinetic energy is generated at a flux of $10^{-2} F_\odot$, and if all of this energy enters the interior, then $\eta \sim 10^{-2}$. Using a spindown time of 3×10^5 years then implies $\omega - \omega_s \sim 2 \times 10^{-5} \, {\rm s}^{-1}$, which is comparable to the synchronous rotation frequency. The implied winds in the interior are then of order $\sim 2000 \, {\rm m \, s}^{-1}$. Even if η is only 10^{-4}, the interior's winds would be $200 \, {\rm m \, s}^{-1}$. The implication is that the interior's spin could be asynchronous by up to a factor of two, depending on the efficiency of energy and momentum transport into the interior.

Any scenario involving different rotation rates of the atmosphere and interior inevitably leads to significant energy dissipation. Since we have considered situations for which the system is in gravitational equilibrium, the energy associated with the flow is provided by the stellar photons. A fraction of the absorbed stellar flux is therefore dissipated at levels other than what would be predicted from radiative transfer. Depending on whether the energy is dissipated at low pressures or deep in the interior, the consequences for the planet's evolution are very different. If energy is dissipated in the high atmosphere, as may be the case for gravity waves, the effect on the evolution will be small. If it is dissipated in the interior, as in the case of a Kelvin–Helmoltz instability, this could potentially be the dominant process governing the planet's evolution.

9.4 Atmospheric Dynamics

I have so far implicitly assumed that Pegasi planets have uniform atmospheres. Because of the strong inhomogeneous stellar irradiation, and the near-synchronous rotation, this hypothesis is in fact probably very far from reality. The consequences of the presence of day/night temperature variations for the evolutions are found to affect only weakly the planet's contraction (Guillot and Showman 2002). Here, I analyse the consequences for the atmosphere of Pegasi planets, on the basis of the articles by Guillot (2001) and Showman and Guillot (2002).

Timescales

Temperature variations across planetary atmospheres are governed by the time required for the atmosphere to absorb the stellar heat, to radiate its heat to space, and by the characteristic advective time scales.

The radiative heating/cooling timescale can be estimated by a ratio between the thermal energy within a given layer and the layer's net radiated flux. In the absence of dynamics, absorbed solar fluxes balance the radiated flux, but dynamics perturbs the temperature profile away from radiative equilibrium. Suppose the radiative equilibrium temperature at a particular location is $T_{\rm rad}$ and the actual temperature is $T_{\rm rad} + \Delta T$. The net flux radiated toward outer space is then $4\sigma T_{\rm rad}^3 \Delta T$ and the radiative timescale is

$$\tau_{\rm rad} \sim \frac{P}{g} \frac{c_p}{4\sigma T^3} . \tag{163}$$

This timescale is thus particularly dependent on the characteristic temperature of the atmosphere. For our giant planets, $T \sim 200\,{\rm K}$, so that the radiative timescale is long, i.e. about a year at 1 bar. This is to be compared, e.g. to the rotation period, which is of the order of 10 hours for Jupiter and Saturn. Their atmospheres are thus found to be relatively uniform. However, as shown in Fig. 37, Pegasi planets have ten times hotter atmospheres, so that $\tau_{\rm rad}$ is of the order of 1 day at photospheric levels, to be compared with their rotation period of ~ 4 days.

The timescale for advection by winds is more difficult to estimate. Guillot (2001) and Showman and Guillot (2002) use a shear instability criterion: assuming that the convective core is locked into synchronous rotation, they assume that at upper levels winds build up with increasing altitude only if they do not exceed the Kelvin–Helmholtz instability criterion (Chandrasekhar

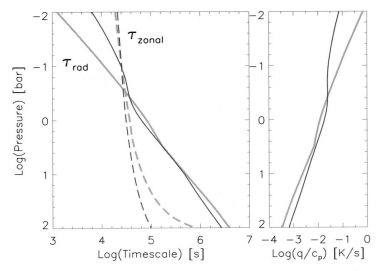

Fig. 37. *Left:* Characteristic time scales as a function of pressure level. τ_{zonal} is the minimal horizontal advection time (*dashed*). τ_{rad} is the timescale necessary to cool/heat a layer of pressure P and temperature T by radiation alone (*solid*). For each case, the *thin black* and *thick gray lines* correspond to the "hot" and "cold" models. *Right:* Approximate cooling/heating rate as a function of pressure. [From Showman and Guillot (2002)]

1961), i.e. if:

$$Ri = \frac{N^2}{(du/dz)^2} > \frac{1}{4}, \quad (164)$$

where Ri is the Richardson number, and $N^2 = (g/H_P)(\nabla_{\text{ad}} - \nabla_T)$ (N is the Brünt–Vaisala frequency). This thus implies a constraint on the wind shear du/dz. The resulting timescale is shown as dashed curves in Fig. 37, for both the "cold" and "hot" cases.

At pressures exceeding 0.1 bar, radiation is slower than the maximal advection by zonal winds, but by less than one order of magnitude. The consequent day/night temperature difference $\Delta T_{\text{day-night}}$ to be expected is:

$$\frac{\Delta T_{\text{day-night}}}{\Delta T_{\text{rad}}} \sim 1 - e^{-\tau_{\text{zonal}}/\tau_{\text{rad}}}, \quad (165)$$

where ΔT_{rad} is the day–night difference in radiative equilibrium temperatures. Rough estimates from Fig. 37 suggest that $\tau_{\text{zonal}}/\tau_{\text{rad}} \sim 0.3$ at 1 bar, implying that $\Delta T_{\text{day-night}}/\Delta T_{\text{rad}} \sim 0.3$. If $\Delta T_{\text{rad}} = 1000\,\text{K}$, this would imply day-night temperature differences of 300 K at 1 bar. Values of $\Delta T_{\text{day-night}}$ even closer to ΔT_{rad} are likely given the fact that slower winds will lead to an even more effective cooling on the night side and heating on the day side.

The small radiative time scale implies that, for the day-night temperature difference to be negligible near the planet's photosphere, atmospheric winds would have to be larger than the maximum winds for the onset of shear instabilities.

Possible Circulation and Atmospheric Chemistry

As discussed by Showman and Guillot (2002), the intense stellar radiation is expected to drive both zonal and meridional winds, but the atmospheric circulation is unknown. However, they note that even if locked into synchronous rotation, the atmospheres of Pegasi planets are characterized by relatively low (~ 0.1) Rossby numbers. This implies that the Coriolis force plays a very important role and that zonal circulation is favored over meridional circulation.

A preliminary numerical simulation with a global circulation model by Showman and Guillot (2002) indicates that a fast superrotating equatorial jet develops, and that the atmosphere is globally superrotating, a situation very similar to that of Venus. This situation is depicted in Fig. 38. It is interesting

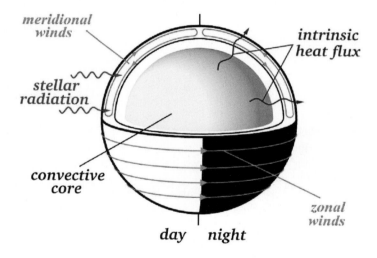

Fig. 38. Conjectured dynamical structure of Pegasi planets: At pressures larger than 100–800 bar, the intrinsic heat flux must be transported by convection. The convective core is at or near synchronous rotation with the star and has small latitudinal and longitudinal temperature variations. At lower pressures a radiative envelope is present. The top part of the atmosphere is penetrated by the stellar light on the day side. The spatial variation in insolation should drive winds that transport heat from the day side to the night side. [From Guillot (2001); Showman and Guillot (2002)]

to notice that this kind of circulation pattern implies that the equator to pole temperature variation is even more pronounced than the day to night one.

However, the consequences for cloud formation and chemistry in the atmospheres of Pegasi planets are still unclear. The solution to that problem depends in fact on whether the heating/cooling is mostly balanced by vertical motions (in which case clouds would tend to form at the substellar point, on the day side), or by horizontal advection. The latter seems to be favored by the simulations and it is instructive to discuss it further.

Let us assume that a superrotating wind advects air roughly on constant pressure levels (negligible vertical advection). In that case, air is cooled on the night side, then it is intensely heated on the day side. As a consequence, any chemical species that condenses on the night side and forms clouds there will evaporate on the day side. The night side should then be relatively cloudy, while the day side would be clear (low albedo). But this circulation has another very important consequence for atmospheric chemistry: most abundant species that condense on the night side are, according to the estimates from Rossow (1978), expected to settle down on short timescales (Guillot 2001). Because, according to our hypothesis, the air is transported on isobars, when it reaches the day side, the condensing species are *undersaturated* everywhere down to the condensation level *on the night side*.

The magnitude of this effect can be estimated as follows: Let us assume that on the day side, the saturation abundance of the condensing species, $x = p/P$ is maximal and equal to $x^\star$ at $P = P^\star_{\rm day}$ (the condensation level on the day side). On the night side, the temperature is lower. Equations (137) and (145) can be used to show that the abundance at saturation on the night side becomes:

$$\ln x(P^\star_{\rm day}) = \ln x^\star - \beta \ln(T_{\rm day}/T_{\rm night}) . \qquad (166)$$

In order to reach condensation, i.e. $x = x^\star$, one has to penetrate deeper into the atmosphere. Equation (146) implies that on the night side,

$$\ln x(P) = \ln x(P^\star_{\rm day}) + (\beta \nabla_T - 1) \ln(P/P^\star_{\rm day}) , \qquad (167)$$

assuming that ∇_T is constant. Using (166), one obtains the condensation pressure on the night side:

$$\frac{P^\star_{\rm night}}{P^\star_{\rm day}} = \left(\frac{T_{\rm day}}{T_{\rm night}}\right)^{\beta/(\beta \nabla_T - 1)} . \qquad (168)$$

Using $\beta \sim 10$, $\nabla_T \sim 0.15$ and $T_{\rm day}/T_{\rm night} \sim 1.2$, one finds $P^\star_{\rm night} \sim 38 P^\star_{\rm day}$, a very significant variation of the condensation pressure. This implies that air flowing on constant pressure levels around the planet would lead to a rapid depletion of any condensing species on the day side, compared to what would be predicted from chemical equilibrium calculations. This can potentially also remove important absorbing gases from the day side, as in the case of TiO, which can be removed by $CaTiO_3$ condensation, or Na, removed by Na_2S

condensation (Lodders 1999). Of course, most of the variation depends on the exponential factor $\beta/(\beta\nabla_T - 1)$, which is infinite in the limit when the atmospheric temperature profile and the condensation profile are parallel to each other. In the discussion, we implicitly assumed $\beta\nabla_T - 1 > 0$; however, when the atmosphere is close to an isotherm, this factor can become negative. In this case the day/night effect is even more severe, as the condensing species is entirely removed from this quasi-isothermal region.

The rapid circulation from the night side to the day side can also lead to a disequilibrium chemistry for non-condensing species when the reaction timescales are longer than the advection timescale ($\sim$ a day). This is for example the case of the N_2 to NH_3 reaction in Gl229B (Saumon et al. 2000b), but many other chemical species should be affected.

Observational Consequences

The structure and evolution of Pegasi planets is much more complex than envisioned when these planets were first discovered. The possibility of dissipation by stellar tides is interesting because it will be directly tested by observations and because this phenomenon is poorly understood even for the planets of our solar system. However, the drawback is that it should be more difficult to infer the planets' global compositions from radii measurements, as first suggested (Guillot 1999b).

However, our understanding of these objects should be greatly increased by the numerous direct or indirect observations that are now possible. With several ground programs (STARE, VULCAN) and space missions (COROT, MONS, MOST, KEPLER, EDDINGTON) aiming at detecting photometric transits of Pegasi planets, there is indeed a good chance that statistically significant information on e.g. the mass radius relationship of Pegasi planets can be gathered.

Measurement of starlight reflected from these planets may allow the albedo to be estimated. Because the star–planet–Earth angle changes throughout the planet's orbit, crude information on the scattering properties of the atmosphere (e.g., isotropic versus forward scattering) may be obtainable. Asymmetries in the reflected flux as the planet approaches and recedes from the transit could give information on the differences of albedo near the leading and trailing terminators, which would help constrain the dynamics. Finally, transit observations of Pegasi planets using high resolution spectroscopy should in the near future yield constraints on the atmospheric temperature, cloud/haze abundance, and winds (Seager and Sasselov 2000; Brown 2001; Hubbard et al. 2001). If these measurements are possible during the ingress and egress, i.e., the phases during which the planets enters and leaves the stellar limb, respectively, asymmetries of the planetary signal should be expected and would indicate zonal heat advection at the terminator.

Remerciements

Merci à toute l'organisation de l'école, et à tout ceux qui m'ont fourni du matériel pour ce cours, écrit difficilement, en pensant à toi petite sœur.

References

Ackerman A.S. & Marley M.S., *ApJ* **556**, 872 (2001).
Acuña M.H., Connerney J.E.P. et Ness N.F., *J. Geophys. Res.* **88**, 8771 (1983).
Alexander D.R. & Ferguson J.W., *ApJ* **437**, 879 (1994).
Allard F. & Hauschildt P.H., *ApJ* **445**, 433 (1995).
Allard F., Hauschildt P.H., Baraffe I. & Chabrier G., *ApJL* **465**, L123 (1996).
Allard F., Hauschildt P.H., Alexander D.R., Tamanai A., & Schweitzer A., *ApJ* **556**, 357 (2001).
Anderson J.D., Campbell J.K., Jacobson R.A., *et al.*, *J. Geophys. Res.* **92**, 14877 (1987).
Bahcall J.N., Pinsonneault M.H., *Rev. Mod. Phys.* **67**, 781 (1995).
Bailer-Jones, C.A.L. & Mundt R., *A&A* **374**, 1071 (2001).
Basri G., *Ann. Rev. Aston. Astrophys.* **38**, 485 (2000).
Barman T., Hauschildt P.H., Allard F., *ApJ* **556**, 885 (2001).
Beule D., Ebeling W., Förster A., Juranek H., Nagel S., Redmer R. & Röpke G., *Phys. Rev. B* **59**, 4177 (1999).
Birnbaum G., *J. Quant. Spectrosc. Radiat. Transfer* **21**, 597 (1979).
Bodenheimer P., Hubickyj O. & Lissauer J.J., *Icarus* **143**, 2 (2000).
Bodenheimer P., Lin D.N.C., Mardling R., *ApJ* **548**, 466 (2001).
Borysow A., Borysow J., & Fu Y., *Icarus* **145**, 601 (2000).
Borysow A., *Icarus* **96**, 169 (1992).
Boss A.P., *ApJ* **503**, 923 (1998).
Boss A.P., *ApJL* **536**, L101 (2000).
Brown T.M., *ApJ* **553**, 1006 (2001).
Brown T.M., Charbonneau D., Gilliland R.L., Noyes R.W., Burrows A., *ApJ* **552**, 699 (2001).
Burgasser A.J. et al., *ApJL* **522**, L65 (1999).
Burrows A. & Liebert J., *Rev. Mod. Phys.* **65**, 301 (1993).
Burrows A., Marley M., Hubbard W.B., Lunine J.I., Guillot T., Saumon D., Freedman R., Sudarsky D., Sharp C., *ApJ* **491**, 856 (1997).
Burrows A., Marley M.S. & Sharp, C.M., *ApJ* **531**, 438 (2000a).
Burrows A., Guillot T., Hubbard W.B., Marley M.S., Saumon D., Lunine J.I. & Sudarsky D., *ApJL* **534**, L97 (2000b).
Busse F.H., *Ann. Rev. Fluid. Mech* **10**, 435 (1978).
Campbell J.K. & Synnott S.P., *Astron. J.* **90**, 364 (1985).
Campbell J.K. & Anderson J.D., *Astron. J.* **97**, 1485 (1989).
Carlson B.E., Lacis A.A., Rossow W.B., *ApJ* **388**, 648 (1992).
Carlson R. et al., *Science* **274**, 385 (1996).

Cavazzoni C., Chiarotti G.L., Scandolo S., Tosatti E., Bernasconi M., & Parrinello M., *Science* **283**, 44 (1999).
Chabrier G., Saumon D., Hubbard W.B., Lunine J.I., *ApJ* **391**, 817 (1992).
Celliers et al., *Phys. Rev. Let.* **84**, 5564 (2000).
Chabrier G. & Baraffe I., *A&A* **327**, 1039 (1997).
Chabrier G. & Baraffe I., *Ann. Rev. Aston. Astrophys.* **38**, 337 (2000).
Chabrier G., Baraffe I., Allard F. & Hauschildt P., *ApJL* **542**, L119 (2000a).
Chabrier G., Baraffe I., Allard F. & Hauschildt P., *ApJ* **542**, 464 (2000b).
Chandrasekhar S., *"Stellar Structure and Evolution"*, The University of Chicago press, Chicago (1939).
Chandrasekhar S., *"Hydrodynamic and hydromagnetic stability"*, Dover, New York (1961).
Charbonneau D., Brown T.M., Latham D.W., Mayor, M., *ApJ* **529**, L45 (2000).
Clayton D.D., *"Principles of stellar evolution and nucleosynthesis"*, McGraw-Hill Book Company, New-York (1968).
Clayton R.N., *Science* **140**, 192 (1963).
Cohen E.R., Taylor B.N., *Rev. Mod. Phys.* **59**, 1121 (1986).
Collins G.W. et al., *Science* **281**, 1179 (1998).
Collins G.W. et al., *Phys. Rev. Let.* **87**, 504 (2001).
Connerney J.P., Acuña M.H. & Ness N.F., *J. Geophys. Res.* **86**, 3623 (1982).
Conrath B.J., Gautier D., Hanel R., Lindal G., Marten A., *ApJ* **282**, 807 (1984).
Conrath B.J., Hanel R.A. & Samuelson R.E. In *Origin and Evolution of Planetary and Satellite Atmospheres* (eds. S.K. Atreya, J.B. Pollack, and M.S. Matthews), Univ. of Arizona Press, Tucson, pp. 513–538 (1989)
Conrath B.J. & Gautier D., *Icarus* **144**, 124 (2000).
Cotton W.R. & Anthes R.A., *"Storm and cloud dynamics"*, Academic Press, Inc., San Diego (1989).
Davies M.E., Abalakin V.K., Lieske J.H., Seidelmann P.K., Sinclair A.T., Sinzi A.M., Smith B.A. & Tjuflin Y.S., *Celestial Mech.* **29**, 309 (1986).
Da Silva L.B. et al., *Phys. Rev. Let.* **78**, 483 (1997).
Emanuel K.A., *"Atmospheric Convection"*, Oxford Univ. Press, New-York (1994).
Fegley B.J. & Lodders K., *Icarus* **110**, 117 (1994).
Fegley B.J. & Lodders K., *ApJL* **472**, L37 (1996).
Folkner W.M., Woo R., Nandi S., *J. Geophys. Res.* **103**, 22831 (1998).
Fontaine G., Graboske H.C. & van Horn H.M., *ApJS* **35**, 293 (1977).
Fowler W.A., Caughlan G.R. & Zimmerman B.A., *Ann. Rev. Aston. Astrophys.* **13**, 69 (1975).
Galli G. et al., *Phys. Rev. B* **61**, 909 (2000).
Gautier D., Owen, T. In *Origin and Evolution of Planetary and Satellite Atmospheres* (eds. S.K. Atreya, J.B. Pollack, and M.S. Matthews), University of Arizona Press, Tucson, pp. 487–512 (1989)
Gautier D., Hersant F., Mousis O. & Lunine J.I., *ApJL* **559**, L183 (2001).

Geballe T.R., Kulkarni S.R., Woodward C.E. & Sloan G.C., *ApJL* **467**, L101 (1996).
Gierasch P.J. & Conrath B.J., *J. Geophys. Res.* **98**, 5459 (1993).
Gladman, *et al.*, *Nature* **412**, 163 (2001).
Goldreich P. & Soter S., *Icarus* **5**, 375 (1966).
Goody R.M. & Yung Y.L., *"Atmospheric radiation"*, Oxford University Press, New York (1989).
Goukenleuque C., Bézard B., Joguet B., Lellouch E., Freedman R., *Icarus* **143**, 308 (2000).
Graboske H.C., Pollack J.B., Grossman A.S. & Olness R.J., *ApJ* **199**, 265 (1975).
Griffith C.A., Yelle R.V. & Marley M.S., *Science* **282**, 2063 (1998).
Griffith C.A. & Yelle R.V., *ApJL* **519**, L85 (1999).
Gudkova T.V. & Zharkov V.N., *Plan. Space. Sci.* **47**, 1201 (1999).
Guillot T., Gautier D., Chabrier G., Mosser B., *Icarus* **112**, 337 (1994a).
Guillot T., Chabrier G., Morel P. & Gautier D., *Icarus* **112**, 354 (1994b).
Guillot T., Chabrier G., Gautier D. & Morel P., *ApJ* **450**, 463 (1995).
Guillot T. & Morel P., *A&AS* **109**, 109 (1995).
Guillot T., *Science* **269**, 1697 (1995).
Guillot T., Burrows A., Hubbard W.B., Lunine J.I. & Saumon D., *ApJ* **459**, L35 (1996).
Guillot T., Gautier D. & Hubbard W.B., *Icarus* **130**, 534 (1997).
Guillot T., *Plan. Space. Sci.* **47**, 1183 (1999a).
Guillot T., *Science* **286**, 72 (1999b).
Guillot T. Atmospheric circulation of hot Jupiters. In *Planetary Systems in the Universe: Observation, Formation and Evolution*, ASP Conf. series, A. Penny et al. eds., in press (2001)
Guillot T. & Showman A., *A&A*, **385**, 156 (2002).
Guillot T., Stevenson D.J., Hubbard W.B. & Saumon D. In *Jupiter*, F. Bagenal *et al.*, eds., Combridge Planetary Science, Vol. 1, pp. 35–37 (2004)
Hayashi C., *PASJ* **13**, 450 (1961).
Hayashi C., *Prog. Theo. Phys. Suppl.* **70**, 35 (1981).
Henry G.W., Marcy G.W., Butler R.P., Vogt, S.S., *ApJ* **529**, L41-L44 (2000).
Hillenbrand L.A. & Carpenter J.M., *ApJ* **540**, 236 (2000).
Hemley R.J., Mao H.K., Finger L.W. et al., *Phys. Rev. B* **42**, 6458 (1990).
Holmes N.C., Ross M. & Nellis W.J., *Phys. Rev. B* **52**, 15835 (1995).
Hubbard W.B., *ApJ* **152**, 745 (1968).
Hubbard W.B. & Lampe M., *ApJS* **18**, 297 (1969).
Hubbard W.B., *Icarus* **30**, 305 (1977).
Hubbard W.B., *Icarus* **52**, 509 (1982).
Hubbard W.B., *"Planetary Interiors"*, Van Nostrand Reinhold Co., Inc., New York (1984).

Hubbard, W.B. In *Origin and Evolution of Planetary and Satellite Atmospheres*, S. K. Atreya, J. B. Pollack, and M. S. Matthews, eds., University of Arizona Press, Tucson, pp. 539–563 (1989)
Hubbard W.B. & Marley M.S., *Icarus* **78**, 102 (1989).
Hubbard W.B., Pearl J.C., Podolak M. & Stevenson D.J., in *Neptune and Triton*, ed. D. P. Cruikshank (Tucson: Univ. of Arizona Press), pp. 109–138 (1995)
Hubbard W.B., Guillot T., Marley M.S., Burrows A., Lunine J.I. & Saumon D.S., *Plan. Space. Sci.* **47**, 1175 (1999).
Hubbard W.B., *Icarus* **137**, 196 (1999).
Hubbard W.B., Fortney J.J., Lunine J.I., Burrows A., Sudarsky D. & Pinto P., *ApJ* **560**, 413 (2001).
Ingersoll A.P., Kanamori H. & Dowling T.E., *Geophys. Res. Lett.* **21**, 1083 (1994).
Ingersoll A.P., Barnet C.D., Beebe R.F. et al., in *Neptune and Triton*, ed. D.P. Cruikshank, University of Arizona Press, Tucson, 613 (1995)
Jeffreys H., *M.N.R.A.S.* **83**, 350 (1923).
Kerley G.I., *Phys. Earth Planet. Interiors* **6**, 78 (1972).
Kippenhahn R. & Weigert A., *"Stellar structure and evolution"*, Springer-Verlag, Berlin (1991).
Kirkpatrick J.D. et al., *ApJ* **519**, 802 (1999).
Klepeis J.E., Schafer K.J., Barbee T.W.III, Ross M., *Science* **254**, 986 (1991).
Knudson M.D., Hanson D.L., Bailey J.E., Hall C.A., Asay J.R. & Anderson W.W., *Phys. Rev. Let.* **87**, 5501 (2001).
Kunde V., Hanel R., Maguire W., Gautier D., Baluteau J.P., Marten A., Chedin A., Husson N. et Scott N., *ApJ* **263**, 443 (1982).
Landau L., Lifschitz E., *"Physique Statistique"*, Editions Mir, Moscow (1976).
Lenosky T.J., Bickham S.R., Kress J.D. & Collins L.A., *Phys. Rev. B* **61**, 1 (2000).
Lenzuni P., Chernoff D.F. et Salpeter E.E., *ApJ Suppl.* **76**, 759 (1991).
Liebert J., Reid I.N., Burrows A., Burgasser A.J., Kirkpatrick J.D. & Gizis J.E., *ApJL* **533**, L155 (2000).
Lin D.N.C., Bodenheimer P., & Richardson D.C., *Nature* **380**, 606 (1996).
Lindal G.F., Wood G.E., Levy G.S., Anderson J.D., Sweetnam D.N., Hotz H.B., Buckles B.J., Holmes D.P., Doms P.E., Eshleman V.R., Tyler G.L., & Croft T.A., *J. Geophys. Res.* **86**, 8721 (1981).
Lindal G.F., Sweetnam D.N., Eshleman V.R., *J. Geophys. Res.* **92**, 14987 (1985).
Lindal G.F., *Astron. J.* **103**, 967 (1992).
Lodders K., *ApJ* **519**, 793 (1999).
Lubow S.H., Tout C.A. & Livio M., *ApJ* **484**, 866 (1997).
Luhman K.L., Rieke G.H., Young E.T., Cotera A.S., Chen H., Rieke M.J., Schneider G. & Thompson R.I., *ApJ* **540**, 1016 (2000).
Lunine J.I., Hubbard W.B., Burrows A., Wang Y.-P. & Garlow K., *ApJ* **338**, 314 (1989).

Marcy G.W., Butler R.P., Williams E., Bildsten L., Graham J.R., Ghez A.M., Jernigan J.G., *ApJ* **481**, 926 (1997).
Marley M.S., Gomez P. & Podolak P., *J. Geophys. Res.* **100**, 23349 (1995).
Marley M.S., Saumon D., Guillot T., Freedman R.S., Hubbard W.B., Burrows A. & Lunine J.I., *Science* **272**, 1919 (1996).
Marley M.S. & McKay C.P., *Icarus* **138**, 268 (1999).
Marley M.S., Gelino C., Stephens D., Lunine J.I. & Freedman R., *ApJ* **513**, 879 (1999).
Martín E.L., Basri G., Delfosse X. & Forveille T., *A&A* **327**, L29 (1997).
Martín E.L., Delfosse X., Basri G., Goldman B., Forveille T. & Zapatero Osorio M.R., *AJ* **118**, 2466 (1999).
Martín E.L., Zapatero Osorio M. R., Barrado y Navascués D., Béjar V.J.S. & Rebolo R., *ApJL* **558**, L117 (2001).
Mayor M. & Queloz D., *Nature* **378**, 355 (1995).
Mayor M., Naef D., Pepe F., Queloz D., Santos N., Udry S., Burnet M. In *Planetary Systems in the Universe: Observation, Formation and Evolution*, IAU Symp. 202, Eds. A. Penny, P. Artymowicz, A.-M. Lagrange and S. Russel ASP Conf. Ser, in press (2001)
Militzer B. & Ceperley D.M., *Phys. Rev. Lett.* **85**, 1890 (2000).
Militzer B. & Ceperley D.M., *Phys. Rev. E* **63**, 6404 (2001).
Montalbán J., D'Antona F., & Mazzitelli I., *A&A* **360**, 935 (2000).
Mosser B., Maillard J.P. & Mékarnia D., *Icarus* **144**, 104 (2000).
Mostovych A.N., Chan Y., Lehecha T., Schmitt A. & Sethian J.D., *Phys. Rev. Let.* **85**, 3870 (2000).
Najita J.R., Tiede G.P. & Carr J.S., *ApJ* **541**, 977 (2000).
Nakajima T., Oppenheimer B.R., Kulkarni S.R., Golimowski D.A., Matthews K. & Durrance S. T., *Nature* **378**, 463 (1995).
Nefedov A.P., Sinel'shchikov V.A. & Usachev A.D., *Physica Scripta* **59**, 432 (1999).
Nellis W.J., Mitchell A.C., van Thiel M., Devine G.J., Trainor R.J., Brown N., *Journal Chemical Physics* **79**, 1480 (1983).
Nellis W.J., Weir S.T., Mitchell A.C., *Phys. Rev. B* **59**, 3434 (1999).
Ness N.F. et al., *Science* **233**, 85 (1986).
Ness N.F. et al., *Science* **246**, 1473 (1989).
Niemann H.B., Atreya S.K., Carignan G.R., Donahue T.M., Haberman J.A., Harpold D.N., Hartle R.E., Hunten D.M., Kaspzrak W.T., Mahaffy P.R., Owen T.C. & Way S.H., *J. Geophys. Res.* **103**, 22831 (1998).
Oppenheimer B.R., Kulkarni S.R., Matthews K. & Nakajima T., *Science* **270**, 1478 (1995).
Oppenheimer B.R., Kulkarni S.R., Matthews K. & van Kerkwijk M.H., *ApJ* **502**, 932 (1998).
Ouyed R., Fundamenski W.R., Cripps G.R., & Sutherland P.G., *ApJ* **501**, 367 (1998).
Owen T., Mahaffy P., Niemann H.B., Atreya S., Donahue T., Bar-Nun A. & de Pater I., *Nature* **402**, 269 (1999).

de Pater I., Massie, S.T., *Icarus* **62**, 143 (1985).
Pearl J.C. & Conrath B.J., *J. Geophys. Res. Suppl.* **96**, 18921 (1991).
Pedlosky J., *"Geophysical fluid dynamics"*, Springer-Verlag, New-York (1979).
Pfaffenzeller O., Hohl D. & Ballone P., *Phys. Rev. Lett.* **74**, 2599 (1995).
Podolak M., Hubbard W.B. & Stevenson D.J., in *Uranus*, eds. J. T. Bergstralh, E. D. Miner, M. S. Matthews (Tucson: Univ. of Arizona Press), pp. 29–61. (1991)
Podolak M., Weizman A. & Marley M.S., *Plan. Space. Sci.* **43**, 1517 (1995).
Podolak M., Podolak J.I. & Marley M.S., *Plan. Space. Sci.* **48**, 143 (2000).
Pollack J.B., McKay C.P. & Christofferson B.M., *Icarus* **64**, 471 (1985).
Pollack J.B., Hollenbach D., Beckwith S., Simonelli D.P., Roush T. & Fong W., *ApJ* **421**, 615 (1994).
Pollack J.B., Hubickyj O., Bodenheimer P., Lissauer J.J., Podolak M. & Greenzweig Y., *Icarus* **124**, 62 (1996).
Potekhin A.Y., Baiko D.A., Haensel P. & Yakovlev D.G., *A&A* **346**, 345 (1999).
Prandtl L., *Z. Angew. Math. Mech.* **5**, 136 (1925).
Ross M., *Phys. Rev. B.* **58**, 669 (1998).
Ross M. & Yang L.H., *Phys. Rev. B* **64**, 134,210 (2001).
Rossow W.B., *Icarus* **36**, 1 (1978).
Roulston, M.S., Stevenson, D.J. Prediction of neon depletion in Jupiter's atmosphere *EOS* **76**, 343 (1995) [abstract].
Salpeter E.E., *ApJ* **181**, L83 (1973).
Sanchez-Lavega A., Colas F., Lecacheux J., Laques P., Parker D., & Miyazaki I., *Nature* **353**, 397 (1991).
Saumon D., Hubbard W.B., Chabrier G. & Van Horn H.M., *ApJ* **391**, 827 (1992).
Saumon D., Chabrier G. & Van Horn H.M., *ApJS* **99**, 713 (1995).
Saumon D., Hubbard W.B., Burrows A., Guillot T., Lunine J.I. & Chabrier G., *ApJ* **460**, 993 (1992).
Saumon D., Hubbard W.B., Burrows A. et. al., *ApJ* **460**, 993 (1996).
Saumon D., Chabrier G., Wagner D.J., & Xie X., *High Pressure Research* **16**, 331 (2000a).
Saumon D., Geballe T.R., Leggett S.K., Marley M.S., Freedman R.S., Lodders K., Fegley B. & Sengupta S.K., *ApJ* **541**, 374 (2000b).
Schmider F.X., Mosser B. & Fossat E., *A&A* **248**, 281 (1991).
Seager S. & Sasselov D.D., *ApJ* **502**, L157 (1998).
Seager S. & Sasselov D.D., *ApJ* **537**, 916 (2000).
Seiff A., Kirk D.B., Knight T.C.D., Young R.E., Mihalov J.D., Young L.A., Milos F.S., Schubert G., Blanchard R.C & Atkinson D., *J. Geophys. Res.* **103**, 22857 (1998).
Sheppard S. & Jewitt D., *Nature* **423**, 261 (2003).

Showman A.P. & Dowling T.E., *Science* **289**, 1737 (2000).
Showman A.P. & Guillot T., *A&A* **385**, 166 (2002).
Smoluchowski R., *Nature* **215**, 691 (1967).
Stevenson D.J. Ph.D. Thesis, Cornell University (1976a)
Stevenson D.J., *Phys. Lett. A* **58**, 282 (1976b).
Stevenson D.J. & Salpeter E.E., *ApJ Suppl.* **35**, 221 (1977a).
Stevenson D.J. & Salpeter E.E., *ApJ Suppl.* **35**, 239 (1977b).
Stevenson D.J., *Ann. Rev. Earth Planet. Sci.* **10**, 257 (1982).
Stevenson D.J., *Rep. Prog. Phys.* **46**, 555 (1983).
Stevenson D.J., *Ann. Rev. Aston. Astrophys.* **29**, 163 (1991).
Strauss M.A. et al., *ApJL* **522**, L61 (1999).
Sudarsky D., Burrows A. & Pinto P., *ApJ* **538**, 885 (2000).
Thompson W.R., & Squyres S.W., *Icarus* **86**, 336 (1990).
Trafton L.M., *ApJ* **147**, 765 (1967).
Trilling D.E., Benz W., Guillot T., Lunine J.I., Hubbard W.B., & Burrows A., *ApJ* **500**, 428 (1998).
Tsuji T., Ohnaka K., Aoki W. & Nakajima T., *A&A* **308**, L29 (1996).
Tyler G.L., Sweetnam D.N., Anderson J.D., et al., *Science* **246**, 1466 (1989).
Vidal-Madjar A., Lecavelier des Etangs A.L., Désert J.-M., Ballester G.E., Ferlet R., Hébrard G. & Mayor M., *Nature* **422**, 143 (2003).
Warwick J.W., Evans D.R., & Roming J.H., et al., *Science* **223**, 102 (1986).
Warwick J.W., Evans D.R. & Peltzer G.R., et. al., *Science* **246**, 1498 (1989).
Weidenschilling S.J., *Astrophys. & Sp. Sci.* **51**, 153 (1977).
Weir S.T, Mitchell A.C. & Nellis W.J., *Phys. Rev. Lett.* **76**, 1860 (1996).
Wigner E. & Huntington H.B., *Phys. Rev. B* **3**, 764 (1935).
Wuchterl G., Guillot T., & Lissauer J.J. In *Protostars & Planets IV*, eds. V. Mannings et al., (Tucson: Univ. of Arizona Press), 1181 (2000)
von Zahn U., Hunten D.M., Lehmacher G., *J. Geophys. Res.* **103**, 22815 (1998).
Zharkov V.N. & Trubitsyn V.P., *"Physics of Planetary Interiors"*, W.B. Hubbard ed., Pachart, Tucson (1978).
Zharkov V.N., *"Interior structure of the Earth and planets"*, Harwood academic pubishers (1986).
Zharkov V.N. et Gudkova T.V., in *"High pressure research: application to Earth and planetary sciences"*, (eds. Y. Syono et M.H. Manghnani), TERRAPUB, Tokyo (1992).

Protostellar Disks and Planet Formation

P. Cassen

1 Introduction

The idea that the planets of the Solar System formed from a "protoplanetary disk" of material swirling about the primitive Sun follows naturally from the observation that the planetary orbital angular momentum vectors are nearly aligned with each other and that of the Sun itself. The existence of such a progenitor disk was implicit in the ideas of Descartes, and has been a common feature of scientific attempts to explain the systematic aspects of the Solar System since then. A corollary of these "nebular" theories is that planetary systems are an ordinary consequence of star formation. Modern astronomy has confirmed the essential aspects of the hypothesis by revealing the common existence of planets around other stars, and disks around young stars.

But the idea that planets form from circumstellar disks carries the further implication that the properties of planetary systems are somehow related to those of their parent disks; that is, an understanding of disk evolution leads naturally to an understanding of the nature of the resultant planetary system. Certainly this premise has been adopted in much of the theoretical work on the formation of the Solar System; indeed, it is the basis upon which much of the content of these lectures is organized. It may be, however, that nature contrives to obscure the conditions of planetary formation to the extent that the disk-planetary system connection is no longer recognizable, at least in some cases. We will discuss some lines of theoretical argument which suggest that this is the case. Certainly the dynamical properties of the extrasolar planetary objects discovered so far are not obviously associated with specific disk properties, or even a disk origin, despite the overwhelming circumstantial evidence that they must have formed from disks. It remains to be seen to what degree these bodies, or the Solar System, are representative of planetary systems in general.

Whether or not the particular properties of a protostellar disk are eventually reflected in the properties of a planetary system, the disk origin of planets is on firm ground. Furthermore, disks are complex and interesting objects in

Cassen P (2006), Protostellar disks and planet formation. In: Mayor M, Queloz D, Udry S and Benz W (eds) Extrasolar planets. Saas-Fee Adv Courses vol 31, pp 369–448
DOI 10.1007/3-540-29216-0_3   © Springer-Verlag Berlin Heidelberg 2006

their own right. Thus the first half of these lectures are devoted to methods of elucidating their properties. The second half deals with theories of planet formation. The literature on these subjects is enormous, so my choice of specific subjects to be treated, as well as the references supplied, are somewhat subjective. I have tried to tie arguments to first principles wherever possible, although references must be relied upon for many details. Inevitably, the derivations of some results are too cumbersome for inclusion here, in which cases the reader is directed to the appropriate references.

2 Observations of Protostellar Disks

2.1 T Tauri Stars

Most of what we know about protostellar disks is derived from observations of T Tauri stars. These are pre-main sequence stars of spectral class G, K and M, originally identified by their observational characteristics: prominent Balmer emission lines, excess ultraviolet (UV) and infrared (IR) emission, variability and evidence for outflows. It is now known that the defining emission characteristics of T Tauri stars are due to the presence of circumstellar *accretion* disks. Lynden-Bell and Pringle (1974) identified the source of IR radiation as dissipation within the disk, due to angular momentum transport and the release of gravitational energy, as material is fed through the disk to the star. The UV radiation was attributed to gas heated to high temperatures by dissipation in a narrow boundary layer between the (rapidly rotating) inner edge of the disk and the surface of the (slowly rotating) star. It turns out that starlight shining on the disk, which is absorbed and re-emitted at longer wavelengths, also contributes substantially to the IR radiation. Also, there is evidence that the UV radiation is due to disk material falling onto the star along stellar magnetic flux tubes, rather than through the viscous boundary layer imagined by Lynden-Bell and Pringle (1974). Nevertheless, the theory developed by these authors (and their predecessor, (Lüst 1952)) forms the basis of the current understanding of protostellar disks, and correctly describes the continuum spectrum a T Tauri star in terms of several components (see Fig. 1): the nearly blackbody radiation from the star itself; a broad IR component from the optically thick part of the disk; microwave (submillimeter-to-millimeter) emission from the more distant, optically thin, parts of the disk; and an optical and UV component from hot gas transferred from the disk to the star.

Figure 2 shows the Hertzprung–Russellure (H–R) diagram for classical T Tauri stars and "weak-line" T Tauri stars (WTTS; premain sequence stars without evidence for disks) in the Taurus–Auriga star-forming region, along with theoretically derived pre-main sequence evolutionary tracks and their isochrons. Note that although some disks apparently last as long as 10^7 years, they appear to be gone by the time a star reaches the main sequence. Also,

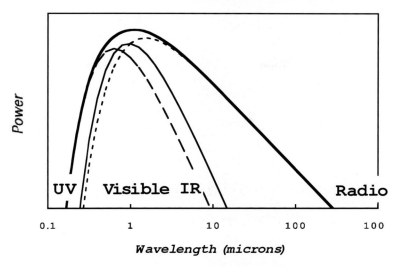

Fig. 1. Accreation disk theory predicts that the continuum spectrum of a pre-main sequence star and disk (*thick solid line*) is the sum of three components: the nearly blackbody radiation from the star (*thin solid line*), a broad component from the disk (*short dashed line*),and a blackbody-like component from hot gas being accreted from the disk to the star (*long dashed line*)

there are some stars that appear to be quite young and yet show no evidence for disks. These stars are of particular interest: either disks never formed; they did form, but were rapidly dissipated; or they rapidly became invisible because the dust in them coagulated to form larger objects (e.g., planets) which are not detected by standard means. These young "weak-line" T Tauri stars deserve more systematic study than they have so far received.

2.2 Interpretation of T Tauri Spectral Energy Distributions

An array of observational methodologies are now used to probe the properties of protostellar disks. High-resolution spectroscopy, optical and near-IR imaging, microwave interferometry (reviewed in chapters of *Protostars and Planets IV*) have all yielded fascinating, detailed information on individual objects. So far, however, the most general information has come from the determination of the spectral energy distributions (SEDs) of a large number star/disk systems, by multiwavelength photometry. The SED is defined as the quantity $\lambda F_\lambda(\lambda)$, where F_λ is the measured flux per unit wavelength λ [or equivalently, in terms of frequency, $\nu F_\nu(\nu)$]. The SED constrains the trends of several properties potentially important for planet formation: disk mass, temperature, accretion rate, lifetime and the degree to which solid particles have coagulated. Therefore, in the remainder of this section, we concentrate on the basics of interpreting SEDs, with an emphasis on the derivation of

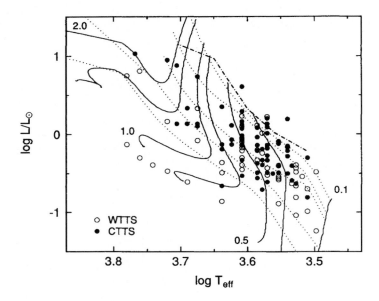

Fig. 2. Positions on the H–R diagram of classical T Tauri stars (*filled circles*) and "weak-line" T Tauri stars (*open circles*) for stars in the Taurus–Auriga star-forming region. The solid lines are theoretical evolutionary tracks, labeled by mass in units of $M_\odot$, from D'Antona and Mazzitelli (1994) The dashed lines are isochrons, corresponding to (*from the top down*): $10^5, 10^6, 3 \times 10^6, 10^7$ and 3×10^7 years, respectively. The dot-dash line is a theoretical upper limit pre-main sequence locations in the diagram (Stahler 1983). (Figure from Kenyon and Hartmann 1985)

fundamental relationships and the approximations commonly employed to obtain useful estimates of disk properties.

We start with some concepts from basic radiative transfer theory (Mihalas 1978). Define the *specific intensity* I_ν, to be the radiant energy dE flowing in direction **k** through an area dA (with normal in direction **n**), in time dt, frequency interval $d\nu$ and solid angle $d\Omega$:

$$dE = I_\nu \mathbf{k} \cdot \mathbf{n} \, dA d\nu d\Omega \ .$$

Then the *flux vector* is defined to be the moment

$$\mathbf{F}_\nu = \int I_\nu \mathbf{k} \, d\Omega \ .$$

Thus the flux component F_ν, coming from a protostellar disk and passing into a telescope pointed at the disk is given by

$$F_\nu = \int (2\pi r dr) \frac{\cos\theta}{d^2} I_\nu \ ,$$

where the integral is to be taken over all disk radii. Interpretation of the measurement of F_ν therefore requires that the quantity I_ν be understood in terms of disk properties. Now the significance of I_ν is that it obeys a conservation law of the form

$$\frac{\partial I_\nu}{\partial t} + c\mathbf{k} \cdot \nabla I_\nu = \text{sources} + \text{sinks} . \tag{1}$$

The first term on the left is negligible in cases where changes in the source and intervening medium are slow compared to the light travel time. In a vacuum, the right hand side is zero. In the presence of matter, the source and sink terms are expressed as

$$\text{sources} = \frac{\varrho\varepsilon_\nu}{4\pi} + (\text{scattering terms})$$

$$\text{sinks} = \varrho\kappa_\nu I_\nu + (\text{scattering terms}) ,$$

where ϱ is the material mass density, ε_ν is the emissivity/mass and κ_ν is the absorption opacity (the latter two being wavelength-dependent). The scattering terms quantify the amount of radiation scattered into and out of the relevant direction and wavelength interval, and we assume them to be unimportant for the present purposes. In the case of *local thermodynamic equilibrium* (LTE), in which the radiative state of material is defined solely by its temperature, the relationship $\varepsilon_\nu = 4\pi\kappa_\nu B_\nu(T)$ holds, where $B_\nu(T)$ is the Planck function:

$$B_\nu(T) = \frac{2h\nu^3/c^2}{\mathrm{e}^{h\nu/kT} - 1} .$$

Thus, (1) can be written

$$\frac{\mathrm{d}I_\nu}{\mathrm{d}\tau_\nu} + I_\nu = S_\nu, \tag{2}$$

where

$$\tau_\nu = \int_0^S \varrho\kappa_\nu \, \mathrm{d}s'$$

is the optical depth along the ray path length s, and the source function S_ν is given by

$$S_\nu = \frac{\varepsilon}{4\pi\kappa_\nu} = B_\nu(T) .$$

The last equality holds for LTE. Equation (1) has the formal solution

$$I_\nu(\tau_\nu) = I_\nu(0)\mathrm{e}^{-\tau_\nu} + \int_0^{\tau_\nu} B_\nu(T) \exp\left[-(\tau_\nu - \tau'_\nu)\right] \mathrm{d}\tau'_\nu .$$

For the radiating disk, we can measure the optical depth along the normal to the disk ($\tau_{\nu\perp}$) and from the midplane, in which case $I_\nu(0) = 0$ and $\tau_\nu = \tau_{\nu\perp}/\cos\theta$. If $B_\nu(T)$ was independent of τ_ν, the integral would yield

$$I_\nu = B_\nu(T)\left(1 - \exp\left(-\tau_{\nu\perp}/\cos\theta\right)\right)$$

and

$$F_\nu = \int (2\pi r\, dr) \frac{\cos\theta}{d^2} B_\nu(T)\left(1 - \exp\left(-\tau_{\nu\perp}/\cos\theta\right)\right) . \qquad (3)$$

Beckwith et al. (1990) used this expression to derive estimates of disk masses and temperature distributions from the SEDs of about 40 T Tauri stars, as described below. Of course the function $B_\nu(T)$ is generally *not* independent of τ_ν, so (3) represents an approximation. The approximation, however, does make sense in the limits of large and small disk optical depth, as long as care is taken in the interpretation of the value of the temperature T in $B_\nu(T)$.

The inner parts of protostellar disks are usually optically thick, even at far IR and millimeter wavelengths. For those regions, the exponential term in (3) can be neglected, and T is clearly to be regarded as a photospheric, or effective radiating temperature, T_e. If one represents its radial distribution as a power law, $T_e = T_{eo}(r/r_o)^{-q}$, a transformation of the variable of integration leads to

$$\nu F_\nu = A(T_{eo})\, \nu^{4-\frac{2}{q}} \left(\frac{\cos\theta}{d^2}\right) \int \frac{x\, dx}{\exp(x^q) - 1} .$$

Thus the slope and magnitude of the SED (the latter modulated by the inclination and distance to the disk) correspond to a particular power law effective temperature distribution.

Simple theoretical arguments led to the expectation that the value of q would be 3/4. First, the power radiated by an accretion disk at any distance was expected to scale as the radial mass flux through an annulus of the disk, times the gravitational energy per unit mass of the annulus. The expression of this power in terms of an effective temperature yields

$$2(2\pi r\, dr)\sigma T_e^4 \propto \dot{M}_d d\left(-\frac{GM}{2r}\right) = \dot{M}_d \frac{GM}{2r^2} dr$$

or $T_e^4 \propto r^{-3/4}$. (This expression turns out to be correct, except for the constant of proportionality; see Sect. 3.1) Second, stellar radiation impinging on a flat, optically thick disk would be absorbed and re-radiated at a temperature T'_e according to

$$T'^4_e = T_*^4 \left(\frac{r_*}{r}\right)^2 \sin\alpha \approx T_*^4 \left(\frac{r_*}{r}\right)^2 \left(\frac{r_*}{r}\right)$$

(where α is the angle upon which starlight impinges the disk), which again implies $T'^4_e \propto r^{-3/4}$. In fact, Beckwith et al. (1990) found q to be greater than 3/4 (Figs.3 and 4), with the most frequent value being close to 1/2, which corresponds to a flat SED in the IR. This observation led to the development of more sophisticated models of disk thermal structure than provided by the simple arguments given above. It was recognized, for instance, that if the surface

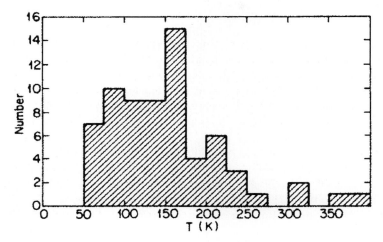

Fig. 3. Histogram of disk temperatures at 1 AU inferred by modeling SEDs, for a sample of T Tauri stars. (Figure from Beckwith et al. 1990)

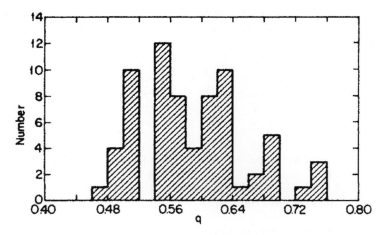

Fig. 4. Histogram of disk temperature power law indices, q, inferred from SEDs, for a sample of T Tauri stars. Most values of q are less than that predicted for accretion or reprocessing from a flat disk, which indicates that SEDs are generally flatter than predicted by the simplest models. (Figure from Beckwith et al. 1990)

of the disk flared upwards with distance from the star, proportionally more stellar radiation would be absorbed and re-emitted at larger distances, which would produce a flatter SED in the IR (Kenyon and Hartmann 1985). In particular, Chiang and Goldreich (1997) analyzed the two-layer model illustrated in Fig. 5.

Dust particles in the upper layer of the disk absorb stellar radiation and emit IR radiation, both outward to space and downward, toward the disk midplane. Because they are smaller than IR wavelengths ($\leq 1\,\mu$m), the particles are inefficient emitters and must become superheated (relative to the gas) in order to attain thermal equilibrium with the radiation they absorb. The lower layers of the disk are heated by the radiation from the dust. The vertical extent of the disk is determined by a balance between pressure and the vertical component of stellar gravity at every radius. In this model, the net effect of this balance is that the thickness of the disk increases faster than linearly with radius, i.e., it flares upward. Both the superheating of the dust and the flaring contribute to flattening the SED relative to that which would correspond to $T'_e \propto r^{-3/4}$, as shown in Fig. 6. The essential features of the Chiang and Goldreich (1997) model are evident in the results of D'Alessio et al. (2001), who analyzed disk structure in much greater detail. This work shows that heating by stellar radiation is the primary determinate of $T_e(r)$ beyond a few AU, and therefore of the slope of the SED at IR and submillimeter wavelengths. Within a few AU, internal

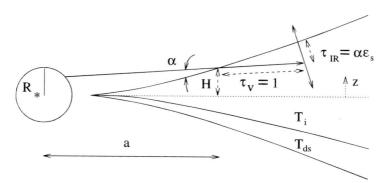

Fig. 5. The two-layer disk model of Chiang and Goldreich (1997), used to explain the flat continuum spectrum of many pre-main stars. Stellar radiation, incident on the disk at distance a, height H, and angle α, is absorbed by dust within one (visible wavelength) optical depth along its path into the upper layers of the disk. The radiation is re-emitted at infrared wavelengths, outward to space and inward toward the midplane. Because the dust particles are too small to be efficient radiators in the infrared, their temperature T_{ds} exceeds that of the surrounding gas, and the temperature T_i of the interior dust (which absorbs and emits the IR radiation from the upper layers with equal efficiency). (Figure from Chiang and Goldreich 1997)

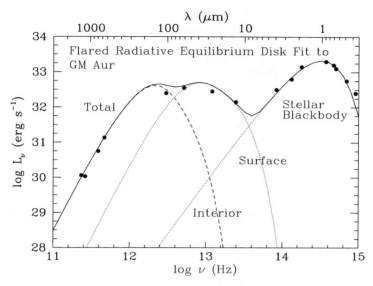

Fig. 6. A fit to the SED of the T Tauri star GM Aur, using the two-layer model. (The SED here is plotted as a function of frequency, with wavelength along the top. Astronomers differ in their preferences of independent variable.) The contribution of the hot surface dust layer dominates throughout the IR, falling below that from the interior only at wavelengths long enough for the disk to be optically thin. (Figure from Chiang and Goldreich 1997)

dissipation associated with disk accretion is important for determining the thermal structure.

At wavelengths where large parts of the disk are optically thin (typically, $\lambda > 1$ millimeter), (3) may be written

$$F_\nu = \frac{1}{d^2} \int (2\pi r \, dr) \frac{2\nu^2 kT}{c^2} \Sigma(r) \kappa_\nu .$$

The quantity $\Sigma(r)$ is the surface density. To obtain this expression, the exponential was approximated in the limit of $\tau_{\nu\perp} \ll 1$, the Planck function was approximated by the Rayleigh–Jeans form $B_\nu \approx 2\nu^2 kT/c^2$, and the substitution $\tau_{\nu\perp} = \Sigma(r)\kappa_\nu$ was made. In protostellar disks, as in the cool interstellar medium, the opacity is determined essentially by the size and composition of dust. Except near resonant absorptions, it can be expressed as a power of frequency, $\kappa_\nu = \kappa_0 (\nu/\nu_0)^\beta$, in which case

$$\nu F_\nu = \frac{A}{d^2} \nu^{3+\beta} \kappa_0 \int (2\pi r \, dr) \, T(r) \, \Sigma(r) .$$

The quantity A is a known constant. In principle, β can be determined by observations alone, through measurements of the same system at different frequencies. The integral then represents a "temperature-weighted" disk mass.

Beckwith et al. (1990) constructed composite models of the SEDs of stars observed at 1.3 and 2.7 mm, by combining their radio data with ground-based optical photometry and IR measurements obtained by the Infrared Astronomical Satellite, and so obtained the disk mass estimates shown in Fig. 7. The masses are mainly in the range 10^{-3}–$10^{-1}\,M_\odot$, and so are typically within an order of magnitude of the minimum mass inferred for the primitive solar nebula, $10^{-2}\,M_\odot$ (Weidenschilling 1977). The lack of correlation of disk mass with age is not understood, but may reflect the fact that it is really *dust* that is being observed, and the mass of dust present in a system could be the fluctuating result of a residual interstellar component, loss by coagulation, and production by collision and fragmentation of larger bodies.

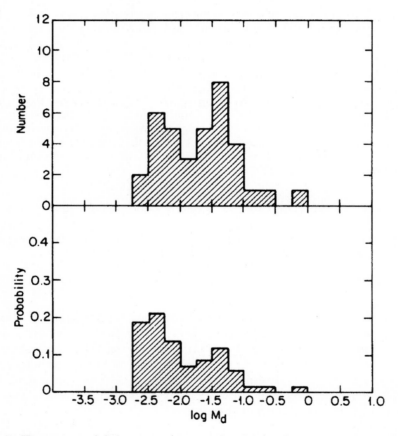

Fig. 7. Histograms of disk masses (measured in $M_\odot$) inferred by modeling SEDs, for a sample of T Tauri stars. The upper panel is that deduced directly from the data; the lower panel is an inferred distribution which accounts for sampling bias. Values of disk masses are subject to uncertainties associated with dust emissivity and abundance. (Figure from Beckwith et al. 1990)

The exponent β is expected to lie in the range $0 < \beta < 2$, with the smaller values favored as particles grow. Comparisons of observed values of β and κ_0 with laboratory measurements and theoretical calculations of absorptivity should then yield information on particle growth in disks (?). There is, in fact, evidence from multiwavelength imaging and SED fits (D'Alessio et al. 2001; Throop et al. 2001) that particle growth to mm-sized pebbles is observed, but compelling, quantitative results are difficult to obtain because of ambiguities in the modeling; see Beckwith et al. (2000) for a discussion.

2.3 Accretion Rates of Protostellar Disks

The accretion rate of a protostellar disk is an important quantity because it determines the amount of energy released *within* the disk, which is the primary determinant of midplane temperatures in optically thick parts of the disk. Optical and UV radiation provide better diagnostics of accretion rates than IR radiation, because the latter tends to be dominated by reprocessing of stellar emission. Figure 8 shows a comparison between the optical spectrum of the T Tauri star BP Tau and the WTTS LkCa7. The notable features of the BP Tau spectrum are the excess emission at shorter wavelengths, the prominent Balmer lines (at 4861, 4340, 4101 Å...; H_α at 6563 Å not shown),

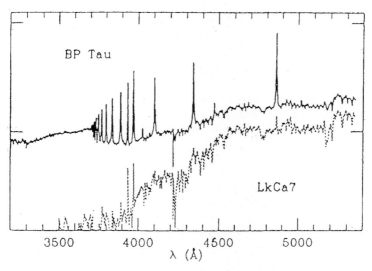

Fig. 8. Optical spectra of the T Tauri star BP Tau (*solid*) and the weak-line T Tauri star LkCa7 (*dotted*). The BP Tau spectrum exhibits excess emission at short wavelengths, prominent Balmer lines and the narrower absorption lines, features which are typical of stars with disks. They are explained by the accretion of hot, infalling gas onto the star. The normal, pre-main sequence photosphere of LkCa7 indicates that this star is not undergoing accretion; its lack of IR excess indicates that it has no disk. (Figure from Hartmann 1998)

and the narrower absorption lines. These features are typical of T Tauri stars, and are attributed to the presence of hot (several thousand K) gas, overlying a normal, pre-main sequence photosphere, such as that of LkCa7. There is evidence that the hot gas is produced by infalling material, channeled along stellar magnetic field lines, from the inner edge of the accretion disk to the surface of the star (Hartmann 1998).

Gullbring et al. (1998) and Calvet and Gullbring (1998) have derived accretion rates for T Tauri stars based on this concept. They assume that the two components (photosphere and hot, infalling gas) contribute to the optical spectrum, so the observed optical and UV luminosity $L_{obs} = L_{photosphere} + L_{accretion}$. The excess continuum emission due to accretion is determined by the subtraction of a "template" spectrum provided by a WTTS of the same spectral type, by the method described by Hartigan et al. (1998). Special procedures must be used to account properly for extinction and UV emission which cannot be observed from the ground. Accretion is then modeled as a one-dimensional flow from a distance R_i (the inner edge of the disk, typically several stellar radii) down a magnetic flux tube which intersects some fraction f of the stellar surface (Fig. 9). The inflow becomes supersonic, and must therefore pass through a shock before hitting the star.

The accretion rate is given by

$$\dot{M}_d = \varrho \left(4\pi r_*^2 f\right) \nu_s ,$$

where ν_s and ϱ are the gas velocity and density at the shock. The former is taken to be the free-fall velocity from R_i:

$$\nu_s^2 = \frac{2GM_*}{r_*} \left(1 - \frac{r_*}{R_i}\right) .$$

The density is determined by the assumption that the shock location occurs at optical depth unity above the star, where pressure balance between the stellar atmosphere and the inflow requires:

$$p_s = \frac{1}{2}\varrho\nu_s^2 = g_* \Sigma_s = g_* \frac{\tau}{\kappa} .$$

Here g_* is the stellar gravity, Σ_s is the surface density of the shock layer, κ is the mean opacity and the optical depth $\tau = 1$. Thus, for known M_*, r_*, g_*, R_i and κ, the flux of energy from the accretion column can be calculated:

$$F_{accretion} = \left(\frac{1}{2}\varrho\nu_s^2\right) \nu_s .$$

The total accretion luminosity is

$$L_{accretion} = \left(4\pi r_*^2 f\right) F_{accretion} ,$$

which then, in principle, determines f. Comparisons of calculated spectra with excess emission determined from observations, and corresponding values of $F_{accretion}$ and f, are shown in Fig. 10. An important result is that accretion apparently occurs on only 1% or less of the stellar surface.

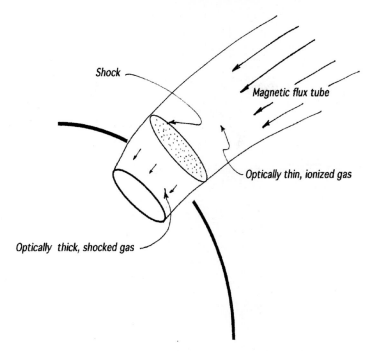

Fig. 9. Accretion rates can be obtained from models in which it is assumed that gas from the inner edge of the disk falls onto the star along stellar magnetic flux tubes. Optically thin gas, flowing supersonically toward the star, passes through a shock wave near the stellar surface. Radiation from the shocked gas heats both the stellar photosphere and the infalling gas above it, which produce excess UV and optical radiation and the characteristic emission features of accreting stars

Accretion rates determined by these methods are shown as a function of stellar age in Fig. 11. Note that a typical accretion rate of $10^{-8}\,M_\odot$/year is consistent with a disk mass of $10^{-2}\,M_\odot$ for a million year old disk. Note also that active (accreting) disks last for many thousands of dynamical (rotation) periods; thus disk evolution is gradual and the processes that cause it are subtle.

2.4 Internal Temperatures of Protostellar Disks

Temperatures within protostellar disks are not directly observable because most disks are optically thick within a few AU of the star, even at long ($\lambda \geq 1\,\mathrm{mm}$) wavelengths. Therefore, some model of the vertical structure, like those referred to above, must be constructed to relate the observed surface fluxes to the internal state. A key parameter in such a model is the rate of energy dissipation within the disk, which is related to the accretion rate, as described below in Sect. 1.3.1. Midplane temperatures for disks around T

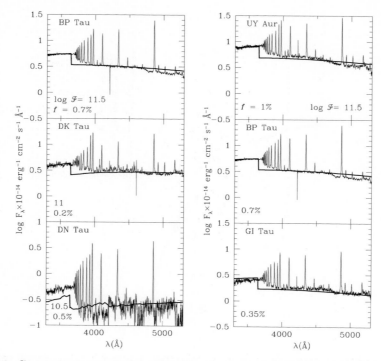

Fig. 10. Comparisons of calculated continuum spectra (*thick line*) with excess emission determined from observations, for a sample of T Tauri stars. Values of $F_{\mathrm{accretion}}$ and f determined from the model illustrated in Fig. 10 are given. Note that $f \leq 0.1$; that is, accretion apparently occurs on only 1% or less of the stellar surface. (Figure from Calvet and Gullbring 1998)

Tauri stars were estimated by Woolum and Cassen (1999), by combining the results described above for disk surface temperatures, masses and accretion rates, with a simple model based on radiative transport through a plane-parallel atmosphere. The observed IR flux was treated as the sum of that due to internally released accretion energy and a reprocessed stellar component. In optically thick parts of the disk, the former largely controls the midplane temperature, while the latter dominates the observed flux. They concluded that midplane temperatures at 1 AU are mainly in the range 200–800 K, for disks with ages of about 1 million years. At the low pressures in these disks ($< 10^{-3}$bars), H_2O would exist as vapor within a few AU of the star (wherever the midplane temperature exceeds 160 K), and so would not be readily incorporated into planetary objects. Icy objects could form beyond 2–3 AU. They also argued that if very young disks were characterized by accretion rates as high as $10^{-6} M_\odot$/year (Fig. 11), midplane temperatures would then be high enough to vaporize even the rock-forming elements (primarily Fe, Mg, Si) at a few AU.

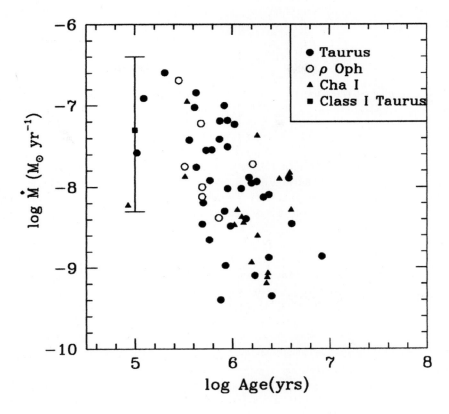

Fig. 11. Mass accretion rates and ages inferred for T Tauri stars in three star-forming regions. The vertical line indicates the mean and dispersion of accretion rates estimated for embedded stars (assumed to be 10^5 years old). (Figure from Calvet et al. 2000)

3 Theory of Disk Structure and Evolution

3.1 Conservation Equations

The equations that govern the structure and evolution astrophysical disks are derived from the equations of mass, momentum and energy conservation, which have the general form

$$\frac{\partial}{\partial t}\begin{pmatrix}\varrho\\ \mathbf{v}\\ e\end{pmatrix} + \nabla\cdot\left[\mathbf{v}\begin{pmatrix}\varrho\\ \mathbf{v}\\ e\end{pmatrix}\right] = \begin{pmatrix}0\\ \text{forces}\\ \text{sources} + \text{transport}\end{pmatrix}.$$

When applied without reference to a specific geometry, they lead to the equations of fluid motion, which are usually expressed as

$$\frac{\partial \varrho}{\partial t} + \nabla \cdot (\varrho \mathbf{v}) = 0 \quad \text{(mass)}$$

$$\varrho \left(\frac{\partial \mathbf{v}}{\partial t} + \mathbf{v} \cdot \nabla \mathbf{v} \right) = -\nabla p + \nabla \cdot \mathbf{w} - \varrho \nabla \Phi \quad \text{(momentum)}$$

$$\varrho \left(\frac{\partial e}{\partial t} + \mathbf{v} \cdot \nabla e \right) = -p \nabla \cdot \mathbf{v} - \nabla \cdot \mathbf{F} + D \quad \text{(energy)} \ .$$

In these equations, $\mathbf{w}$ is the non-diagonal part of the stress tensor, w_{ij} being the stress on a j-facing surface in the i-th direction; Φ is the gravitational potential; $\mathbf{F}$ is the energy flux vector due to radiation, conduction, or other means; and D is the rate of energy dissipation associated with stresses:

$$D = w_{ij} \frac{\partial v_i}{\partial w_j} \ .$$

Magnetic fields have been ignored in these equations, as they will be in the rest of this section; but they can be important and will be discussed later.

The equations of disk structure and evolution are most illuminatingly derived by application of the conservation equations in their general form to a control volume such as that shown in Fig. 12. Variations in ϕ are not considered. Because the disk is assumed to be thin, it is practical to consider the values of z-integrated quantities, such as the surface density:

$$\Sigma = \int_{-\infty}^{+\infty} \varrho \, \mathrm{d}z \ .$$

Thus, the conservation of mass equation, integrated over z, is:

$$\frac{\partial \Sigma}{\partial t} + \frac{1}{r} \frac{\partial \dot{m}}{\partial r} = 0, \tag{4}$$

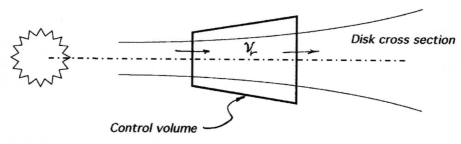

Fig. 12. Control volume for the derivation of conservation equations for a disk with radially flowing material and no azimuthal gradients. The radial velocity v_r is taken positive outward by convention, but will generally be inward in the inner parts of the disk and outward in the outer parts

where

$$\dot{m} = \int_{-\infty}^{+\infty} \varrho v_r r \, dz$$

is $1/2\pi$ times the radial mass flux through the disk (the "reduced" mass flux). In what follows, it is assumed that the radial velocity v_r varies little over the disk thickness, so $\dot{m} = \varrho v_r r$.

To a high degree of accuracy, the vertical momentum balance is hydrostatic, just as in a planetary atmosphere. Pressure forces are balanced by gravity:

$$0 = -\frac{\partial p}{\partial z} + \varrho \frac{\partial \Phi}{\partial z}.$$

The gravitational potential consists of two terms, one due to the vertical component of stellar gravity and one due to the self-gravity of the disk. The latter is negligible for stable, unperturbed disks, so

$$\frac{\partial \Phi}{\partial z} = \frac{\partial \Phi_*}{\partial z} = -\frac{GM_*}{(r^2+z^2)}\frac{z}{r} \approx -\frac{GM_*}{r^2}\frac{z}{r} = -\Omega_K^2 z.$$

The quantity $\Omega_K^2 = GM_*/r^3$ is the square of the Keplerian frequency, i.e., the angular velocity of a freely orbiting object in a circular orbit at distance r from the star. Thus,

$$\frac{1}{\varrho}\frac{\partial p}{\partial z} = -\Omega_K^2 z. \tag{5}$$

If we characterize the distance z by the thickness of the disk measured, say, in terms of a scale height h, and note that the quantity p/ϱ is, to order unity, the sound speed c_s, we find that the relative thickness $h/r = c_s/r\Omega_K = c_s/V_K \ll 1$, for a thin disk. ($V_K$ is the Keplerian velocity.) Disk models typically yield values of h/r in the range 10^{-2}–10^{-1}.

The radial momentum equation is:

$$\frac{\partial (v_r \Sigma)}{\partial t} + \frac{1}{r}\frac{\partial (v_r \dot{m})}{\partial r} = \Sigma\left(\frac{\partial \Phi}{\partial r} + r\Omega^2\right) - \frac{\partial P}{\partial r},$$

where $P \equiv \int_{-\infty}^{+\infty} p \, dz$ and $\partial \Phi/\partial r = -GM_*/r^2$. It is readily seen that, as long as the characteristic evolution time r/v_r is much longer than the orbital period $2\pi/\Omega$, the terms on the left are much smaller than the first term on the right-hand side. Also, the pressure gradient is a factor of $(h/r)^2$ smaller than the first term on the right. Thus, to this order, $\Omega = \Omega_k$; the orbital motion is Keplerian. (It turns out that the small radial pressure gradient is important when considering the fate of solid objects in the disk, as discussed in Sect. 1.4.1.).

Disk evolution is addressed directly by the angular momentum equation, which is:

$$\frac{\partial(\Sigma j)}{\partial t} + \frac{1}{r}\frac{\partial(\dot{m}j)}{\partial r} = \frac{1}{r}\frac{\partial(r^2 W_{r\phi})}{\partial r} + \frac{1}{2\pi}\frac{\partial T}{\partial r}.$$

Here, j is the angular momentum/mass $= r^2\Omega_K = rV_K = (GM_*r)^{1/2}$, T is any externally applied torque, and

$$W_{\phi r} = \int_{-\infty}^{+\infty} w_{\phi r}\, dz\ .$$

The mass conservation equation (4) can be used to solve for the reduced mass flux:

$$\dot{m} = \frac{r}{j}\frac{\partial}{\partial r}\left(\frac{T}{\pi} + 2r^2 W_{\phi r}\right). \tag{6}$$

This equation and (4) specify the rearrangement of surface density in terms of the forces that affect angular momentum (we omit the external torque term from now on, for brevity):

$$\frac{\partial \Sigma}{\partial t} = -\frac{2}{r\sqrt{GM_*}}\frac{\partial}{\partial r}\left[r^{1/2}\frac{\partial}{\partial r}\left(r^2 W_{\phi r}\right)\right].$$

Now it is usually supposed that $W_{\phi r}$ is a *Newtonian stress*; that is, it is linearly proportional to the rate of strain (as is true for many viscous fluids). If this is the case,

$$W_{\phi r} = \Sigma \nu r \frac{\partial \Omega}{\partial r},$$

where we follow standard notation and use ν for the kinematic viscosity (not to be confused with frequency, in Sect. 1.2). The disk evolution equation is then

$$\frac{\partial \Sigma}{\partial t} = \frac{3}{r}\frac{\partial}{\partial r}\left[r^{1/2}\frac{\partial}{\partial r}\left(r^{1/2}\Sigma \nu\right)\right]. \tag{7}$$

This equation has the form of a diffusion equation for Σ, and the problem of disk evolution has been essentially reduced to one of determining the proper expression for ν. For instance, if ν were given as any function of r, the equation would be linear, and could be solved by any of a number of standard techniques (Lynden-Bell and Pringle 1974). In general, solutions for disks which conserve overall angular momentum (but can lose energy) have the properties shown schematically in Fig. 13. Material in the inner parts of the disk lose angular momentum and spiral in to the central star, while material in the outer disk gains angular momentum and expands outward. This very

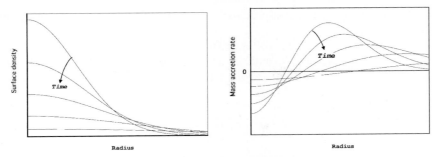

Fig. 13. Schematic representation of the variations of surface density (*left*) and accretion rate (*right*), as functions of time and radius. Surface density decreases in the inner parts of the disk as the disk spreads. The mass accretion rate is negative (inwards) in the inner parts, nearly independent of radius and diminishing in magnitude with time. There is a stagnation radius, at which the radial motion changes direction; the location of the stagnation radius moves outward with time

fundamental behavior of accretion disks can be demonstrated by the application of simple conservation arguments, and is independent of the particular mechanism(s) of angular momentum transport (see Lynden-Bell and Pringle 1974).

What is the value of ν indicated by observations? The viscous evolution time is given by r_d^2/ν, where r_d is a characteristic disk size, say, 100 AU. A disk age of 1 million years then yields a kinematic viscosity of about 10^{15} cm^2/sec, far greater than the ordinary molecular viscosity of hydrogen gas, which is about 10^6 cm^2/sec for conditions appropriate for these disks. It is therefore commonly supposed that turbulence in the disks enhances the viscosity far above the molecular value. The turbulent kinematic viscosity (or eddy diffusivity, as atmospheric scientists call it) can be expressed as the product of a turbulent velocity and a mixing length. If the latter is as large as the scale height of the disk, relatively modest turbulent velocities, about 10^3 cm/sec would suffice to provide the inferred value of the viscosity.

The concept of turbulent viscosity gives rise to a prescription originally postulated by Shakura and Sunyaev (1973), in another context. They assumed

$$\nu = \alpha c_s h$$

where the parameter $0 < \alpha < 1$, as expected if the scale height and sound speed are upper limits to the mixing length and turbulent velocity, respectively. This widely adopted formulation shifts the burden of quantifying angular momentum transport from ν to α. If one adopts expected values of c_s and h (say 1 km/sec and 0.1 AU, respectively), a value of α between 10^{-3} and 10^{-2} would provide the necessary viscosity. Useful as such prescriptions may be for some purposes, one should bear in mind that the clear inadequacy of molecular viscosity renders even the assumption of Newtonian viscosity suspect. For this reason, it is prudent to recognize the more basic forms of the evolution (4) and (6).

The energy lost from an evolving disk contributes to the observed radiation and provides diagnostic information, as discussed in Sect. 1.2. If the dissipated energy is expressed in terms of an effective radiating temperature T_e, the energy conservation equation is

$$\frac{\partial (\Sigma e)}{\partial t} + \frac{1}{r}\frac{\partial (\dot{m} e)}{\partial r} = \frac{1}{r}\frac{\partial (rW_{\phi r}v_\phi)}{\partial r} - 2\sigma T_e^4 + (\text{external irradiation}) \,. \quad (8)$$

The thermal energy is $(h/r)^2$ times smaller than the kinetic and gravitational energies, so the pressure work term has been ignored, and a good approximation for the energy/mass is

$$e = \frac{v_\phi^2}{2} + \Phi = -\frac{GM_*}{2r} \,.$$

The last term in (8) may include energy deposition from the star, the background radiation field, and radiation from other parts of the disk itself. Ignoring it for the moment, and using the mass conservation equation to eliminate terms in (8), one finds

$$2\sigma T_e^4 = rW_{\phi r}\frac{\partial \Omega_k}{\partial r} = -\frac{3}{2}W_{\phi r}\Omega_K \,.$$

This formula can be expressed in terms of the disk mass flux by noting that solutions of the disk evolution equation generally produce an r-independent $\dot{m}$ over some portion of the inner disk. In that region, (5) can be integrated to obtain

$$W_{\phi r} = \frac{\dot{m} j}{r^2} \,.$$

Therefore

$$\sigma T_e^4 = \frac{3\dot{M}_d GM}{8\pi r^3} \,,$$

where $\dot{M}_d = -2\pi \dot{m}$. Note that this frequently encountered formula for the distribution of internally generated, radiated energy does not depend on the nature of the stresses causing angular momentum transport. Its validity does require that energy be dissipated "locally"; that is, that radial transport of energy is negligible compared to vertical transport, and that the disk material behave as a Newtonian fluid. And, as discussed above, it is not the sole source of the observed radiated energy, which can be dominated by external irradiation which is absorbed and re-emitted by the disk.

Finally, it is important to have a description of the vertical structure of the disk, especially when considering the accumulation of solid material into planets. Equation (5) gives the mechanical balance. The thermal state must be ascertained by an energy equation of the form

$$\frac{\partial F_z}{\partial z} = -(\text{dissipation}) + (\text{external irradiation}),$$

where F_z is the flux of energy in the z direction and the terms on the right must now be specified as functions of z. Although F_z may have radiative and convective components, the latter is usually unimportant for heat transport in disks (Cassen 1993; D'Alessio et al. 1999). Several other factors, however, do complicate vertical structure models. First, without a detailed model for the stresses responsible for angular momentum transport, some assumption must be made regarding the vertical distribution of dissipation. Usually it is assumed that the dissipation rate is proportional to the local mass density, although this need not be the case in a real disk. Second, the evaporation, condensation and coagulation of dust and ice, which are the major contributors to the opacity (Pollack et al. 1994; Henning and Stognienko 1996), must be solved for self-consistently. Third, the effect of the external radiation can be a complicated function disk radial structure (Bell 1999).

The nature of the most useful approximations and assumptions employed to determine vertical structure may depend on the particular objective. Studies directed toward the astronomical appearance of disks must properly account for the external illumination, but may simplify the issues of opacity structure and internal dissipation, as in, for instance, Chiang and Goldreich (1997). Studies directed toward understanding the thermal conditions of planet-building may simplify the treatment of external irradiation, which has a minor effect on midplane temperatures where disks are optically thick, but must account for the vertical distributions of dissipation and opacity (e.g., Cassen 2001).

A simplification that is usually valid is to treat the vertical structure as "quasi-steady". That is, the time for even an optically thick disk to adjust to new heating and cooling conditions,

$$t_{\text{thermal}} = \frac{\Sigma c_s^2}{2\sigma T_e^2}$$

is usually much shorter than other evolutionary timescales (e.g., the characteristic dust coagulation time or the time over which the local accretion rates changes). There are conditions, however, when even this simplification is invalid (e.g., during outbursts). See the papers by D'Alessio et al. (1998, 1999, 2001), Bell and Lin (1994) and Bell et al. (1997) and Cassen (2001) for further applications of vertical structure modeling.

3.2 Turbulence in Disks

The problem of disk evolution is that of determining the precise mechanisms of angular momentum transport. What are the processes that produce the torques and stresses that appear in (6) and how are they quantified? Instability leading to turbulence is the most frequently invoked phenomenon for providing the required viscosity, but the nature of the instability and the consequences of the turbulence remain controversial, despite assertions that the problem has been solved (e.g., Balbus and Hawley (1991, 2000). The problem begins with the realization that a Keplerian disk, although possessing a strong radial gradient of azimuthal velocity (shear), does not suffer the kind of shear-induced instability known to produce turbulence in other situations. An annulus of disk material displaced outward from radius r_1 to r_2, while conserving its angular momentum, experiences a centrifugal force j_1^2/r_2^3 at its new location. The centrifugal force required to maintain it in equilibrium is j_2^2/r_2^3. So, as long as $j_1 < j_2$, as it is in a Keplerian disk, the material experiences a net force which restores it toward its original location. The disk satisfies Rayleigh's criterion for stability,

$$\frac{\mathrm{d}j^2}{\mathrm{d}r} > 0$$

and some other source of turbulence must be sought.

Other instabilities exist, of course. For instance, it was proposed by Lin and Papaloizou (1980) that convective instability, which stirs a fluid when radiation alone would require a superadiabatic temperature gradient, could provide the required radial mixing of angular momentum. They envisioned a *feedback loop*, in which internally dissipated energy in an optically thick disk would drive convection; the turbulence so produced would give rise to $W_{\phi r}$ stresses; the stresses would result in the net outward transfer of angular momentum from the (more rapidly rotating) inner annuli to the (less rapidly) rotating outer annuli; the loss of angular momentum by the inner annuli would release gravitational energy, which would be the ultimate source of the dissipated energy driving convection. It is just such a feedback process that is required to overcome the inherent stability conferred by rotation.

But strong arguments have been presented against the existence of *any* such feedback process, in the absence of magnetic forces Balbus et al. (1996). The issue can be addressed by examining the equations of turbulent motion, derived from the mass and momentum equations by separating variables into an *average* part and a *fluctuating* part. For instance, for the velocity component v_i,

$$v_i = \langle v_i \rangle + u_i$$
$$\langle v_i \rangle = \frac{1}{2\pi \Delta r} \int v_i \, \mathrm{d}\phi \, \mathrm{d}r \, \mathrm{d}z,$$

where the r-integration is taken over some suitably small interval. In particular, one can derive a set of energy equations by multiplying each component of the momentum equation by its respective velocity component, and dropping terms smaller than second order in the fluctuating quantities (and other terms assumed to be small). The results for the r and ϕ directions, derived by Balbus et al. (1996), are:

$$\frac{\partial}{\partial t}\left\langle\frac{\varrho u_r^2}{2}\right\rangle = 2\Omega\left\langle\varrho u_r u_\phi\right\rangle - \left\langle u_r\frac{\partial p}{\partial r}\right\rangle - \left\langle\varrho\nu|\nabla u_r|^2\right\rangle$$

$$\frac{\partial}{\partial t}\left\langle\frac{\varrho u_\phi^2}{2}\right\rangle = -\frac{\langle\varrho u_r u_\phi\rangle}{r}\frac{\mathrm{d}\left(r^2\Omega\right)}{\mathrm{d}r} - \left\langle\frac{u_\phi}{r}\frac{\partial p}{\partial \phi}\right\rangle - \left\langle\varrho\nu|\nabla u_\phi|^2\right\rangle . \quad (9)$$

These equations do not represent the energy conservation law, being derived from the momentum equations alone, but are relations that must hold between the mechanical energies associated with turbulent fluctuations and the mean flow, the latter being represented by Ω and its derivative. The last term on the right is the energy dissipated by viscosity; it always reduces the energy of the fluctuating part of the flow (the terms whose derivatives appear on the left). The first term on right prescribes the interaction of the turbulent stress, $\langle\varrho u_r u_\phi\rangle$, with the mean flow. The point stressed by Balbus et al. (1996) is that, in Keplerian disks (or any disk that is stable by Rayleigh's criterion) this interaction provides a *negative* feedback for the energy of ϕ-fluctuations if the stress is such as to transport angular momentum outward, i.e., $\langle\varrho u_r u_\phi\rangle > 0$. But it is necessary for turbulence to transport angular momentum outward in an accretion disk, because it is the loss of angular momentum that allows material to flow inward. Thus, turbulence that allows accretion appears to be self-defeating, no matter what the source of the turbulence. (True, the radial fluctuations are not damped, but *correlated* azimuthal fluctuations are necessary to produce any turbulent stress.) What about the ϕ-pressure gradient term? Balbus et al. (1996) argue that it cannot isotropize the turbulence enough to overcome the damping effect of the positive angular momentum gradient (while pointing out that long-range correlations in pressure fluctuations associated with *organized waves* can provide the desired effect, but would not be considered to be turbulence).

The conclusion that purely hydrodynamic turbulence cannot be self-sustaining in Keplerian disks is supported by numerical calculations of the evolution of a turbulent field in a "local" patch of a disk. In these calculations (e.g., Balbus et al. (1996) and Hawley et al. (1999), in which an appropriate form of periodic boundary conditions are imposed on a sheared, rotating fluid, instabilities are not manifested and an initially turbulent field decays. Furthermore, calculations in which convective turbulence is forced by imposition of an *ad hoc* heat source produce stresses which induce *inward* angular momentum transport, consistent with the arguments given above.

These results indicate that protostellar disk evolution must be driven by either magnetic instability or the action of waves, topics discussed below. And yet the issue should not be considered settled; not all researchers are ready to rule out the possibility of hydrodynamic turbulence in the unequivocal manner that its detractors have. For instance, calculations by Klahr and Bodenheimer (2000) indicate that *baroclinic* instabilities lead to sustained positive turbulent stresses and consequent outward angular momentum transport (see also Sheehan et al. (1999) and Li et al. (2001)). Baroclinic instability, well-known in planetary atmospheres, can occur when surfaces of constant pressure do not coincide with surfaces of constant density, i.e., when

$$\nabla p \times \nabla \varrho \neq 0 \ .$$

This condition is precluded in any calculation for which a barotropic relation between pressure and density, $p \sim \varrho^\gamma$, is assumed, for then

$$\nabla p \times \nabla \varrho \sim \nabla \varrho^\gamma \times \nabla \varrho = 0$$

and surfaces of constant p and ϱ do coincide. But it is virtually inevitable in an optically thick protostellar disk, where radial entropy gradients prevent the simple proportionality represented by the constant in the above equation. Local calculations, which assume a barotropic relation or do not account for a radial entropy gradient, and which exhibit decaying turbulence or negative turbulent stresses, do not allow the possibility of baroclinic instability. Now a common feature of the nonlinear development of baroclinic instabilities is the generation of relatively long-lived, organized structures, such as vortices, jet streams and spiral shock waves. Associated with these structures are non-local, correlated fluctuations which contribute to the transport of angular momentum. Related instabilities, which depend on the existence of a locally steep pressure gradient have similar properties (e.g., Lovelace et al. (1999) and Li et al. (2001). The full implications of these sources of turbulence (and possibly others; real disks are complicated structures) remain to be worked out (see the comments on Rossby waves, below).

In whatever way the issue of non-magnetic turbulence is resolved, Balbus and Hawley (1991, 1998) have shown that magnetically coupled protostellar disks are inevitably turbulent in a manner that produces outward angular momentum transport at the level required to drive the observed activity. Additional stress terms appear in (9) which can act as sources of azimuthal fluctuations. The instability that produces the turbulence relies on the fact that adjacent annuli are magnetically tethered, so they act like spring-coupled, rotating masses. If the spring constant is very strong, the instability is suppressed, but for a weak spring (magnetic field), perturbations in displacement grow. The reader is referred to the above references for a formal stability analysis. Here, we develop a useful analogy provided by Balbus and Hawley (1998) that gives some insight into the nature of the instability, and can be used to describe its essential properties.

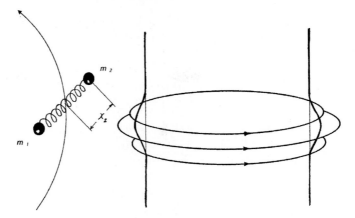

Fig. 14. Two masses, attached by a spring and displaced from a common orbit about a star (*left*), obey the same equations as orbiting, perfectly conducting fluid annuli threaded by a vertical magnetic fluid (*right*). Both systems are unstable for a range of values of the effective spring constant. This unstable range encompasses the only physically realistic values for the magnetic fluid

Consider two masses (identical for convenience) in orbit about a star, but attached by a spring with spring constant f_s (Fig. 14a) The equation of motion of either mass, in the rotating frame, is

$$\frac{d^2 \mathbf{r}}{d^2 t} = \mathbf{\Omega} \times (\mathbf{\Omega} \times \mathbf{r}) + 2\mathbf{\Omega} \times \frac{d\mathbf{r}}{dt} = -2f_s \mathbf{x} + \mathbf{g},$$

where $\mathbf{x}$ is the displacement from the equilibrium orbit. Let $\mathbf{r} = \mathbf{r}_0 + \mathbf{x}$ and expand $\mathbf{g}$ about its value at $\mathbf{r}_0$ to obtain the set

$$\ddot{x}_r - 2\Omega \dot{x}_\phi = x_r \left(-2f_s + 3\Omega^2 \right)$$
$$\ddot{x}_\phi + 2\Omega \dot{x}_r = -2f_s x_\phi .$$

The effective spring constant for the r motion is $2f_s - 3\Omega^2$, which must be positive for a restoring force. Now it turns out that these equations are exactly equivalent to those for the small displacement of annuli in a disk of perfectly conducting gas, threaded by a vertical magnetic field $\mathbf{B}$ (Fig. 14b), with the correspondence

$$2f_s - 3\Omega^2 \rightarrow r\frac{d\Omega^2}{dr} + (\mathbf{k} \cdot \mathbf{u_a})^2 .$$

The last term is the scalar product of the wave vector of the disturbance and the Alfvén velocity, the latter defined by

$$\mathbf{u_a} \equiv \frac{\mathbf{B}}{\sqrt{4\pi \varrho}} .$$

Note that since $\mathrm{d}\Omega^2/\mathrm{d}r < 0$ in a Keplerian disk, stability requires

$$(\mathbf{k}\cdot\mathbf{u_a})^2 > -r\frac{\mathrm{d}\Omega^2}{\mathrm{d}r} \Rightarrow u_a^2 > \left|\frac{r}{k^2}\frac{\mathrm{d}\Omega^2}{\mathrm{d}r}\right|.$$

But k must be large enough to allow a wavelength to fit within the thickness of the disk: $k > \pi/h$. So the right hand inequality can be extended to

$$u_a^2 > \left|\frac{r}{k^2}\frac{\mathrm{d}\Omega^2}{\mathrm{d}r}\right| > \left|\frac{rh^2}{\pi^2}\frac{\mathrm{d}\Omega^2}{\mathrm{d}r}\right| = \frac{3\Omega^2 h^2}{\pi^2} \approx \frac{3c_s^2}{\pi^2}.$$

But this condition for stability requires that the magnetic field pressure dominate the thermal pressure, a condition difficult to achieve in disks because magnetic buoyancy effects tend to expel such strong fields. Therefore, magnetic disks are unstable; in fact, they are unstable even in the limit of vanishingly weak field. Numerical simulations confirm the instability and indicate that the resulting turbulence would be sufficient to provide the inferred accretion rates of protostellar disks Hawley et al. (1995).

Can we conclude that magnetic turbulence is the main process responsible for protostellar disk evolution? For this to be the case, the disk gas must be well-coupled to the magnetic field. This condition is quantified by the requirements that the magnetic Reynolds number be larger than unity and that collisional ion-neutral momentum exchange occur rapidly compared to, say, the orbital period. Disks are dense enough in most places to insure that the latter condition is fulfilled, but the former condition requires a level of ionization which, although not very high (ionization fraction $\approx 10^{-13}$), is still difficult to attain in their cold, dusty interiors. One can identify four sources of ionization: (1) galactic cosmic rays, (2) stellar energetic particles and x-rays, (3) radioactive nuclides and (4) thermal (collisional) excitation. Galactic cosmic rays penetrate no more than about 10^2 gm/cm^2 of material, and so would be largely excluded from the inner disks ($r \leq 1$ AU) where the surface density is estimated to typically exceed 10^3 gm/cm^2. (The current galactic cosmic ray flux is obviously incapable of significant ionization of most of the approximately 10^3 gm/cm^2 of terrestrial atmosphere.) Furthermore, one expects that the intense stellar wind associated with young stars would attenuate the flux of such particles to levels well below that currently experienced by the solar system. Stellar particles and x-rays, although abundant, penetrate only about 1 gm/cm^2, and so are expected to ionize only a very thin "ionodisk" at high altitudes. Similarly, the most abundant energetic particles from radioactive nuclides have very limited ranges. Only close to the star, perhaps within a few tenths of an AU, is the disk temperature expected to be high enough to evaporate most of the dust, a condition that is probably required to maintain the ionization level necessary for magnetically induced turbulence. Thus the issue of turbulent angular momentum transport throughout the disk remains open. For a recent discussion of relevant issues by the advocates of *exclusively* magnetic turbulence, see Balbus and Hawley (2000).

3.3 Waves in Disks

It was mentioned above that organized, non-axisymmetric structures transport angular momentum. These structures frequently have the form of waves, which may be thought of as coherent perturbations of the flow through which the fluid medium flows. An astounding variety of fluid waves have been identified, many of them familiar in our everyday experience; one should expect to find many of them in disks. Waves usually result from the interaction of some disturbing force (or instability) and the natural restoring forces present in the system, necessary for the existence of an equilibrium. The restoring forces can be associated with the natural frequencies of a system. Some of the important natural frequencies of a disk are its orbital frequency Ω, the epicyclic frequency (associated with the Coriolis force and characteristic of radial oscillations), and the Brünt-Väisälä frequency (associated with pressure forces and characteristic of vertical oscillations). In a thin disk, these frequencies are all of comparable magnitude, so one might expect that there are complicated interactions among various wave types.

The nature of a wave is quantitatively described by a dispersion relation, which is a relation between the frequency of oscillation of the wave, ω, and the wave number, $k = 2\pi/\lambda$ (λ is the wavelength): $\omega = \omega(k;\text{parameters})$. From the dispersion relation, one can determine the rate at which energy is carried by the wave ($\partial \omega / \partial k$), the speed at which the wave pattern moves (ω/κ), places where the waves can and cannot propagate, and so forth. The usual method of deriving a dispersion relation involves three steps: (1) all variables are expressed as the sum of their equilibrium values and small perturbations which oscillate in time and space; (2) the fluid equations are linearized by retaining only terms first-order in the perturbations and using the fact that the equilibrium values satisfy the steady-state equations exactly; and (3) spatial derivatives in the equilibrium values are considered negligible compared to the spatial oscillations of the perturbations (WKB approximation). These steps result in a set of linear, *algebraic* equations (algebraic because the spatial and temporal derivatives of oscillating quantities are proportional to the quantities themselves). If there are no explicit forcing functions, the equations are homogeneous and have solutions (the "free wave" solutions) only if there is a specific relation between ω (specifying the temporal oscillation) and k (specifying the spatial oscillation); this is the dispersion relation.

In Keplerian disks, the strong variation in rotation rate with radius tends to shear disturbances, so that waves commonly have a spiral pattern. The equation of a spiral is $\phi = \psi(r)$; the equation of a spiral rotating at frequency ω is $\phi = \psi(r) + \omega t$; and many such spirals, with the same shape and frequency, are described by $m\phi = \psi(r) + \omega t$, where m is an integer. Therefore, in analyzing disk waves, it is expeditious to express the oscillating perturbations in the form:

$$x = X e^{i[\omega t - m\phi + \psi(r)]} .$$

where X is the amplitude of the perturbation. The radial wavenumber is given by $k = \mathrm{d}\psi/\mathrm{d}r$ and the pattern speed is $\Omega_p = \partial\phi/\partial t = \omega/m$. A positive wavenumber corresponds to a *leading* wave (phase increasing with r) and a negative wave number corresponds to a *trailing* wave. It turns out that only trailing waves carry angular momentum in ways that are physically sustainable.

Waves in disks have been studied using many different approximations, assumptions and techniques, with the consequence that it is not always easy to relate the results of one study to those of another. There are approximations based on disk geometry: 1-D (axisymmetric), 2-D (r,ϕ), and "shearing sheet" (or local Cartesian). In the last approximation, the (r,ϕ) coordinate system is replaced by a local, rotating (x,y) system, and terms associated with the curvature of streamlines are neglected. The "tightly wound" (short radial wavelength) assumption exploits the spiral geometry of the wave and is usually equivalent to the WKBJ approximation. Physical forces that are frequently (but not always) neglected are viscous, magnetic and disk self-gravitational. The following approximations of the equation of state are often encountered: isothermal, polytropic and Boussinesq (in which density variations due to dynamic pressure are neglected, but not those in the base state). In particular, it is often assumed for mathematical convenience that the surface density (rather than the volume density) is a power of the vertically integrated pressure, a kind of polytropic approximation unjustifiable by any simple physical assumption. Finally, analyses can be Lagrangian, in which the fluid displacement is used as a dependent variable, or Eulerian, in which fluid velocities are the primary dependent variables. I will identify the specific assumptions used in the analyses described below.

Spiral density waves can be considered to be the primary wave-form in astrophysical disks. They were originally studied in connection with galactic structure (Lin and Shu 1964), but observational confirmation of the theory was obtained in the structure of Saturn's rings. A standard analysis (Shu 1992) considers the disk to be infinitely thin, so that perturbations are restricted to the (r,ϕ) plane and the volume density $\varrho = \Sigma\delta(z)$, where δ is the Dirac function. It is assumed that the vertically integrated pressure is a function of only the surface density: $P = P(\Sigma)$. Viscous stresses and dissipation are ignored, but the self-gravity of the disk is included. For this purpose, the standard conservation equations of Sect. 3.1 are supplemented by Poisson's equation for the gravitational potential:

$$\frac{1}{r}\frac{\partial}{\partial r}\left(r\frac{\partial \Phi}{\partial z}\right) + \frac{1}{r^2}\frac{\partial^2 \Phi}{\partial r^2} + \frac{\partial^2 \Phi}{\partial z^2} = 4\pi G \delta(z) + \varrho_*\ .$$

Here, ϱ_* represents the contribution to the stellar potential. Shu (1992) shows how this equation can be simplified and solved in the WKB approximation to yield a simple relation between the amplitudes of the surface density and gravitational potential perturbations. The following dispersion relation can then be derived from the linearized conservation equations:

$$(\omega - m\Omega)^2 = \Omega^2 + k^2 c_s^2 - 2\pi G |k| \Sigma_0 . \tag{10}$$

Subscript 0 refers to the unperturbed state. (Here and in the following it is assumed that the orbital and epicyclic frequencies are the same, as they nearly are whenever the disk mass is small compared to the stellar mass. More general relations can be derived, in which κ appears instead of Ω.) The derivation of (10) involves some subtle issues of ordering. It must be recognized that the perturbation to the surface density is intrinsically larger than the velocity, displacement and potential perturbations. This effect is illustrated geometrically in Fig. 15, where a density wave has been constructed by representing the perturbed streamlines of the disk by a set of nested ellipses, each ellipse being rotated slightly with respect to its neighbor. Note how small displacement perturbations (eccentricities) produce large surface density perturbations (tightly bunched streamlines).

The dispersion relation yields an axisymmetric ($m = 0$) stability criterion. Recognize that $\omega^2 > 0$ (i.e., ω real) is required for oscillating (not

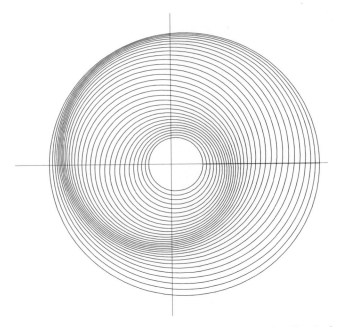

Fig. 15. Motions in a spiral density wave can be represented by elliptical streamlines with radially varying phases. Even a small displacement of the streamlines from circular can produce a large surface density perturbation, as seen in the tightly bunched streamlines

exponentially growing) perturbations, and find from (10) (with $m = 0$) that the condition

$$Q \equiv \frac{c_s \Omega}{\pi G \Sigma} > 1 \qquad (11)$$

is required for stability; i.e., $\omega^2 > 0$ positive for all values of k. This important inequality is commonly known as Toomre's stability criterion (Toomre 1964), although Toomre derived a slightly different relation for a galactic disk of stars, and it has been derived by in other forms by a number of people (perhaps first by Safronov (1960)). (The analysis of non-axisymmetric perturbations is more complicated, but it is found that values of Q only somewhat greater than unity are required for stability.) It is seen that the effects of pressure and rotation are stabilizing, while the effect of gravity is destabilizing. Estimates of the temperatures and surface densities obtained by the methods described in Sect. 1.2 usually indicate that protostellar disks are stable by this criterion. It is quite possible, however, that a disk could become unstable at some time during its evolution. This might happen during the formation of the disk, if mass builds up faster than it is accreted by the star, or at later times if the outer part of the disk (for instance) becomes sufficiently cool. The question of what happens to an unstable disk is of great importance to the issue of giant planet formation, and will discussed in Sect. 1.6.2.

The dispersion relation also reveals other important properties of the waves. These are conveniently described in terms of the nondimensional wavenumber and frequency defined by

$$k' = \frac{k 2\pi G \Sigma_0}{\Omega^2}$$
$$\omega' = \frac{\omega - m\Omega}{\Omega} .$$

The radius at which $\omega' = 0$ is the *corotation* radius, where the pattern speed $\Omega_p = \omega/m$ matches the orbital frequency (which, for our Keplerian protostellar disks is the natural frequency for radial oscillations). The quantity ω' thus measures the "distance" from corotation in frequency space. The locations at which $\omega' = \pm 1$ also have special significance. This condition defines the *Lindblad resonances*, where the orbital frequency is exactly an integer times the difference in the orbital and wave pattern frequencies:

$$\Omega = \pm m \left(\Omega_p - \Omega \right) .$$

The resonant radii are found from the Keplerian rotation law:

$$r_L = \left(\frac{m \pm 1}{m} \right)^{1/3} r_{\text{corotation}} \ (m = 1, 2, 3, ...) . \qquad (12)$$

The plus and minus signs correspond to *outer* and *inner* Lindblad resonances, respectively. At these locations, the disturbance caused by the wave is in phase with the local natural frequency of oscillation, and one expects that disturbances would interact strongly with the disk material there.

The dispersion relation, solved for k', then becomes

$$|k'| = \frac{2}{Q^2}\left[1 \pm \sqrt{1 - Q^2\left(1 - \omega'^2\right)}\right] . \tag{13}$$

The requirement that the radical be real excludes wave propagation from within a certain distance of corotation:

$$\omega'^2 > 1 - \frac{1}{Q^2} .$$

Furthermore, the fact that the right-hand side of (13) must be positive indicates that waves associated with the minus sign ("long" waves) cannot propagate where $\omega'^2 > 1$. For very stable disks ($Q \gg 1$), the range of long wave propagation is very limited. These waves are governed primarily by rotational and gravitational forces. The propagation of "short" waves (for the plus sign), which are like acoustic waves, is not so restricted. The situation is illustrated in Fig. 16. Spiral density waves can be excited by instability if $Q < 1$ somewhere in the disk, or by an object embedded in, or external to, the disk, such as a planet or stellar companion. In fact, they can be excited by any number of disturbances. Their importance for disk evolution lies in their ability to transport angular momentum. One can get some idea about how this works by examining the fate of energy, E_w, and angular momentum, J_w, carried by

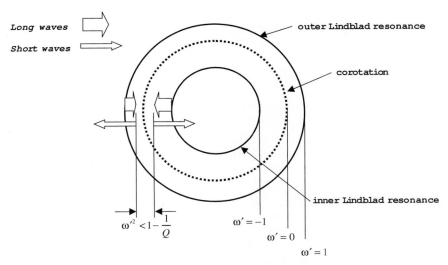

Fig. 16. Long (gravity-like) waves of a given spiral mode (m) are restricted to propagate in a limited range between the inner and outer Lindblad resonances, the boundaries of which exclude the corotation radius. Short (acoustic-like) waves can propagate within the inner Lindblad resonance and beyond the outer Lindblad resonance, but are also excluded by the same boundaries near corotation

a wave. In the absence of dissipation, wave energy and angular momentum are conserved, in the sense that they obey conservation equations of the form:

$$\frac{\partial (E_w, J_w)}{\partial t} + \frac{1}{r}\frac{\partial [rc_{gr}(E_w, J_w)]}{\partial r} = 0,$$

where c_{gr} is the group velocity, and the quantities (E_w, J_w) represent the azimuthally averaged energy and angular momentum, per unit surface area, associated with the wave. They are, of course, functions of the wave parameters ω, k, m and the radius r. Their specific functional dependencies need not concern us now (see Shu (1992) for formulae), except to note that quite generally, $J_w = E_w/\Omega_p$. In the presence of dissipation, waves exchange energy with the disk material, and the wave conservation equations take the form

$$\frac{\partial E_w}{\partial t} + \frac{1}{r}\frac{\partial [rc_{gr}E_w]}{\partial r} = -D - \frac{\partial E_d}{\partial t} + \frac{1}{r}\frac{\partial [rv_r E_d]}{\partial r}$$

$$\frac{\partial J_w}{\partial t} + \frac{1}{r}\frac{\partial [rc_{gr}J_w]}{\partial r} = -\frac{\partial J_d}{\partial t} + \frac{1}{r}\frac{\partial [rv_r J_d]}{\partial r},$$

where D is the appropriately averaged dissipation rate and

$$E_d = -\frac{GM_*\Sigma}{2r}$$

$$J_d = \Sigma\sqrt{GM_*r}\ .$$

From these equations, with the aid of the azimuthally averaged mass conservation equation, and assuming that the disk quantities do not change in the time of wave propagation, one can derive an explicit formula for the radial mass flux induced by angular momentum deposition from the attenuated waves:

$$2\pi \Sigma r v_r = \frac{4\pi D}{\Omega(\Omega_p - \Omega)}\ .$$

This relation demonstrates that the mass flux is proportional to the dissipation, and that it is inward inside corotation and outward outside of corotation. The relation does not hold at corotation because it was derived from the free wave dispersion relation, which must be violated at the corotation resonance. We return to the connection between waves and resonances in the discussion of planet-disk interactions (Sect. 1.7.1).

What about the *vertical* structure of these waves? The 2-D analysis just described cannot address this question, of course. In fact, a complete 3-D analysis has yet to be accomplished. Interesting results have been obtained, however, by means judicious approximations (Lubow and Pringle (1993); Korycansky and Pringle (1995); Lubow and Ogilvie (1998); Ogilvie (1998)). The strategy employed is the following. The conservation equations, in the shearing sheet approximation, including z and r dependencies, are linearized in the usual way, but only axisymmetric waves are considered. The WKB approximation is used for r, but not z dependencies, so r takes the role of a parameter.

This procedure reduces the problem to a set of 1-D (in z) differential equations (eigenvalue problems), with coefficients dependent on r. These equations are solved for the z structure of the waves. The radial propagation properties are diagnosed by solving the equations at different r and constraining the solutions to obey conservation of wave energy. It is then argued that the results are applicable to non-axisymmetric waves, with the replacement of ω by $\omega - m\Omega$, as long as the waves are tightly wound in the sense that $m \ll kr$. Self-gravity is not included, so only the counterparts of the short (acoustic) modes described above are present. Finally, it is assumed that the fluid obeys a polytropic equation of state: $p \propto \varrho^\gamma$, as might be the case for an optically thick disk. This point is important for the following reason: the vertical extent of a locally polytropic disk is finite. That is, there is an altitude at which temperature, density and pressure fall to zero, which defines the thickness of the disk. To see this, solve the vertical hydrostatic equation (5) to find that the state variables are proportional to a power of the quantity $(1 - z^2/H^2)$, where

$$H^2 = 2\gamma p_0 / (\gamma - 1) \Omega^2 \varrho_0$$

and subscript 0 refers to values at $z = 0$. This kind of vertical structure has the effect of a waveguide, and strongly affects the nature of waves and their propagation properties, as described below.

Resolving the vertical disk structure reveals a multiplicity of modes, each with its own characteristic vertical oscillations, some of which are associated with buoyancy restoring forces. Some modes have counterparts in stellar oscillations, and are therefore designated accordingly. The p-waves are compressible, with the main restoring force provided by pressure. The g-waves are essentially gravity waves, and are incompressible, with buoyancy providing the restoring force. Waves of the third mode, designated r-waves, have no counterpart in (non-rotating) stars; they are inertial waves, with angular momentum and buoyancy providing the restoring forces. Each of these modes possess z-symmetric and antisymmetric components, and each is represented by a series of waves with n nodes in the z direction. In addition, there are two *fundamental* (designated f) modes (z- symmetric and antisymmetric), which turn out to be particularly important because they carry almost all of the angular momentum contained in waves stimulated at resonances. They are the dominant 3-D counterparts of the short, 2-D spiral density waves.

Like their 2-D counterparts, the radial propagation of these waves is constrained, as shown in Fig. 17. Note that vertical resonances (where radial oscillations are in phase with vertical oscillations) now play a role in restricting the radial propagation of the p-waves. The waves also become vertically constrained, in some cases severely so, as they propagate away from resonances. The f, p and g modes become confined to within a wavelength of the surfaces of the disk; the r modes become restricted to a distance of about $(\lambda H)^{1/2}$ from the midplane. An example of the confinement of the f mode is shown in Fig. 18. This effect is due to the waveguide character of disks with finite

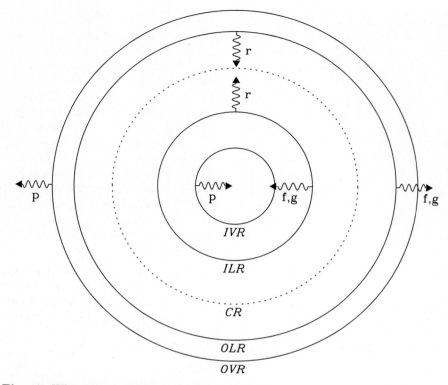

Fig. 17. When the vertical structures of spiral density waves are resolved, variety of additional modes are discovered. These also have restricted ranges of propagation, some of which are bounded by the locations of vertical resonances, as well as the Lindblad resonances. Shown are the boundaries of allowed propagation for the $r-$, $p-$, $g-$ and $f-$ modes discussed by Lubow and Ogilvie (1998). I and O refer to inner and outer, LR and VR are Lindblad and vertical resonances, respectively, and CR is corotation. (Figure from Lubow and Ogilvie 1998)

vertical extent. In fact, it disappears in disks that are vertically isothermal, for such disks formally extend to infinite z, as demonstrated by the appropriate solution of (5). So what about a more realistic disk, which might resemble the polytropic model except near the surfaces, where an optically thin, isothermal atmosphere should exist? This situation was analyzed by Ogilvie and Lubow (1999), who found that wave confinement still occurred, but was less severe.

Now recall that the angular momentum carried by a wave is deposited in the disk according to how the wave energy is dissipated. The fact that the energy of spiral density waves tends to be concentrated toward the disk surfaces could have a profound effect on their non-linear behavior and the ultimate manner in which they are dissipated. They might form shocks or "break" in other ways. The problem deserves further attention, probably most fruitfully by the application of high resolution numerical simulations.

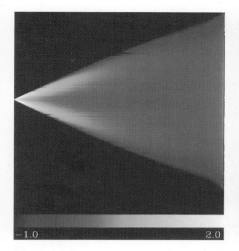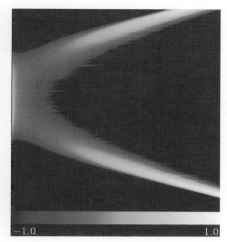

Fig. 18. A representation of the distribution in disk cross-section of the wave energy associated with the fundamental, $m = 2$ mode, launched at an inner (**a**) and outer (**b**) Lindblad resonance. The scale refers the logarithm to the base 10 of a dimensionless energy. The plot illustrates the channeling effect imposed by vertical structure. (Figure from Lubow and Ogilvie 1998)

Before leaving the subject, something should be said about another kind of wave, the Rossby wave. Rossby waves have the property of propagating *vorticity*, which is a measure of the local shear in a flow, defined as the curl of the velocity. These waves are well-studied in the context of planetary atmospheres and have been shown to be potentially important in protostellar disks. Indeed, the baroclinic instability (and related instabilities) referred to in Sect. 3.2 is known to stimulate such waves, which may then couple with other modes in complicated ways. The Rossby wave dispersion relation, as derived by Sheehan et al. (1999), is

$$(\omega - m\Omega)^2 = \Omega^2 + k^2 c_s^2 + m^2 c_s^2/r^2 \ . \tag{14}$$

This looks very much like (10), except that the gravitational term has disappeared (because self-gravity was not included in the analysis), replaced by the last term on the right. This term was neglected in the spiral density dispersion relation because it was smaller than other terms, for short wavelengths ($kr >> 1$) in thin disks. But Rossby waves have long wavelengths and are revealed only when this term is retained. For this reason, radial gradients of the base state should not be suppressed, i.e., the WKBJ approximation is not valid. [In other respects, the assumptions leading to (14) are the same as those for (10).]

In planetary atmospheres, Rossby waves are known to be responsible for the extraction of energy from external sources (such as solar radiation or internally generated heat) and its transfer to organized motions. For instance, large storms and jet streams in the terrestrial atmosphere are manifestations of Rossby waves, as are the vortices and high speed belts on Jupiter. The role of these waves in disk angular momentum transport has yet to be determined. Their effectiveness will depend on the existence of appropriate instabilities and the establishment of a feedback loop that permits energy extraction from the disk itself, as discussed above.

This ends our discussion of disks as astrophysical objects independent of their potential for forming planets. The remaining sections focus on the theory of how they produce planetary systems.

4 Dust-Gas Dynamics

4.1 Drift and Settling Velocities in the Absence of Turbulence

In a cool protostellar disk of the same composition as our Sun, about 0.4 % of the mass is in the form of rock-forming solids (mainly iron and magnesium silicates). If it is cold enough (less than about 160 K), another 1.5 % exists as H_2O ice. Initially, these solids are in the form of sub-micron particles, or "dust", inherited from the interstellar medium, or newly condensed, if and where the disk was once hot. Small dust particles are well-coupled to the gas, and so follow the overall gas motions closely. While dispersed, their main effect is as a source of opacity; they have no direct effect on the gas dynamics, because they comprise such a small mass fraction. Occasionally they collide with each other, stick together (see ?, and references therein), and gradually accumulate, reducing the opacity and altering the nature of their dynamical interaction with the gas. So begins, it is believed, the process of rocky planet formation.

The dynamical interaction of gas and solids is described by the equations of motion for the solid particles, which may be written (e.g. Dubrulle et al. 1995)

$$\frac{dv_{rp}}{dt} = g_r + \frac{j_p^2}{r^3} - \frac{v_{rp} - v_r}{t_r}$$

$$\frac{dj_p}{dt} = -\frac{j_p - j}{t_f}$$

$$\frac{dv_{zp}}{dt} = g_z - \frac{v_{zp} - v_z}{t_f} . \qquad (15)$$

Subscript p refers to the particles. The last term on the right hand side of each equation is the frictional force between the gas and the particles. The "friction time" t_f is the characteristic time in which a particle of given size and mass exchanges momentum with the gas (sometimes referred to as the "stopping time"). The way it is calculated depends on whether the mean free path in the gas is larger or smaller than the particle and whether the initial relative velocity is super- or sub-sonic. The mean free path for gas at 1 AU in a typical protostellar disk is a few tens of centimeters. For smaller particles moving subsonically through such a gas,

$$t_f = \frac{\varrho_p r_p}{\varrho c_s},$$

where ϱ and r_p are the particle's density and radius, respectively. This formula can be derived by equating the frictional force to the rate at which thermal gas molecules transfer momentum to a solid, spherical particle and solving the resulting momentum equation. (Of course, the particles are likely to be irregularly shaped rather than round, in which case r_p must be interpreted to be an appropriate dimension. For instance, it might be the characteristic size of the individual particles making up a fractal structure.) For particles larger than the mean free path, the friction time depends on the relative velocity itself:

$$t_f = \frac{8\varrho_p}{3\varrho C_d (v_p - v)}.$$

Here, C_d is the drag coefficient, which also depends on the relative velocity as well as the particle size and shape, and gas viscosity; however, its value is generally of order unity. In most of the following, we will be concerned with particles smaller than the mean free path.

The first important point to be learned from these equations is the following: solid particles experience an azimuthal drag due to the fact that the gas motion is slightly sub-Keplerian, and this drag can cause a substantial radial drift as particles lose angular momentum. In Sect. 3.1, we used the gas radial momentum equation to derive the fact that, to order, $(h/r)^2$, $\Omega = \Omega_K$. But if the pressure gradient term is retained in that equation, one finds for the gas

$$(r\Omega)^2 = (r\Omega_K)^2 + \frac{r}{\varrho}\frac{\partial p}{\partial r}$$

or

$$j^2 = j_K^2 + \frac{r^3}{\varrho}\frac{\partial p}{\partial r}.$$

With this expression, the solid particle radial momentum equation (the first of 15) yields

$$v_{rp} - v_r = \frac{t_f}{\varrho}\frac{\partial p}{\partial r}.$$

Since the pressure gradient is usually negative, the effect produces an inward drift. Weidenschilling (1977) calculated the drift rate in a model disk, for a variety particle sizes and gas densities, for a disk in which $v_r = 0$. His results are shown in Fig. 19, in which it is seen that very small particles are well-coupled to the gas and have small drift rates, and large solid objects

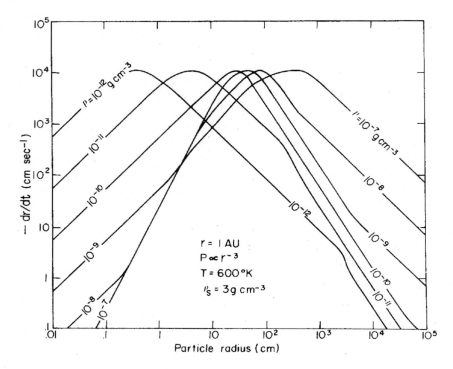

Fig. 19. Particle drift rates due to loss of angular momentum by gas drag, as a function of particle size and gas density, for typical disk conditions. Small particles have small drift rates because they are easily carried along by the gas; large objects have small drift rates because drag forces are small compared to inertial forces. Intermediate-sized particles, in this case about a meter in size, drift rapidly. (Figure from Weidenschilling 1977)

are decoupled from the gas and also have small drift rates. Intermediate-sized particles, however, can drift rapidly. The peak drift rates in Fig. 19, 10^4 cm/sec, would deliver a particle from 1 AU to the Sun in only 100 years. The survival of planetary-sized objects therefore implies that growth through the critical size range (about 1 meter, for a typical gas density) was rapid, or that there were sustained, systematic outward gas velocities which were larger than the drift rate, or that collective effects (for which formula derived for individual, isolated particles do not apply) protected the particles from drift loss.

The solid particle equations also quantify the rate at which vertical settling and concentration at the midplane occurs. For the gas velocity $v_z = 0$, and writing $v_{zp} = \mathrm{d}z/\mathrm{d}t$ and $\mathrm{d}v_{zp}/\mathrm{d}t = \mathrm{d}^2z/\mathrm{d}t^2$, one finds

$$\frac{\mathrm{d}^2 z}{\mathrm{d}t^2} + \frac{1}{t_f}\frac{\mathrm{d}z}{\mathrm{d}t} + z\Omega^2 = 0 \,.$$

This is the equation of a damped harmonic oscillator, which is critically damped for $t_f = 1/2\Omega$. Particles larger than $r_p = \Sigma/4\varrho_p$ follow orbits which oscillate through the midplane; smaller particles gradually spiral toward the midplane. If $t_f \ll 1/\Omega$, the terminal velocity $v_{zp} = -z\Omega^2 t_f$ is quickly attained. Suppose $\Omega = 2 \times 10^{-7}\,\text{sec}^{-1}$, the orbital frequency of the Earth, and $\varrho_0 = 10^{-9}\,\text{gm/cm}^3$, a typical value at 1 AU. Then, if we measure the quantity $\varrho_p r_p$ in cgs units, we find $t_f \approx 5 \times 10^3\,(\varrho_p r_p)$ sec, $v_{zp} \approx 200\,(\varrho_p r_p)$ cm/sec, and the settling time $= h/v_{zp} \approx c_s/v_{zp}\Omega \approx 10^3/\Omega \varrho_p r_p$ sec. The density ϱ_p is of order unity, so one sees immediately that micron-sized particles would take millions of years to settle, and the least breeze in the vertical direction would inhibit any concentration by settling. They must grow first. On the other hand, a golf-ball sized rock would settle, in the absence of vertical gas velocities, in 10^3–10^4 years.

4.2 Particle Growth and Trajectories in the Absence of Turbulence

Differential velocities among particles of different sizes cause them to collide, stick and grow. The rate of growth of a particle and the path that it follows as it grows can be determined by integration of the set (15), augmented by an equation for the growth rate. The latter is given by

$$\frac{dm_p}{ds} = \pi r_p^2 \varrho_s,$$

where m_p is the particle mass (assumed here to be $4\pi r_p^3 \varrho_p/3$), s is the path length along the trajectory, and ϱ_s is the volume density of solids encountered along the trajectory. This expression is valid if all solids encountered by the growing particle stick to it, as is expected when relative velocities between particles are less than about 1 m/sec (Dominik and Tielens 1997; ?). If this is not the case, the right-hand side must be appropriately reduced by some efficiency factor. If relative velocities are too high, fragmentation occurs, rather than growth; see the discussion in Wurm et al. (2001). Note also that the density of solids, ϱ_s, is an evolving quantity whose value depends on how all of the particles are settling. Nakagawa et al. (1986) calculated growth and trajectories by assuming that at early times ϱ_s was well represented by its initial value, a constant fraction of the gas density. At later times, as settling proceeded, it was approximated by Σ_s/z, where Σ_s is the initial column density of solids. They also adopted typical forms for the distribution of gas density, and assumed that there were no gas velocities. Their results are shown schematically in Fig. 20, for a particle starting at altitude Z_0. There is little radial drift until the particle has settled and grown. At Z_1, it is large enough to drift radially, but rapidly accumulates smaller (background) particles, and begins to spiral vertically again at Z_3. But by this time, it has entered a region (at Z_2) in which settling has caused the dust density to be comparable to the gas density. Further details are given in Fig. 21. Most of the settling time is spent at high altitudes, while the particle grows. The total

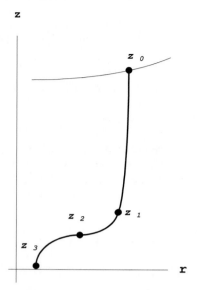

Fig. 20. Schematic representation of the trajectory of a particle settling and growing in a non-turbulent disk. The initial position of the particle is at altitude Z_0. It settles mainly vertically until Z_1, where it has become large enough to drift radially. It then rapidly accumulates smaller particles, whereupon, at Z_2, it enters a region where the dust density is comparable to the gas density. After Z_3 it spirals vertically toward the midplane. Small amounts of turbulence, however, can prevent the formation of the dense dust layer; see Fig. 23 and the discussion in Sect. 4.3. (Figure adapted from Nakagawa et al. 1986)

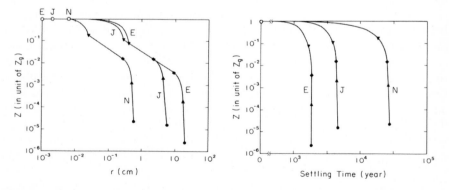

Fig. 21. Quantitative results for settling and growth in a non-turbulent disk: altitude vs. particle size (a), and altitude vs. time. The quantity z_g is the disk scale height. The curves labeled E, J and N refer to locations in a disk at the orbits of Earth, Jupiter and Neptune, respectively. The solid symbols correspond to the points z_1, z_2 and z_3 of Fig. 21, and the endpoint of the calculation. (Figure from Nakagawa et al. 1986)

radial drift experienced by these particles is limited to a small fraction of an AU, because rapid radial drift is associated with rapid growth through the size of maximum drift.

4.3 The Effect of Turbulence on Particle Settling

Although the calculation just described is useful for understanding the fundamental processes and timescales, even a small degree of turbulence could have a substantial impact on some of the quantitative details. The rather small values of the settling velocity estimated above motivate the following question: What values of the turbulence parameter α would prevent (or inhibit) settling? Suppose we calculate

$$\left|\frac{v_{zp}}{v_{turb}}\right| = \frac{z\Omega^2 t_f}{\alpha c_s h/l_{turb}} = \frac{2}{\alpha \Sigma}\frac{zl_{turb}}{h^2}\varrho_p r_p \ .$$

Here, l_{turb} is the turbulent mixing length. Realizing that $zl_{turb}/h^2 < 1$, one concludes that

$$\left|\frac{v_{zp}}{v_{turb}}\right| < \frac{2}{\alpha \Sigma}\varrho_p r_p = r_p \frac{2 \times 10^{-3}}{\alpha}$$

for $\varrho_p = 1\,\mathrm{g/cm^3}$ and a typical value (at 1 AU) of $\Sigma = 10^3\,\mathrm{gm/cm^3}$. Turbulent velocities would therefore exceed the settling velocities of small grains ($r_p \ll 1\,\mathrm{cm}$) even for values of α well below that required to evolve a disk in a million years (10^{-3}–10^{-2}). Therefore, even if turbulence is not the primary agent for transporting angular momentum in disks, it could still be important for controlling the initial distribution and evolution of the solid component.

It is often useful to characterize the turbulent field by an ensemble of transient *eddies* (or vortices), with some wavelength distribution of energy, usually expressed as a power law in the wavenumber k:

$$E_k = \frac{v_k^2}{k_0}\left(\frac{k}{k_0}\right)^{-a} \ .$$

The characteristic eddy velocity and size are v_k and $2\pi/k$, respectively. In random turbulence, eddies typically last only a single turnover time, so their lifetime t_k is given by $1/kv_k$. The motion of a solid particle in a turbulent field is then a diffusive process, as its velocity is altered by random encounters with turbulent eddies. The velocity perturbations depend on the particle size and the gas properties, as characterized by t_f. For instance, if $t_f > t_k$, the eddy disappears before the particle is entrained in it. If the particle has a systematic velocity v_p, and if the "eddy crossing time" $t_{cross} \equiv 1/kv_p$ is greater than t_k, the particle can be entrained, but will also encounter smaller eddies within the eddy of size $2\pi/k$. If $t_{cross} < t_f$, the particle passes through the eddy without being entrained. Such considerations must be accounted for when deriving a diffusion coefficient κ_p for the particles (Volk et al. 1980). The diffusion coefficient can then be represented as an integral of the effects of each eddy size. For particles settling in a turbulent disk, Dubrulle et al. (1995) derive

$$\kappa_p = \kappa_0 \left[\int_{k_0}^{\infty} \frac{E_k(k)\,dk}{k^2} \right]^{1/2},$$

where the coefficient κ_0 represents the effects of the largest eddies, which dominate the transport, and is a function of t_f, v_{zp} and k_0 This diffusion coefficient κ_p is to be used in the equation governing the diffusive settling of particles:

$$\frac{\partial \varrho_s}{\partial t} + \frac{\partial (\varrho_s v_{zp})}{\partial z} = \frac{\partial}{\partial z}\left[\varrho \kappa_p \frac{\partial (\varrho_s/\varrho)}{\partial z}\right].$$

Dubrulle et al. (1995) consider a disk with an isothermal vertical structure and calculate the evolutions of initially uniform dust distributions, of a single particle size, for various values of α. They find that stationary solutions, shown in Fig. 22 (for $\alpha = 2 \times 10^{-3}$), are attained in a few times the timescale

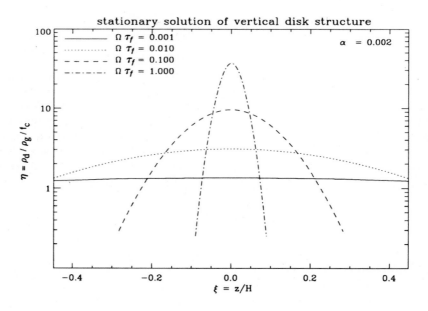

Fig. 22. Steady state, vertical concentrations of solid particles in a turbulent disk with turbulence parameter $\alpha = 0.002$. The quantity plotted on the vertical axis is the enhancement in the concentration of solids above the initial, uniform value. Particles are assumed to be of uniform size. The curves correspond to particle sizes of 250 cm (*dot-dashed*), 25 cm (*dashed*), 2.5 cm (*dotted*) and 0.25 cm (*solid*), for values of the disk surface density of 10^3 gm/cm² and particle density 2 gm/cm³. (Figure from Dubrulle et al. 1995)

$t_f/(t_f\Omega)^2$. (These solutions can be scaled to other values of the parameters by preserving $\alpha/t_f\Omega$.) Strong concentrations about the midplane are only attained by particles considerably larger than 1 cm. The theory predicts that only large rocks ($r_p > 2.5$ m) could settle to a layer in which the density of solids approaches that of the gas (but the analysis breaks down for such large objects, which could oscillate about the midplane). Even if turbulence was vanishingly small in the ambient disk, a dense layer of solids would produce local turbulence by virtue of the velocity difference between it and the gas (Weidenschilling and Cuzzi 1993), so the formation of such a layer appears to be self-inhibiting. The conclusion is that substantial particle growth must occur before a dense, dusty (or rocky) layer can form at the midplane.

4.4 The Initial Stages of Accumulation

The conclusion of the last section has important consequences for the initiation of planet-building. If dense particle layers could form by settling, as in the calculation by Nakagawa et al. (1986), they might form clumps due to a gravitational instability in the layer (Safronov 1969; Goldreich and Ward 1973). The instability would be like that which occurs in a gas disk when the criterion (11) is violated. This is an appealing prospect, because the most primitive meteorites, which are objects representative of the earliest stage of accumulation, appear to be made by the indiscriminant collection of nebular solids, as might occur in such gravitational clumping. But the results described above indicate that gravitationally unstable layers are not readily formed; accumulation must proceed snowball fashion, as individual particles and collections of particles collide. Calculations of this kind of accumulation in the presence of turbulence have been performed (see Wasson 1985) and indicate that it could produce rocky objects on a timescale of 10^4 years. But the effects of fragmentation are either ignored or represented by untested hypotheses; furthermore, the detailed interactions of particles with the turbulent field are approximated by averages which may obscure important physics.

An attempt to address these issues more rigorously was undertaken by Cuzzi et al. (1996, 2001), with interesting results. They noted that the instantaneous distribution of solid particles entrained in a turbulent flow tends to be inhomogeneous; particles are transiently concentrated at stagnant regions of the flow, between eddies. Moreover, there is a preferred particle size for concentration, selected by the condition $t_f = t_\eta$, where $\eta = k_{max}$ and k_{max} is the wavenumber of the smallest eddies. (This kind of concentration is not the same as that which is sometimes seen at the center of vortices, which only occurs when long-lived, large scale eddies are present.) The expected concentrations and preferred sizes depend on the turbulent characteristics.

Now the properties of the smallest eddies can be found by using the well-known Kolomogorov rules of steady turbulence (Tennekes and Lumley 1972). The starting point is the observation that, in steady, 3-D turbulence, energy does not accumulate at any wavenumber despite the fact that it is being continuously transferred from larger to smaller eddies. Thus, $\dot{E}_k$, the rate at which energy is transferred at wavenumber k, must be independent of k:

$$\dot{E}_k = \frac{v_k^2}{t_k} = \frac{v_k^2}{1/kv_k} = v_k^3 k = v_0^3 k_0 = v_\eta^3 k_\eta .$$

Subscript 0 refers to the largest wavelength. Introduce the *Reynolds number*, a dimensionless combination that measures the ratio of inertial forces to viscous forces:

$$R_e = \frac{\text{velocity} \times \text{length}}{\text{kinematic viscosity}} .$$

The large scale turbulence is characterized by

$$R_e = v_0/k_0 \nu_{mol},$$

where ν_{mol} is the molecular viscosity. The smallest scale is that where energy is dissipated by molecular viscosity, so

$$R_{e\eta} = v_\eta/k_\eta \nu_{mol} .$$

From these relations, and the invariance of $v_k^3 k$, one finds

$$\eta/k_0 = R_e^{3/4}$$
$$v_\eta/v_0 = R_e^{-1/4}$$
$$t_\eta/t_0 = R_e^{-1/2} .$$

These expressions give the turbulent characteristics at the dissipation scale in terms of the large scale turbulent characteristics. So what is the value of R_e in a protostellar disk? It can be expressed in terms of the turbulence parameter α as follows:

$$R_e = \frac{v_0}{k_0 \nu_{mol}} = \frac{\alpha c_s h}{\nu_{mol}} \approx \frac{\alpha h}{l_{mfp}} \approx \alpha \times 10^{11} .$$

Here we have used the fact that $\nu_{mol} \approx c_s l_{mpf}$, and chosen representative values of $l_{mpf} = 10$ cm and $h = 10^{12}$ cm to obtain the numerical factor. The preferred size for concentration is given by the condition $t_f = t_\eta = R_e^{-1/2} t_0$, so for this we need to know the lifetime of the largest eddies. It seems safe to assume that, in random turbulence, this lifetime would be comparable to an orbital period. With this assumption, the expression for preferred particle size becomes

$$t_f = \frac{\varrho_p r_p}{\varrho c_s} = \frac{1}{R_e^{1/2} \Omega}$$

or

$$r_p = \frac{\Sigma}{\varrho_p R_e^{1/2}} = \alpha^{-1/2} 10^{-3} .$$

The numerical factor in the last equality corresponds to $\Sigma = 10^3$ gm/cm^2 and $\varrho_p = 3$gm/cm^3. Values of α commonly expected for disk evolution, or derived from numerical simulations of angular momentum transport, (say, 10^{-4}–10^{-2}) produce concentrations of particles in the size range 10^{-2}–10^{-1}. Cuzzi et al. (2001) have emphasized that this is the range of sizes of *chondrules*, the abundant and ubiquitous igneous pebbles found in primitive meteorites. In fact, the *distribution* of sizes they find in concentrations from numerical simulations matches that of chondrules in meteorites remarkably well (Fig. 23). Also, from such simulations, and a fractal description of the turbulent properties as a function of R_e, they predict large concentration factors for the high

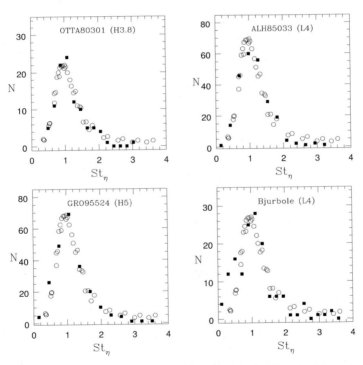

Fig. 23. Size distributions of chondrules from four meteorites (*solid squares*), compared with predicted size distributions from numerical simulations of turbulent concentration (*open circles*). The quantity plotted on the horizontal axis is the Stokes number, a non-dimensional parameter that measures (in this case) the ratio of the friction time, t_f, to the frequency of the smallest turbulent eddies; it is proportional to the product of particle size and density. Thus the plots show, in effect, particle size vs. density. (Figure from Cuzzi et al. 2001)

Reynolds numbers of protostellar disks (see their Fig. 4, where concentrations are expressed as the fraction of total particles expected to exceed a given enhancement in spatial density, as a function of R_e).

From this analysis, Cuzzi et al. (2001), noting that settling is inhibited by turbulence, propose that the initial stages of rocky body accumulation occurred by the formation of chondrule precursors by sticking collisions, their subsequent transformation into chondrules (an enigmatic process, not understood despite much attention and applied creativity), and concentration in the manner described above. However, it must be emphasized that turbulent concentrations are transient; particles actually flow through the regions of concentration, so some further development must be postulated to attain larger solid objects. Perhaps concentrations are high enough to promote gravitational instability or some other positive feedback due to heavy mass loading on the gas by solids. Incidentally, I note that an equivalent analysis of systematic inhomogeneities produced by waves has not been performed.

Other mechanisms for the concentration of solid material at the earliest stage of planet formation have also been suggested. An example is the two-phase (gas-dust) fluid instability proposed by Goodman and Pindor (2000), which produces radial fluctuations in a dust layer at the midplane. Again, the perturbations must grow to the point where gravitational instability sets in, if the process is to lead to the formation of long-lived solid objects.

The fact is, we do not yet know exactly how the first rocks form from the dusty component of protostellar disks. This is the case despite the fact that we actually have samples of these rocks, in the form of meteorites, for our own Solar System. The most primitive of the meteorites look like sediments, the result of a gentle accumulation of disparate components, the ensemble of which escaped familiar planetary equilibrating processes (heat and pressure). Our lack of understanding of how this happened does not prevent us from analyzing the later stages of planet formation, which, in some ways, is a less complicated problem.

5 Growth of Planetesimals to Planets

5.1 Basics of Collisional Planet-building; Runaway Growth

For the present purposes, we define a planetesimal to be a body large enough to gravitationally perturb its neighbors, but smaller than a full-grown planet, say Moon-sized. The first part of this definition can be quantified by saying that a planetesimal produces gravitationally induced velocity perturbations larger than typical drift or settling velocities, perhaps 10^3 cm/sec. The velocity perturbations are comparable to the escape velocity, so we are talking about rocky objects of roughly a kilometer in radius or greater. Somehow these objects formed, but, as indicated in the preceding discussion, gasdynamic effects are sufficiently complex that the mechanisms of growth to this stage remain poorly described. Further growth is dominated by the gravitational interactions among bodies, so the nature of the problem changes (although gasdynamic effects are not necessarily negligible).

The number of kilometer-sized objects required to build the terrestrial planets is about 10^{11}. Even with modern computers and sophisticated algorithms for integrating the equations of motion, this is far too many bodies to treat by any direct numerical means; statistical methods are required. Despite the difficulty of the problem, a great deal of progress has been made in understanding planetesimal growth in the past two decades. The reasons, I believe, are to be found not only in the advances in computational capability and the technical creativity of researchers, but in a cooperative effort in which a combination of numerical and statistical treatments have been continuously tested against each other to resolve discrepancies.

The statistical approach to understanding the coalescence of many small objects into a few large ones starts with the coagulation equation:

$$\frac{dn_k}{dt} = \frac{1}{2} \sum_{i+j=k} A_{ij} n_i n_j - n_k \sum_{i=1}^{\infty} A_{ik} n_i \ .$$

Here, n_k denotes the number of objects with mass m_k, and A_{ij} is the probability of collision (and merger) of bodies with masses m_i and m_j. The first term on the right is the gain of bodies with mass m_k and the second term is the loss of bodies with m_k that merge to make bigger bodies. All of the physics is in the collision term A_{ij}, which usually depends in complicated ways on the relative velocities of colliding bodies, their masses, number densities and so forth. Thus one must understand both the detailed physics of collisions (including the effective cross-section for collision) and the dynamical evolution of the ensemble, in order to calculate relative velocities and mass distributions.

Let us start with a description of collisional growth in the simplest situation. If all collisions result in merger, one can say

$$\frac{dm_p}{dt} = \left(\pi r_p^2\right) v_{rel} \varrho_s F_g \ .$$

The first factor on the right is the geometric cross-section of the particle, v_{rel} is the mean relative velocity of the ensemble of particles, ϱ_s is their spatial density and F_g is a gravitational enhancement factor, which represents an increase in effective cross-section due to the gravitational bending of particle paths toward the growing object. In the simplest case, the growing body sweeps up smaller particles (with much smaller masses) by two-body interactions. It is readily shown by solution of this "scattering" problem that

$$F_g = 1 + \left(v_{esc}^2 / v_{rel}^2\right) \ ,$$

where the escape velocity is given by

$$v_{esc}^2 = \frac{2 G m_p}{r_p} = \frac{8}{3} \pi r_p^2 \varrho_p \ .$$

Thus

$$\frac{dm_p}{dt} = \left(\pi r_p^2\right) v_{rel} \varrho_s \left(1 + \frac{8\pi r_p^2 \varrho_p}{3V_{rel}^2}\right).$$

Note that the dependence of growth rate on the radius of the growing particle depends strongly on the relative magnitudes of v_{esc} and v_{rel}. If $v_{esc} \ll v_{rel}$, $dm_p/dt \propto r_p^2$; in the opposite limit, $dm_p/dt \propto r_p^4$. In general, $dm_p/dt \propto r_p^y$, where $2 \leq y \leq 4$. This dependence turns out to be extremely important in determining the nature of solutions to the coagulation equation and the physics of planetary growth.

To see this, consider the growth of two large objects sweeping up smaller objects by two-body interactions. Calculate the time dependence of their mass ratio:

$$\frac{d}{dt}\left(\frac{m_1}{m_2}\right) = \left[\left(\frac{r_1}{r_2}\right)^{y-3} - 1\right].$$

The "+" term on the right indicates a positive definite factor, the details of which are unimportant for the present point. What is important, is that the ratio m_1/m_2 continuously grows if $y > 3$ (as for $v_{esc} \gg v_{rel}$). The possibility of this "runaway growth" of one (or a few) body(ies) with respect to others has a substantial effect on how the planetesimal growth problem must be solved. For instance, because the coagulation equation involves probabilities and deals with mean quantities (as do all statistical treatments), particular care must be used in evaluating such quantities which can be skewed in unphysical ways if the distribution of masses (or other properties) becomes discontinuous. (In fact, there are other aspects of the coagulation equation that are potentially unphysical; see Wetherill (1990), for a discussion.) Because runaway growth occurs when relative velocities are smaller than escape velocities, it is important to have an accurate representation of the dynamical state of the entire ensemble. The central problems of planetesimal growth are therefore those of obtaining real collisional cross-sections and outcomes (including the effects of fragmentation), and describing the real velocity evolution of the system.

5.2 Three-Body Effects on Collision Cross-Section

Greenzweig and Lissauer (1990) used numerical integrations of the restricted three-body problem to determine the gravitational enhancement factor, F_g, for an object sweeping up massless "test" particles in orbit about a star. Averages of the outcomes of many integrations, starting from different initial conditions, were used to evaluate

$$F_g = F_g\left[i,\, e,\, (r_p/r_{hill})\right],$$

where i and e are inclination and eccentricity of a test particle, respectively, and r_{hill} is the radius of the "Hill sphere":

$$r_{hill} = \left(\frac{m_p}{3M_*}\right)^{1/3} a_p \; .$$

The quantity a_p is the semi-major axis of the growing planetesimal. (The Hill sphere is a measure of the extent of the gravitational influence of a secondary compared to the gravitational influence of the primary. Its radius is defined to be the distance from the secondary, in the direction of the primary, at which the potential is a minimum, in a frame rotating with the secondary; that is, the distance to the inner Lagrange point. It arises as a natural parameter in the "shearing sheet" coordinate system.) By performing many integrations and relating the eccentricities and inclinations to relative velocities, the results shown in Fig. 24 were obtained. Scaling relations allow the results to be applied to general values of m_p and a_p. Although the two-body formula for F_g indicates divergence as relative velocities vanish, three-body effects limit its value. It is noteworthy that the orbits of test particles near the growing planetesimal can be extremely complicated (see Fig. 1 of Greenzweig and Lissauer 1990). These orbits should give pause to anyone contemplating a brute force, numerical attack on the entire planetesimal growth problem.

Armed with the details of these encounters, one can say something about the *limits* to runaway growth. Runaway slows when a planetesimal has

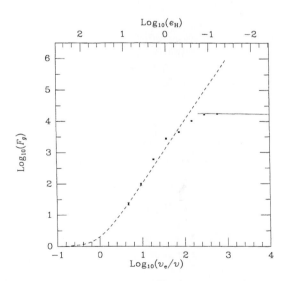

Fig. 24. The gravitational enhancement in collision cross-section, including three-body effects, as a function of the ratio of escape velocity to planetesimal velocity dispersion (or planetesimal eccentricity). The dashed line indicates the two-body approximation. (Figure from Lissauer 1993)

substantially cleared its neighborhood, after its own random velocity has diminished so that the boundaries of its neighborhood are reasonably stable. From numerical integrations like those referred to above, this occurs after test particles are depleted from a zone Δr of a few r_{hill} in width about the planetesimal. The mass contained in such an annulus therefore represents the mass attained after runaway:

$$m_{\text{runaway}} \approx (\text{few times}) 2\pi a_p \Delta r \Sigma_s \approx \frac{\left(10\pi a_p^2 \Sigma_s\right)^{3/2}}{(3M_*)^{1/2}}, \quad (16)$$

where Σ_s is the surface density of solids. Note that this is not the upper limit to the mass that may ultimately be attained; it is merely the mass that corresponds to the end of the runaway phase, after which the growth rate diminishes substantially. We note here that, for typical parameters, m_{runaway} in the terrestrial planet region is about $0.04 M_\oplus$, or 3 lunar masses.

5.3 Evolution of the Velocity Distribution

One of the most successful techniques for describing the evolution of the planetesimal velocity distribution is based on classical methods of statistical mechanics (Stewart and Kaula 1980; Lissauer and Stewart 1993). It employs the collisional Boltzmann equation, or more specifically a Fokker-Planck equation, in which encounters are regarded as instantaneous and local, and motions between encounters are effectively freely orbiting. (These techniques have also been used in stellar dynamics.) As will be seen, distant, non-local encounters are also important, so modifications must be made to incorporate their effects. The basic premise is that the planetesimal ensemble can be described by as a perturbation to a Boltzmann-like distribution function, $f(\mathbf{r}, \mathbf{v})$, expressed in terms of inclination and eccentricity as $f(i, e)$, where the number density of planetesimals is

$$n = \int f \, d^3\mathbf{v} \, .$$

Evolution obeys the collisional Boltzmann equation

$$\frac{\partial f}{\partial t} + \frac{d\mathbf{r}}{dt} \frac{\partial f}{\partial \mathbf{r}} + \frac{d\mathbf{v}}{dt} \frac{\partial f}{\partial \mathbf{v}} = \left. \frac{\delta f}{\delta t} \right|_{encounter}. \quad (17)$$

It is assumed that encounters cause perturbations to an equilibrium distribution, $f = f_0 + f_1$, where $f_1 \ll f_0$ and

$$f_0(i, e) = \frac{4\Sigma_s}{m_p} \frac{ei}{\langle e^2 \rangle \langle i^2 \rangle} \exp\left(-\frac{e^2}{\langle e^2 \rangle} - \frac{i^2}{\langle i^2 \rangle}\right) \, . \quad (18)$$

Equation (17) is linearized, using the fact that, because f_0 is an integral of free orbital motion, it satisfies the homogeneous equation. The result is a linear, inhomogeneous equation for f_1.

It remains to specify the encounter term on the right hand side of (17). The reader is referred to Wetherill and Stewart (1989) for mathematical details; here I only describe the physical mechanisms that are included. The term is usually divided into five components:

1. Gravitational "viscous stirring", due to close encounters between orbiting bodies converts ordered, orbital energy to random kinetic energy. This component is always positive, in the sense that it increases relative velocities.
2. Inelastic collisions also produce "viscous stirring", which increases random energy at the expense of ordered, orbital motion. Cross-sections for collision are provided by data such as that shown in Fig. 24.
3. Inelastic collisions can, of course, also extract energy from the random component, which reduces relative velocities.
4. The effect of "dynamical friction" is also included in the encounter term, although this effect really depends on the long-range, collective interaction of large numbers of small objects with their larger counterparts. It acts to drive the system toward a state of energy equipartition, in which different masses have the same energy. Thus, it reduces the relative velocities of the more massive bodies, and is a thereby a critical driver of runaway accretion.
5. Gas drag extracts energy from both the ordered orbital motion and the random velocities, but it can actually increase the relative velocities of disparate masses by differential drag.

It is useful to examine steady state velocity distributions (the mean relative velocity as a function of mass) derived by setting the sum of encounter terms equal to zero. An example is shown in Fig. 25. The index q defines the power law mass distribution; the value $q = 2$ gives an equal mass in each logarithmic mass interval. When mass is distributed rather smoothly among the bodies ($q < 2$), relative velocities are relatively uniform. For steeper power laws, and particularly for $q > 2$, all relative velocities are less, but dynamical friction caused by the relatively large number of smaller bodies dramatically decreases those of the largest bodies. This, of course, promotes runaway accretion. (See the discussion in Lissauer and Stewart 1993).

5.4 Calculating Planetesimal Growth

Modern computer codes that calculate the evolutionary development of an ensemble planetesimals employ complicated algorithms that have been tuned extensively by comparison with direct numerical simulations of more restricted (and therefore more manageable) problems, careful comparison with exact solutions of the coagulation equation (Wetherill 1990), and comparisons among different methods (e.g., Stewart and Ida (2000) and Inaba et al. (2001)). Many issues, not only of physics, but of numerical stability and accuracy must be

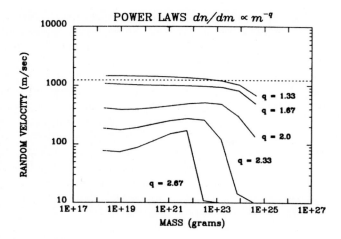

Fig. 25. Equilibrium velocity dispersions as a function of planetesimal mass, for different mass power law distributions. Larger objects tend to have lower random velocities. As the mass power law steepens, the random velocities of the largest objects are significantly depressed, a result of the equipartitioning effect of dynamical friction. (Figure from Lissauer and Stewart 1993)

confronted. To give some flavor of what is involved in a comprehensive calculation, we describe, with unfortunate but necessary brevity, the calculation of Wetherill and Stewart (1993).

As in all such statistical treatments, a "box" of planetesimals is considered to be representative of the ensemble evolution at a given distance from the star. Wetherill and Stewart (1993) start with a box at 1 AU, of radial width $\Delta a = 0.17$ AU, containing approximately 10^9 equally massive objects (i.e., all objects reside in a single mass bin, $m_1 = 4.8 \times 10^{18}$ gm). At $t = 0$, the distribution of velocities is given by (18), from which mean horizontal and vertical relative velocities, and collision probabilities can be found. The number of mergers, and the change in relative velocities in timestep Δt due to encounters, is then calculated from $\delta f/\delta t|_{encounter}$. The number of mass bins in increased to two; one containing objects with initial mass m_1, and one containing objects with mass m_2 ($= 2m_1$, if all collisions resulted in perfect mergers). Fragmentation is accounted for by creating a multi-mass bin, into which fragments are distributed according to a fragmentation size distribution law. The smallest of these fragments are particularly vulnerable to the gas drag term in $\delta f/\delta t|_{encounter}$, and may be lost from the system (dragged into the Sun). The populations and mean relative velocities are calculated for each mass bin, and each is advanced through Δt, the number of mass bins being increased as bodies merge. But soon the number of objects in the largest mass bins becomes rather small, challenging the validity of mean quantities calculated for those bins. So an algorithm for identifying and treating such runaways must be implemented. Wetherill and Stewart (1993) find that these runaways tend to be on circular, coplanar orbits, and so become isolated from each other (but not from smaller objects). They define a "gravitational interaction radius" $R_g = R_g\left(r_{Hill}, a, e\right)$ and regard as *isolated* the largest N bodies, where N is defined by the condition that

$$\sum_N R_g \geq \Delta a \ .$$

These objects are not permitted to collide with each other, but they do interact with objects in bins of smaller mass. For more details the reader should consult Wetherill and Stewart (1993) and subsequent literature.

A representative of the (Wetherill and Stewart 1993) results are shown as successive cumulative mass distributions in Fig. 26. After 10^3 years, 52 bodies

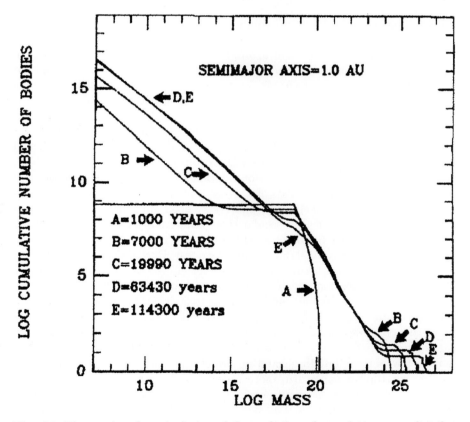

Fig. 26. The results of a calculation of the evolution of cumulative mass distribution in the terrestrial planet. These calculations considered planetesimals in a box, 0.17 AU on a side, located at 1 AU, using special methods to treat the runaway bodies. (Figure from Wetherill and Stewart 1993)

with $m > 30m_1$ have formed; after 7×10^3 years years, 50 bodies have become isolated (the largest being bigger than 10^3 km in radius); by about 20×10^3 years, fragmentation has begun to increase the small body population noticeably; and by 1.2×10^5 years there are 7 runaway bodies containing roughly half the mass of the system, the largest body being the size of Mercury. The results of another calculation is shown in Fig. 27 (Weidenschilling et al. 1997). This simulation is based on similar techniques, but treats the isolation of the largest objects directly by explicitly calculating the properties of individual runaway bodies, and including 100 radial zones, encompassing $a = 0.5$–1 AU. Their results are similar in essential ways to those of Wetherill and Stewart (1993).

5.5 The Final Stage of Accumulation; Rocky Planets in the Terrestrial Planet Region

Planetesimal growth simulations indicate that a final stage of rocky planet accumulation begins when several (or perhaps a few tens) of lunar or Mars-sized protoplanets, which have grown by runaway accretion, become relatively dynamically isolated. Secular resonances among these objects and/or externally produced perturbations (i.e., those produced by giant planets) then must increase their eccentricities and inclinations to the point where their orbits cross and further collisional growth ensues. Modern computeralgorithms (e.g., Sugimoto et al. (1990), Wisdom and Holman (1991), Saha and Tremaine (1992), Saha and Tremaine (1994), Makino et al. (1997), Duncan et al. (1998) and Chambers (1999)) can integrate the orbits of several tens to hundreds of individual gravitating objects for hundreds of millions of years, and so can be applied to the problem of this final stage. Monte Carlo techniques (e.g., Wetherill (1992, 1994, 1996)) have also been applied with considerable success, although they neglect secular effects. Inevitably, such calculations are performed specifically for the terrestrial planets and asteroids, as they provide the only currently available quantitative test. Here, I only summarize the main features of current results.

A typical calculation which begins with several tens of "protoplanets" in the terrestrial planet region usually evolves to a situation in which only a few planets remain (Fig. 28). The entire process may take a few times 10^8 years. A few conclusions about the process are robust: (1) If there are no external perturbers, mass and angular momentum are conserved to a high degree (only a few percent lost in ejected bodies or collisions with the star). In the presence of Jupiter or Saturn, more objects are lost, the amount depending on how much material originally resided close to the regions most affected by the giant planet resonances (i.e., beyond about 2 AU). (2) After a few hundred million years, a few terrestrial-mass planets remain (if the appropriate mass was present to begin with), although the final orbital configurations and individual masses are stochastically determined. Thus, specific quantitative predictions about the precise nature of the final planets cannot be made.

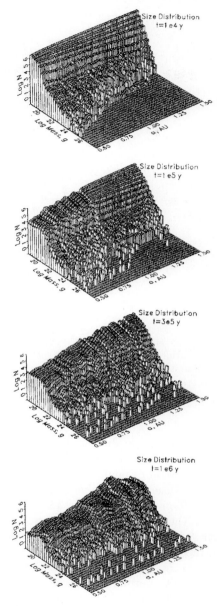

Fig. 27. The results of another calculation of the evolution of cumulative mass distribution in the terrestrial planet region. These calculations considered planetesimals between 0.5 and 1.5 AU, and allowed for the interactions between a continuum distribution of small bodies and a population of discrete runaway objects in individual orbits. There is qualitative agreement with the results shown in Fig. 27. (Figure from Weidenschilling et al. 1997)

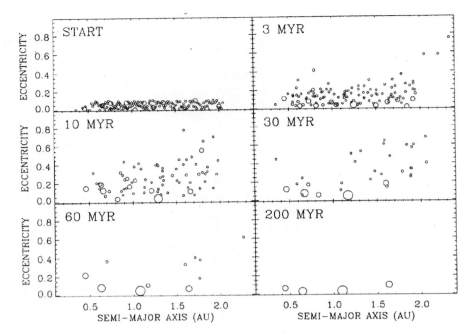

Fig. 28. Results of an N-body simulation (starting with about 150 bodies) of the final stage of planetary formation in the inner Solar System. The size of the symbols are proportional to the planetesimal radii. This particular simulation produces a reasonable facsimile of the terrestrial planets. (Figure from Chambers (2001))

(3) Giant collisions were inevitable during the last stages of rocky planet formation. (4) There is some radial mixing of material, but not complete homogenization, during the period of final assembly.

Some problems remain. Because it is believed that the giant planets formed in less than a hundred million years, their perturbations on terrestrial planet formation are included in the most recent calculations. Yet, when this is done, there is a tendency for N-body simulations to produce eccentricities and inclinations higher than those of the present terrestrial planets. Sometimes this results in only one or two massive planets being formed in the simulations, because more eccentric orbits permit more mergers. The results appear to be somewhat sensitive to the initial mass distribution, and it might be that the configuration of the terrestrial planet system reflects initial conditions not well-represented by the calculations so far. But it is also clear that the locations of the giant planets, and the timing of their growths can have important consequences. See Chambers and Wetherill (1998) and Chambers (2001) for recent discussions of these issues.

6 The Formation of Gas Giant Planets

6.1 Atmospheric Capture or Gravitational Collapse?

Although Jupiter and Saturn contain a greater mass of hydrogen and helium than rock-forming and icy material, they are nevertheless enhanced in the heavy elements relative to solar abundances. This fact excludes gravitational collapse from the protoplanetary disk, a compositionally indiscriminant process, as the sole mechanism by which these planets formed. Moreover, Uranus and Neptune, which have managed to retain large amounts of hydrogen and helium, are still made mostly of the heavier elements. Thus a great deal of attention has been directed toward understanding the growth of planets, by the means described in Sect. 5.5, to the point where they can capture a massive atmosphere. To be sure, one can imagine that these planets formed by gravitational collapse followed by some other process (e.g., preferential loss of gases, or gain of rocky planetesimals) which resulted in their present compositions. Furthermore, what appear to be gas giant planets have been detected around other stars, and we do not know whether or not their compositions are the same as that of their parent star; these might have formed by gravitational collapse. So two distinct modes of giant planet formation may occur; both are interesting.

6.2 Giant Planet Formation by Atmospheric Capture

Let us begin by asking: What do calculations predict for planetesimal growth at 5 AU and beyond? Growth rate is proportional to collision frequency, which is less at 5 AU than at 1 AU, because both orbital frequencies and (presumably) the surface density of solids decrease with distance from the star. Note, however, that settling in a non-turbulent nebula would still be expected to proceed on a 10^3–10^4 year timescale (see Fig. 21). As a planetesimal grows, it is capable of retaining an increasing mass of gas in its atmosphere. In fact, upon reaching a critical mass, it can induce the collapse of all the gas available in its neighborhood, limited only by a finite reservoir, or by dynamical effects associated with rotation (see Sect. 7.1). The planetesimal becomes the core of a gas giant planet.

This mechanism of gas giant planet formation is frequently called the "core instability" model, although it can be caused by evolution to a state in which there is no static equilibrium. The phenomenon is a consequence of the simultaneous constraints imposed by hydrostatic equilibrium and radiative transfer. A simple illustration is given by Stevenson (1982), although his argument involves some sleight-of-hand. Hydrostatic equilibrium requires that the atmosphere above the core satisfy

$$\frac{1}{\varrho}\frac{dp}{dr} = -\frac{GM_r}{r^2},$$

where M_r is the mass contained within radius r. Radiative transport in the optically thick atmosphere is governed by

$$\frac{16\sigma T^3}{3\kappa\varrho}\frac{\mathrm{d}T}{\mathrm{d}r} = -\frac{L}{4\pi r^2},$$

where the luminosity L (supplied by core accretion) and opacity κ are assumed constant, for the sake of example. As long as M_r is dominated by the core mass (and is therefore nearly constant), these equations are satisfied by state variables with the dependencies

$$\varrho \propto 1/r^3, \quad p \propto 1/r^4, \quad T \propto 1/r\ .$$

But when the core becomes large enough to attract an envelope which itself begins to contribute substantially to the total mass, no physically realistic solutions exist; the gravitational force within the atmosphere cannot be balanced by the pressure gradient and the atmosphere collapses. Figure 29 shows the critical core masses calculated by Mizuno (1980), along with an those derived from an analytic expression derived by Stevenson (1982). Note that the critical core mass has a strong dependency on opacity; it also depends on the core mass accretion rate.

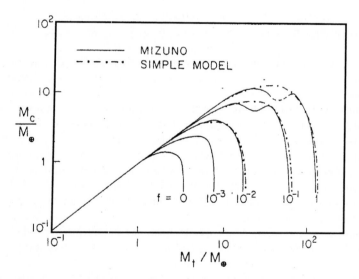

Fig. 29. Core mass as a function of total (core plus atmosphere) mass, for different (normalized) concentrations of dust in the atmosphere (or, equivalently, opacity normalized to a standard, solar composition). The *solid lines* are from calculations of Mizuno (1980); the *dot-dashed line* is derived by a simple model of Stevenson (1982). Portions of the curves with negative slope are unstable; no static equilibrium solutions exist to the right of the vertical portions of the curves. (Figure from Stevenson 1982)

What limits the mass once the nebular gas begins to collapse around the core? Undoubtedly, rotational effects become important at some point. Gravitational interactions with the disk, i.e., gap clearing Lin et al. (1996), may act to diminish Lubow and Artymowicz (2000) or terminate atmospheric accumulation. Or, it may be that the mass of available gas diminishes with time, so later-forming planets accrete less gas Shu et al. (1993). It has also been proposed that gas accumulation can be terminated by a hydrodynamic instability associated with accretion, if nebular densities are sufficiently low (Wuchterl et al. 2000), and references therein). There is currently no consensus on this important question.

The mass of Jupiter's core is uncertain. It could be as large as $15\,M_\oplus$, but data constraining its size also permit models with *no* core (Guillot et al. 1997). (If the latter were the case, the atmospheric capture model would be invalid, of course.) Can a core of $15\,M_\oplus$ grow before the gas of the protoplanetary disk has disappeared? Recall that the lifetimes of protostellar disks do not appear to exceed 10^7 years. Recall also that planetesimal growth slows considerably after the runaway phase, and that even in the terrestrial planet region (where planetesimal growth is faster) full-sized planets are attained only after 10^7–10^8 years. Thus it seems necessary that the critical core mass for Jupiter be attained before runaway ceases. The runaway mass is given by (16), which demands that $\Sigma_s \geq 10\,\text{gm/cm}^2$ for $m_c = 15\,M_\oplus$. This surface density, if spread over the Jovian zone, would yield a mass of solids of about $50\,M_\oplus$, substantially greater than that which now resides in Jupiter. We shall return to the question of what happened to this "extra mass" in a moment, but supposing it to have existed, a critical mass core, consistent with the limits imposed by models of Jupiter's composition and structure, could have accumulated within about 10^6 years.

Pollack et al. (1996) have constructed quantitative models of the formation of Jupiter and Saturn, based on the atmospheric capture idea. The study incorporates a model for the growth of planetesimals and a calculation of the hydrostatic structure of the gaseous envelope, including its interaction with accreting planetesimals. They consider a zone in the disk centered on Jupiter's orbit (assumed not to change), with a width $\Delta a = \Delta a\left(r_{Hill}, \langle e^2 \rangle\right)$. That is, Δa grows as Jupiter's Hill sphere grows, and as the eccentricity dispersion of the accreting planetesimals increases. The planetesimals are assumed to be distributed instantaneously uniform throughout Δa, but their number diminishes as they are consumed by Jupiter (or increases as Δa expands). The fates of the planetesimals as they encounter Jupiter's atmosphere are calculated. They may have a grazing encounter, entering and leaving the atmosphere; they may be trapped by gas drag and disintegrate in the atmosphere; or they may plunge all the way to the core. Their kinetic energy is deposited accordingly. Calculations are performed for an ensemble of impact parameters. Statistical averages are then used as inputs to the atmospheric envelope growth calculation. The envelope is supported by the luminosity of accretion, and gas is supplied to it from the disk in accordance with the growth of the Hill sphere.

As might be suspected from the discussion above, opacity in the envelope is a key variable. Pollack et al. Pollack et al. (1996) assume that it corresponds to cosmic abundances.

Results of this study for the growth of Jupiter are shown in Fig. 30. (Calculations for Saturn were also performed.) Runaway growth of the core to about $12\,M_\oplus$ occurs in less than one million years, after which it becomes dynamically isolated and core growth slows. Atmospheric mass grows faster until, at about 8 million years, it exceeds the core mass and induces dynamical collapse. The authors point out that the quantitative aspects of this solution depend rather sensitively on the assumed surface density of solids. If Σ_s is much greater than $10\,\mathrm{gm/cm^2}$, the fraction of heavy elements (i.e., the core mass) is too large to be consistent with models of Jupiter. If Σ_s is much less than $10\,\mathrm{gm/cm^2}$, the core grows too slowly. Thus the model is essentially "tuned" to provide the maximum core mass allowed by the data, so that it can form within the inferred lifetime of the nebula. Variations on model parameters are discussed in Pollack et al. (1996).

What of the "extra mass" in the Jupiter region implied by this value of the surface density? Theory predicts that a few other potential giant planet cores would form from this material (e.g., Kokubo and Ida 1998). Thommes et al. (1999), noting the difficulty with which planetesimal accumulation theories have in explaining the formation of Uranus and Neptune (bodies become too isolated; growth is too slow) suggested that these planets formed between Jupiter and Saturn and were gravitationally scattered to their present orbital locations during the growth of Jupiter and Saturn. Their orbits were circularized, according to Thommes et al. (1999), by interaction with the residual

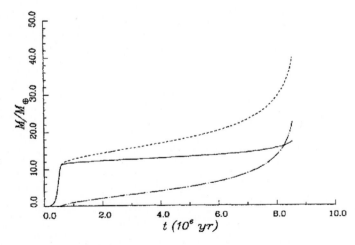

Fig. 30. Growth of Jupiter as a function of time, according to calculations of Pollack et al. (1996). Shown are the core mass (*solid line*), atmosphere mass (*dot-dashed line*) and total mass (*dotted line*).(Figure from Lissauer 1993)

planetesimal disk. This is an appealing hypothesis, potentially solving a few puzzles at once.

At this point, one should realize that the planetesimal accumulation theory of planet formation has attained considerable success in explaining the overall dynamical configuration of the Solar System, the only planetary system available for meaningful tests. But there remain difficulties and uncertainties concerning both the terrestrial and gas giant planets (and the asteroids, which I have not discussed). Moreover, the existence of extrasolar giant planets close to their parent stars is not readily explained by gradual planetesimal accumulation alone. We now turn to the possibility of planetary formation by gravitational instability.

6.3 Giant Planet Formation by Gravitational Collapse

The idea that planets could form, like stars, by direct gravitational collapse of gas and dust together, is an appealing one, particularly in the light of the discoveries of extrasolar planets. Gravitational collapse would likely be fast and efficient, taking only several orbital periods to isolate the planetary mass. This is an obvious advantage, as the atmospheric capture process, as applied to Jupiter, seems barely able to satisfy the requirement that the process be complete before the disk gas disappears. Most of the extrasolar planets exceed the mass of Jupiter and have non-circular orbits. It is not known how effective atmospheric capture might be in producing such objects, but if they formed rapidly by gravitational instability, one can readily imagine situations in which multiple protoplanets scatter each other into irregular orbits, with occasional mergers increasing the largest masses. Indeed, it has been argued (without notable acceptance, I should add) that the masses and distribution of eccentricities of extrasolar companions indicate that they represent a low mass extension of the stellar population, and that they therefore probably formed as stars do, by gravitational collapse, and should not even be called planets (Stepinski and Black 2001). It should be understood, however, that here we are talking about gravitational collapse *within a protostellar disk*, which represents quite a different situation than the collapse of a molecular cloud core.

Gravitational instability occurs if the criterion of (11) is violated. It is possible to estimate Q from the diagnostics outlined in Sect. 1.2.2 and theoretical models. Figure 31 shows the radial distribution of Q calculated from the models of D'Alessio et al. (1998) for typical T Tauri stars. It is seen that the disks are quite stable over all radii less than about 10 AU, but that it is only the effect of stellar radiation that stabilizes them beyond that distance. Thus, if some heating event in the inner disk caused it to swell vertically and shadow the outer disk (for instance), an instability might develop beyond 10 AU. Or it might be that disks were substantially more massive during their early histories, thereby lowering Q, as in the models of Boss (1997, 1998). Given the model results shown in Fig. 31, it is reasonable to suspect that instances of gravitational instability do occur.

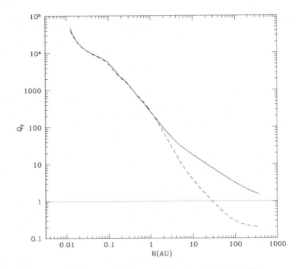

Fig. 31. The radial distribution of the gravitational stability parameter Q (*solid line*) for a typical T Tauri star, calculated from the models of D'Alessio et al. (1998). The dashed line indicates the value that would result if the disk was not illuminated by stellar radiation. Disks appear to be stabilized by stellar radiation over radii less than about 10 AU. They might be unstable at an earlier, more massive, stage, or if shadowed by a hot inner region.(Figure from D'Alessio et al. 1998)

In Sect. 3, I stated that gravitational instability produces spiral density waves. This fact has been confirmed in many numerical simulations, and is not surprising because almost *all* disturbances in highly sheared disks produce spiral structure. But the manifestations of instability are difficult to assess by direct observation. The waves seen vividly in Saturn's rings can usually be attributed to the disturbance of a satellite rather than instability, and the precise origin of galactic spiral structure is complicated by overlying processes such as star formation and molecular cloud evolution. Thus, so far, we have had to rely on the numerical simulations to determine the ultimate fate of unstable disks.

There are two fundamental problems that plague gas disk simulations, unique among numerical astrophysical problems. One source of difficulty is the intrinsic dynamic range of global phenomena in Keplerian disks, as defined by the variation of orbital period over the radial extent of the disk. Gravitational disturbances close to the star develop rapidly and can affect things far away, where the much slower orbital period can retard response. Thus a useful simulation must be sustained, and resolve a large range of orbital periods. This problem has been addressed in galactic simulations by the development of variable timestep techniques and tree codes, in which integrations are performed non-uniformly over the disk; equivalent techniques have not been developed for gas disks. The second problem stems from the fact that small changes in kinetic and gravitational energy can cause large changes in internal energy, because the latter quantity is a factor of $(h/r)^2$ smaller than the former. And, as we shall see, the response of a disk's internal energy to instability plays a large role in determining the nonlinear development of the instability.

The first problem is somewhat mitigated in simulations of gas giant planet formation by gravitational instability, because the instability and interesting effects are presumed to occur in the outer disk, with minimal influence of the (highly stable) inner disk. Thus the inner disk has been effectively removed from some calculations (e.g., Boss (2000) and Pickett et al. (2000b)), although it remains to be demonstrated rigorously that this is an acceptable procedure. Up until recently, the calculations have also been performed under simplifying assumptions regarding the disposition of internal energy. It has been established that if an unstable disk remains everywhere at its initial temperature (as might be possible under optically thin conditions), instabilities proceed violently to a state dominated by strong density waves and high density contrasts (Boss 1997, 2000; Nelson et al. 1998; Pickett et al. 1998), as in Fig. 32. Boss (1997) argued that some of the density perturbations were, in fact, gravitationally bound and therefore could be regarded as potential protoplanets. But it has also been shown that disks which respond *adiabatically* to instability-generated disturbances sustain much milder density waves (Pickett et al. 2000a; Boss 2000; Nelson et al. 2000). That is, the energy loss rate determines the effect of the instability: isothermal disks, in which energy is easily lost (and gained, in gas undergoing expansion) remain unstable and evolve violently; adiabatic disks tend to heat up and become more stable.

It is clear that processes which transport and dissipate energy must be accurately represented if we are to determine, from numerical simulations, whether or not planets can form by gravitational instability and, if so, what their characteristics would be. I can't help mentioning that these calculations, as difficult (and therefore subject to error of assumption or algorithm) as the calculations of planetesimal growth, have been undertaken by relatively few researchers. A satisfactory theoretical understanding would be hastened, no doubt, by the kind of cooperative attention that has so advanced the planetesimal growth problem. Of course, determination of the heavy element abundances of extrasolar giant planets could provide a definitive test: compositional identity of planet and parent star would heavily favor gravitational collapse.

7 Planet Migration

7.1 Tidal Interaction and Angular Momentum Exchange Between Planet and Disk

The discovery of large planets in orbits very close to their parent stars has focussed attention on a theoretical result that had been previously disregarded

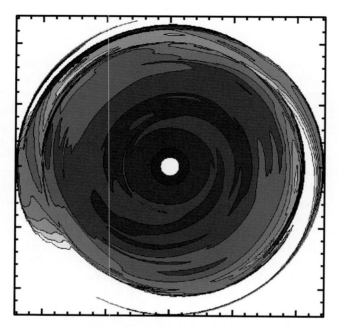

Fig. 32. A representation of equatorial density in a gravitationally unstable protostellar disk. In this numerical simulation, the evolution was assumed to be isothermal; that is, its temperature was fixed everywhere to its initial value. The contour/gray-scale interval is 0.5 in the log of the fractional density perturbation; it spans 4.5 orders of magnitude. An important unresolved question for giant planet formation is whether the high density (*dark*) arcs seen in this simulation eventually evolve into discrete, gravitationally coherent objects (protoplanets), or whether they are transient entities, like the many waveforms that dissolve into the background. (Figure from Pickett et al. 2000b)

in most theories of planet formation. The result states that angular momentum can be exchanged between a planetary body (or other stellar companion) and a circumstellar disk, due to tidal interaction, such that the net torque on the planetary body causes rapid orbital evolution (Goldreich and Tremaine 1979). Thus, even when a solid body has grown too large to be coupled to the gas by viscous drag, it may still be coupled by gravitational forces strong enough to affect its orbital motion.

An intuitive derivation of the tidal torque was given by Lin and Papaloizou (1979), by means of an "impulse approximation", in which the interaction is considered to arise from the local gravitational deflection of fluid parcels in

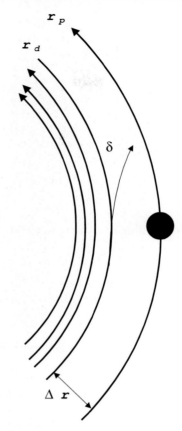

Fig. 33. A planet gravitationally deflects gas near the edge of a disk, changing its angular momentum. If close encounters between the planet and a given fluid element occur in a phase-coherent way, angular momentum is systematically exchanged between planet and disk. This tidal interaction can cause radial migration of the planet

the disk by the planet (Fig. 33). The fluid is deflected through an angle δ as it passes the planet, located Δr away:

$$\delta = \frac{\Delta v_r}{v_{rel}} = \frac{2Gm_p}{\Delta r v_{rel}^2} \; .$$

Here v_{rel} is the relative velocity between disk gas and planet and m_p is the mass of the planet. The change in the gas angular momentum (per unit mass) is

$$\Delta j = -v_{rel} r_d \left(1 - \cos\delta\right) \approx -v_{rel} r_d \frac{\delta^2}{2} = -\frac{2G^2 m_p^2 r_d}{(\Delta r)^2 v_{rel}^3} \; .$$

This interaction occurs every time the planet comes near the particular fluid under consideration, or every $\Delta t = 2\pi/|\Omega - \Omega_p|$, so the rate of angular momentum exchange is given by

$$\frac{\Delta j}{\Delta t} = \frac{\Delta j}{2\pi/|\Omega - \Omega_p|} = -\frac{G^2 m_p^2 |\Omega - \Omega_p|}{\pi r_p^2 (\Delta r)^2 (\Omega - \Omega_p)^3}.$$

We have used the fact that $v_{rel} = r_d \Omega - r_p \Omega_p$. To find the *total* angular momentum exchanged between planet and disk, this expression should be integrated over the whole disk:

$$\frac{dJ}{dt} = \int_{\Delta r_d}^{\infty} \Sigma \frac{\Delta j}{\Delta t} 2\pi r \, d(\Delta r).$$

But it is assumed that the torques nearest the planet dominate, so Ω may be expanded

$$\Omega^2 = \left(\Omega_p + \left.\frac{d\Omega}{dr}\right|_p \Delta r + ...\right)^2.$$

The result is

$$\frac{dJ}{dt} = T = -\frac{8}{27} \left(\frac{r_p}{\Delta r}\right)^3 \left(\frac{m_p}{M_*}\right)^2 \Omega_p^2 \Sigma r_p^4. \tag{19}$$

Remarkably, this formula for the torque on the disk close to the planet is almost correct, requiring only a modest adjustment of the constant to match a more rigorous derivation. Note that, according to this formula, the torque diverges as $\Delta r \to 0$, a fact which leads to the notion that the planet actually clears a gap around itself, by extracting angular momentum from the gas within its orbit and giving angular momentum to the gas outside of its orbit. The size of the gap is presumably set by the Δr at which the tidal torques are balanced by some other (e.g., viscous) torque. It is then the difference between the inner and outer torques that determines the net torque on the planet, i.e., how fast and in what direction the planet's orbit will change.

The problem with this derivation is that it obscures the real physics of the interaction, which is rich in hydrodynamical phenomena, and is really one of resonant (or near-resonant) interaction and wave excitation (Goldreich and Tremaine 1979, 1980). These aspects are revealed only by solving the full fluid equations, as in Sect. 3 for waves, but now including the forcing by the gravitational potential of the planet. The linearized equations that lead to the dispersion relation (9) then have inhomogeneous "driving" terms which represent the Fourier decomposition of this potential. Solutions of the inhomogeneous equations can be matched to the homogeneous, "wave" solutions to obtain a full description of the excitation of the disk caused by the planet (Goldreich and Tremaine 1979; Yuan and Cheng 1989). The torque is indeed concentrated near the planet, but at the special locations identified previously as the Lindblad resonances (Sect. 3). These local disturbances excite spiral density waves which, behaving as previously described, propagate away from the resonances carrying angular momentum to be deposited wherever the waves damp. The effects associated with each resonance must then be summed to obtain the total angular momentum exchange rate. The torque exerted on the disk *near a single resonance* (designated as the mth Lindblad resonance) is given by Ward (1997a); cf. Artymowicz (1993) as:

$$T_m = -\frac{\pi^2 m \Sigma \psi_m^2}{r\dfrac{\mathrm{d}D_*}{\mathrm{d}r}} , \qquad (20)$$

where the terms have the following definitions:

$$\psi_m = \frac{r\dfrac{\mathrm{d}\phi_m}{\mathrm{d}r} + 2m\phi_m f}{\sqrt{1 + 4\left(mc_s/\Omega r\right)^2}}$$

$$\phi_m = -\frac{Gm_p}{r_p} b_{1/2}^m \left(r/r_p\right)$$

$$b_{1/2}^m = \frac{2}{\pi} \int_0^\pi \frac{\cos m\theta \, \mathrm{d}\theta}{\sqrt{1 - 2\left(r/r_p\right)\cos\theta + \left(r/r_p\right)^2}}$$

$$f = m\left(\Omega - \Omega_p\right)/\Omega$$

$$D_* = \Omega^2 - m^2\left(\Omega - \Omega_p\right)^2 + \left(mc_s/r\right)^2 .$$

The quantity $b_{1/2}^m$ is the coefficient of the mth term in the Fourier expansion of the planet's gravitational potential (in the frame of the center of the disk). Examination of these terms shows that the numerator of (20) contains the intrinsic strength of the forcing term and the denominator contains the distance (measured in frequency space) from the resonance. Artymowicz (1993) showed that the terms proportional to $mc_s/r\Omega$, which had been neglected in previous analyses because this factor was considered to be small, of order the disk thickness ratio, are important in limiting the maximum value of torque. Recall that the impulse approximation indicates that the torque diverges as the distance to the planet Δr approaches zero. But note that the highest order resonances also occur closest to the planet, so the relevant m becomes large as Δr becomes small; thus terms proportional to $mc_s/r\Omega$ are not necessarily negligible. These terms reflect the displacement of the resonant locations due to the non-zero pressure gradient in the disk.

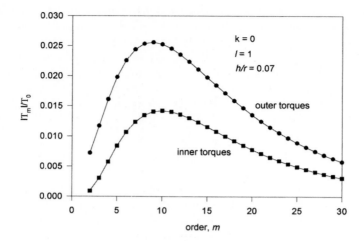

Fig. 34. Normalized torque as a function of mode number m, for typical values of disk parameters. In this figure, k refers to the power law index for the base state surface density (i.e., $k = 0$ for constant surface density) and l is the (negative) power law index for the base state temperature. Outer torques are systematically higher than inner torques, an asymmetry which produces inward planetary migration. (Figure from Ward 1997a)

Ward (1997a) calculated T_m from (20) for a typical disk configuration and obtained the results shown in Fig. 34. A maximum value of T_m occurs for intermediate values of m. Thus the highest order Lindblad resonances are of diminishing importance and there exists a "torque cut-off" which limits the total torque even when there is no gap. Most importantly, note that the torques associated with outer Lindblad resonances systematically exceed those of the inner Lindblad resonances. This imbalance is due to intrinsic characteristics of Keplerian disks; outer resonances are slightly closer to the planet than their inner counterparts, and are less diminished by pressure effects, among other factors (Ward 1997a). The result is that the dominance of the outer resonances produces a net negative torque on the planet, which induces inward migration.

The total torque can be found by the following trick. Calculate the *torque density* by assuming that the torque is distributed more or less smoothly between Lindblad resonances:

$$\frac{dT}{dr} = \frac{T_m}{\Delta r_{Lm}} \ .$$

The distance r_{Lm} is the separation of successive resonances; see (12). The total torque is then found as before, from the integral

$$T = \int_{\Delta r}^{\infty} \frac{dT}{dr}\, dr \ .$$

The trick is rigorously justified if the waves excited by the resonant interaction are damped immediately, i.e., at or very close to the resonance. Even if they are not, the *total* torque is usually accurately represented, because the effect of propagating waves is simply to redistribute the torque deposited near the resonance. If the terms proportional to $mc_s/r\Omega$ are ignored, a formula essentially the same as (19) is found, except with the coefficient 0.84 instead of $8/27 = 0.296$ (Goldreich and Tremaine 1980). Including the torque cut-off effect yields a different coefficient, which depends on the details of the disk base state.

7.2 Rates of Orbital Evolution

What do the results described above predict for the rate of planetary migration? To answer this question, one must calculate both the total torque *and* the response of the disk, in a self-consistent manner. Note that the torque is proportional to the local value of the surface density (19), which changes as the disk material is redistributed in response to the torque. Thus one cannot simply use the base state disk conditions to calculate the torque on the planet. Ward (1997a) found an illuminating solution to this problem, in which the planet migrates inward at constant radial velocity, accompanied by a steady, wave-like disturbance in the disk density distribution (Fig. 35). In essence,

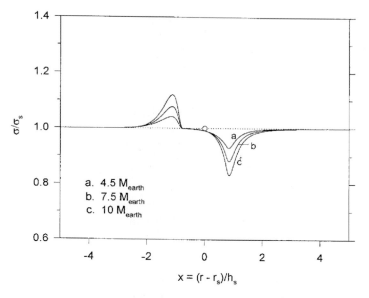

Fig. 35. The surface density perturbation produced by a migrating planet that is not large enough to clear a gap (Type I drift), for three planetary masses and typical disk parameters. The horizontal axis is the distance from the planet (normalized by the scale height) in a frame moving with the planet. (Figure from Ward 1997b)

the solution represents a situation in which the disk surface density is locally disturbed in just the way necessary for the planet-induced torque to produce a constant mass flux in the frame of the radially migrating planet. For an inviscid disk, the planet's radial velocity is given approximately by

$$v_p = .25 c_s \frac{m_p}{M_* Q} \frac{r^3}{h^3},$$

where Q is Toomre's stability parameter (11). For typical parameter values ($c_s = 1\,\text{km/sec}$, $h/r = 0.1$, $Q = 10$, $M_* = 2 \times 10^{33}\,\text{gm}$), one finds $v_p = 7.5\,(m_p/M_\oplus)$ cm/sec. Thus a $1\,M_\oplus$ planet would migrate from 1 AU to the Sun in about 6×10^4 years and a $10\,M_\oplus$ planetary core would migrate from 5 AU to the Sun in about 3×10^4 years. This mobility (called Type I migration, when the planet is not massive enough to clear a gap around itself) can therefore result in large radial displacements of planetary objects in times comparable to or less than their formation times.

It is found, however, that there is a critical mass, m_c, above which a gap in the disk forms around the planet and Type I migration ceases. Because viscosity acts to spread material into the gap, the critical mass increases with disk viscosity; it is given by Ward (1997b)

$$m_c \approx \text{(constant)}\, M_* \alpha \left(\frac{M_*}{\Sigma r^2}\right) \left(\frac{c_s}{r\Omega}\right)^2,$$

where the constant depends on specifics of the disk properties, and is generally of order unity. For masses exceeding m_c, radial flow past the planet is inhibited and the planet becomes "trapped" in the gap. The planet and gap then move with the local disk gas, and drift at the rate

$$v_p \approx \frac{\nu}{r} \approx \frac{\alpha c_s h}{r} = \frac{\alpha c_s^2}{r\Omega}.$$

This drift is called Type II migration, and is slower than Type I migration, although it is clearly significant on timescales relevant to the evolution of the disk. Note that Type II migration could be outward, if the planet was located in the expanding part of the disk. Ultimately, however, most of the disk moves inward, and the planet would be carried with it toward the star. Figure 36 summarizes the drift rates for Type I and II migration, as a function of planet mass, for a typical disk configuration (Ward 1997a).

Although there is no obvious evidence for tidal drift in the Solar System, it is a likely for the Jovian-sized planets found very close to other stars. That is, it seems implausible that these planets formed so close to their parent stars (Bodenheimer et al. 2000), so one might suppose that they formed further away and then migrated inward to their present locations (Lin et al. 1996). Why might some systems suffer extensive tidal drift while others exhibit no effects? Trilling et al. (1998) addressed this question by calculating the orbital

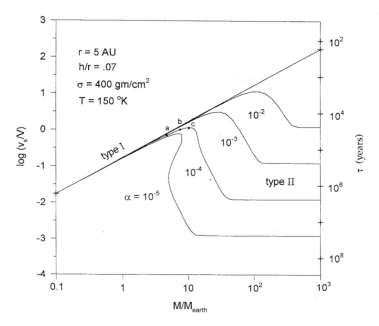

Fig. 36. Inward migration velocity as a function of planetary mass and turbulent viscosity parameter α, showing the transition from Type I to Type II drift. The velocity is normalized by $2\,(M_\oplus/M_*)\,(\pi\,r^2\,\Sigma/M_*)\,(V_K/c_s)^3\,V_K$. The time shown on the right-hand scale is the characteristic orbital decay time from 5 AU. (Figure from Ward 1997b)

evolutions of planets subject to Type II migration in a circumstellar disk. The planets were considered to be fully formed at the beginning of the calculation, and located at a specified distance from a solar mass star, usually about 5 AU. Subsequent orbital migration due to disk tides was calculated using the impulse approximation. If and when a planet approached the star, it suffered additional torques due to mass exchange with the star (Roche lobe overflow) and tidal interaction with the star, both of which tend to increase the orbital angular momentum of the planet and thereafter oppose the disk tidal torques. These torques depend on the size and internal structure of the planet, which were also calculated. Finally, it was assumed that the disk disappeared after a time interval of, say, 10^7, after which tidal interaction with the disk ceased.

Representative results from Trilling et al. (1998), for planets with initial masses between 1 and 5 times that of Jupiter, are shown in Fig. 37. For the case shown (disk mass $1.1 \times 10^{-2}\,M_\odot$, $\alpha = 1 \times 10^{-3}$), planets with initial masses less than 3.36 Jupiter masses migrate toward the star and ultimately lose most or all of their mass by Roche lobe overflow. Planets with initial masses greater than 3.41 Jupiter masses lose no mass, and the most massive do not migrate far from their initial location before the disk disappears.

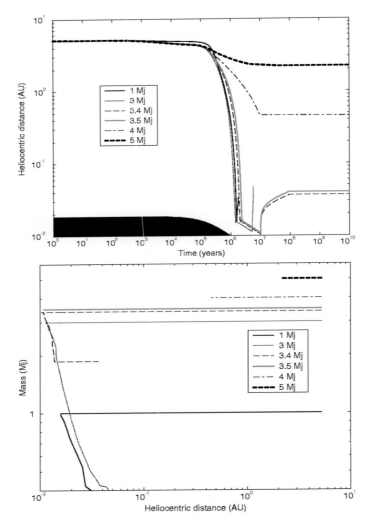

Fig. 37. (*upper*) Orbital radius as a function of time for planets of various initial masses, starting at 5 AU in a typical disk, and (*lower*) planetary mass versus orbital radius for the same planets. Masses are in units of Jupiter's mass. The dark region in the *upper* figure represents the radial extent of the central star. Outward motion is caused by mass loss and tidal interaction with the star. (Figure from Trilling et al. 1998)

Planets with initial masses between 3.36 and 3.41 Jupiter masses migrate to distances at which they lose mass to the star, but are saved by the disappearance of the disk. Trilling et al. (1998) suggest that the few percent of the planets that fall into the last category may account for the extrasolar giant planets found close to other stars, although the relevant mass range would vary with model parameters (disk mass and lifetime, initial planetary location, etc.).

Note that the analysis of Trilling et al. (1998) is not consistent, in detail, with the solutions found by Ward (1997a) First, Type I migration, which is consequential during the planet building stage, being more rapid than Type II, cannot be rigorously treated in the impulse approximation and is therefore ignored in Trilling et al. (1998). Also, according to Ward's analysis, Type II migration rates are mass-independent, because gap formation in a viscously evolving disk produces density perturbations which adjust the torque imbalance on the planet to that which causes the planet migration to match the movement of disk material. In contrast, according to Fig. 37, larger planets migrate slower than smaller ones, a result attributed by Trilling et al. (1998) to the larger gaps cleared by the larger planets. In fact, the final word on planet migration has yet to be spoken. Not all of the potentially relevant feedback effects have been examined (e.g., the effects of waves on thermal state, turbulent viscosity, dissipation, etc.). The interactions of multiple migrating planets have yet to be fully assessed. For instance, differential Type I migration of planets of differing masses could affect planetary growth rates (Ward and Hahn 1995). Numerical simulations by Artymowicz and Lubow (1996) show that gap formation may not be complete, in the sense that material can be accreted through the gap, which, in turn, would affect the migration rate. See Ward (1997a) and Lin et al. (2000) and Ward and Hahn (2001) for discussions of some of these issues.

7.3 Modeling the Formation of the Solar System

The Solar System will continue to provide the main test case for theories of planet formation until the orbital configurations of other systems have been determined to a much greater degree of completeness than they are presently known. Thus our quantitative theories of particle accumulation, collisional growth, giant planet formation, planetary migration, etc. must, in the end, satisfactorily explain the dynamical state of the Solar System, which we know to high accuracy. As discussed in these lectures, much progress has been made, but the problem is of such complexity that a completely satisfactory model does not yet exist, and may indeed be unattainable without further specific constraints.

In fact, aside from dynamical factors, there exists a wealth of information about the composition and physical state of the Solar System's individual members, which must also be consistent with our theories. For instance, the outstanding characteristic of Solar System planetary composition is the extreme volatile depletion of the terrestrial planets compared to the gas giants. The former lack not only their cosmic share of the light gases hydrogen and helium and noble gases, but are also vastly deficient in water and carbon- and nitrogen-containing compounds. This state is commonly attributed to the relatively small masses of the terrestrial planets and the higher temperatures they have experienced by virtue of their proximity to the Sun, but a quantitative integration of this idea with a general theory of planet formation is lacking.

Among the most intriguing objects of the Solar System are the meteorites: rocky fragments of asteroids, as well as some from the Moon and Mars, whose orbits have brought them to Earth. Of the many kinds of meteorites, the most common (the "chondritic" meteorites) are primitive, in the sense that they have escaped the equilibrating processes (sustained high temperatures and pressures) that are characteristic of terrestrial rocks. Thus they are composed of an unequilibrated mix of pebbles and fine-grained material. The outstanding features of chondritic meteorites are the following:

1. They are as old as the Solar System. The ages of individual components, as determined by nucleochronological methods, are typically within several million years of 4.56 Gyr (Wadhwa and Russell 2000, and references therein). The oldest known components (calcium-aluminum rich inclusions, or CAIs) are commonly taken to be the first solid objects formed in the Solar System, with ages as great as 4.566 ± 0.002 Gyr. Despite the ambiguities encountered in interpreting radiometric ages, there is no doubt that the chondritic meteorites have retained their essential character since their formation in earliest Solar System history.
2. Their composition, as measured by the relative abundances of elements, is the same as the Sun's, within a factor of two or so, for all but the most volatile elements. Although there are modest systematic differences in composition, these meteorites formed from "solar material", presumably the protoplanetary disk itself, and have experienced little compositional modification since then.
3. They are composed largely of igneous (melted and solidified) pebbles, the "chondrules" from which the chondritic meteorites derive their name. These chondrules exist within a matrix of fine-grained material to form the bulk rock, which, laboratory studies have shown, could not have experienced the high temperatures required to melt the individual chondrules. Moreover, there also exist within the meteorites, material with isotopic compositions which identify them as *pre-solar*, and which apparently survived the formation of the Solar System intact, to be incorporated and preserved in these primitive rocks. For the detailed properties of chondritic meteorites and pre-solar material, see Bernatowicz and Zinner (1997), Kerridge and Matthews (1988), and Hewins et al. (1996).
4. They contain unequivocal evidence for the existence of short-lived radionuclides (with half-lives less than a few million years) in the early Solar System (Goswami and Vanhala 2000). These radionuclides were either made in stellar sources or the interstellar medium shortly before the Solar System formed, or they were produced by energetic events within the Solar System itself.

The first three properties listed above identify the chondritic meteorites as early products of the planet-building epoch. They have therefore been regarded as key elements in understanding planet formation in the Solar System. However, attempts to explain their properties have led to widely divergent hypotheses. I will mention three approaches to the problem, to illustrate the diversity of ideas regarding the origin of the primitive meteorites and their radically different consequences for planetary formation theories.

A traditional approach has been to relate the bulk compositions of planets and chondritic meteorites to the thermal state of the protoplanetary disk at the time of their formation (e.g., Lewis (1974), Cameron (1978), Wasson (1985) and Cassen (2001)). In this case, it is assumed that thermal gradients reflect, directly or indirectly, position in the disk, and that a memory of these gradients has been retained in the final product. This approach focuses on bulk properties and usually has little to say regarding the details of physical state, chondrule formation, or isotopic properties.

A contrasting theory has been extensively developed by Shu et al. (1996, 2001), which postulates extensive redistribution of planetary material by inward drift within the disk and subsequent outward transport by the protosolar wind. The theory takes advantage of the energetic environment near the Sun (within 0.1 AU) to account for the observed thermal processing of chondrules and CAIs (as well as the production of short-lived radionuclides). Because of the high abundance of chondrules in chondritic meteorites (up to 80% by volume), a large amount of material must be recycled from close to the Sun back to the terrestrial planet region and the asteroid belt. Therefore, in this theory, one would expect that the final bulk compositions of planetary material retained little memory of its initial distribution in the protoplanetary disk.

Yet another idea (currently unpopular among meteoriticists) is that the chondritic meteorites are the products of vaporizing collisions among massive (perhaps lunar-sized), volatile-rich planetesimals of chondritic composition. Collisional production of chondrules has been dismissed for various reasons (Grossman 1988; Bos 1996), but I am partial to it because (1) the chondritic meteorites reflect a range of physical and chemical conditions (pressures, temperatures and oxidation states) which might be expected in such collisions, and which are not particularly characteristic of those we believe prevailed in the solar disk, (2) the time scales inferred for chondrule heating and cooling (minutes to days) (Jones et al. 2000) are commensurate with those expected to result from such large collisions, (3) large amounts of material might be locally processed and rapidly re-accumulated, thus preserving the individual characteristics of various meteorite classes and (4) such collisions are predicted by planetary formation dynamics. A consequence of this hypothesis would be that the detailed properties of the chondritic meteorites would have little to do with disk properties or any astronomically observable phenomena. But until a rigorous model for the remnants of such collisions is developed, the hypothesis will remain in speculative limbo.

7.4 Concluding Comments

I began these lectures by noting that, despite the unequivocal genetic association of circumstellar disks with planetary systems, there was reason to be skeptical of the premise that the nature of a planetary system could be predicted (or even modeled from) the properties of its progenitor disk. This skepticism might be provoked just by the existence of planets with highly eccentric orbits. Indeed, an argument can be made that final planetary configurations are determined only by the dynamical laws governing many-body systems, regardless of the origins of the bodies. I mentioned that the distribution of inferred extrasolar planet eccentricities appears to be similar to the distribution stellar binary eccentricities (Stepinski and Black 2001), which could be interpreted to mean that dynamics alone determines the orbits of gravitating objects, regardless of origins. Also, it must be remembered that the first extrasolar planets discovered were two Earth-sized objects in regular orbits about a pulsar (Wolszczan and Frail 1992), a planetary system of orthodox mass and orbital configuration, but surely with an origin and history dramatically different from other systems. One might speculate that we will eventually recognize classes of planetary systems, distinguished by distinct (or a continuous distribution of) dynamical histories, solely determined by the range and variety of stable (i.e., long-lived) states.

The idea that any connection with a primordial disk is effectively obscured is supported by numerous theoretical concepts, besides the gravitational scattering of large objects: the radial mobility of small solid objects in disks due to gas drag, the orbital evolution of planets due to tidal interactions with a gaseous disk, the potential recycling of material by a stellar wind and so forth. And yet the dynamical regularity found in our own Solar System, as well as its enigmatic but systematic compositional regularities, seem to provide sound evidence for inheritance from a disk. It still seems useful, therefore, to pursue the idea that the properties of individual planets, as well as the system they comprise, can be traced to their disk origins.

Acknowledgments

I thank Didier Queloz for inviting me to give these lectures, which I found challenging beyond all expectations. My thanks also to Didier, Michel Mayor, Willy Benz, Stephane Udry and Elisbeth Teichmann for being such gracious hosts and expert organizers, and making our stay in Grimentz so enjoyable. And thanks to Dotty Woolum, who provided invaluable support throughout.

References

P. Artymowicz: ApJ. **419**, 155 (1993)
P. Artymowicz, S. Lubow: ApJLett. **467**, L77 (1996)

S.A. Balbus, J.F. Hawley: ApJ **376**, 214 (1991)
S.A. Balbus, J.F. Hawley: Rev. Mod. Phys. **70**, (1998)
S.A. Balbus, J.F. Hawley: *From Dust to Terrestrial Planets*, (W. Benz, R. Kallenbach, and G.W. Lugmair, Kluwer Academic Publishers 39, 2000)
S.A. Balbus, J.F. Hawley, J.M. Stone: ApJ **467**, 76 (1996)
S.V.W. Beckwith, A.I. Sargent, R.S. Chini, R. Güsten: AJ **99**, 924 (1990)
S.V.W. Beckwith, Th. Henning, Y. Nakagawa:*Protostars and Planets IV*, (V. Mannings, A.P. Boss and S.S. Russell, University of Arizona Press Tucson, 533, 2000)
K.R. Bell: ApJ, **526**, 411, (1999)
K.R. Bell, D.N. C. Lin: ApJ **427**, 987, (1994)
K.R. Bell, P. Cassen, H.H. Klahr, Th. Henning: ApJ. **486**, 372, (1997)
J. Blum: *From Dust to Terrestrial Planets*, (W. Benz, R. Kallenbach, and G. W. Lugmair, Kluwer Academic Publishers, 265, 2000)
P. Bodenheimer, O. Hubickyj, J.J. Lissauer: Icarus **143**, 2, (2000)
A.P. Bos: *Chondrules and the Protoplanetary Disk*, (R.H. Hewins, R. H. Jones and E.R. D. Scott, Cambridge, 257, 1996)
A.P. Boss: Science **276**, 1836, (1997)
A.P. Boss: ApJ **503**, 923, (1998)
A.P. Boss: ApJLett. **536**, L101, (2000)
N. Calvet, G. Gullbring: ApJ, **509**, 802, (1998)
N. Calvet, L. Hartmann, S.E. Strom: *Protostars and Planets IV*, (eds. V. Mannings, A.P. Boss and S.S. Russell, University of Arizona Press, Tucson, 453, 2000)
A.G.W. Cameron: Moon and Planet, 18,5, (1978)
P. Cassen: *LPSC XXIV*, 261, (1993)
P. Cassen: Met. and Planet. Sci. **36**, 671, (2001)
J.E. Chambers: Icarus **152**, 205, (2001)
J.E. Chambers, G.W. Wetherill: Icarus **136**, 304, (1998)
J.E. Chambers: MNRAS **304**, 793, (1999)
E I. Chiang, P. Goldreich: Astrophys. J. **490**, 368, (1997)
J.N. Cuzzi, A.R. Dobrovolskis, R.C. Hogan: *Chondrules and the Protoplanetary Disk*, (R.H. Hewins, R.H. Jones, & E.R. D. Scott, Cambridge Univ. Press, 35, 1996)
J.N. Cuzzi, R.C. Hogan, J.M. Paque, A.R. Dobrovolskis: ApJ **546**, 496, (2001)
P. D'Alessio, J. Cantó , N. Calvet, S. Lizano: ApJ **500**, 411, (1998)
P. D'Alessio, N. Calvet, L. Hartmann, S. Lizano, J. Cantó: ApJ **523**, 893, (1999)
P. D'Alessio, N. Calvet, L. Hartmann: ApJ **553**, 321, (2001)
F. D'Antona, I. Mazzitelli: ApJS **90**, 467, (1994)
C. Dominik, A.G. G. Tielens: ApJ **480**, 647, (1997)
B. Dubrulle, G. Morfill, M. Sterzik: Icarus, **114**, 237, (1995)
M.J. Duncan, H.F. Levison, M.H. Lee: ApJ **116**, 2067, (1998)
P. Goldreich, S. Tremaine: ApJ **233**, 857, (1979)
P. Goldreich, S. Tremaine: ApJ **241**, 425, (1980)
P. Goldreich, W.R. Ward: ApJ **183**, 1051, (1973)

J. Goodman, B. Pindor: Icarus **148**, 537, (2000)

J.N. Goswami, H.A. T. Vanhala: *Protostars and Planets IV* (V. Mannings, A.P. Boss and S.S. Russell University of Arizona Press, Tucson, 963,2000)

Y. Greenzweig, J.J. Lissauer: Icarus, **87**, 40, (1990)

J.N. Grossman: *Meteorites and the Early Solar System*, (J.F. Kerridge and M.S. Matthews, University of Arizona Press, Tucson, 680, 1988)

T. Guillot, D. Gautier, W.B. Hubbard: Icarus **130**, 534, (1997)

E. Gullbring, L. Hartmann, C. Briceño, N. Calvet: ApJ, **492**, 323, (1998)

L. Hartmann: *Accretion processes in Star Formation*, (Cambridge University Press 1998)

P. Hartigan, L. Hartmann, S.J. Kenyon, R. Hewett, J. Stauffer: ApJS **70**, 899, (1998)

J.F. Hawley, S.A. Balbus, W.F. Winters: ApJ **518**, 394, (1999)

J.F. Hawley, C. Gammie, S.A. Balbus: ApJ **440**, 742, (1995)

Th. Henning, R. Stognienko: A&A **11**, 291-303, (1996)

S. Inaba, H. Tanaka, K. Nakazawa, G.W. Wetherill, E. Kokubo: Icarus **149**, 235, (2001)

R.H. Jones, T. Lee, H.C. Connally Jr., S.G. Love, H. Shang: *Protostars and Planets IV* (eds. V. Mannings, A.P. Boss and S.S. Russell University of Arizona Press, Tucson, 927, 2000)

S. Kenyon, L. Hartmann: ApJs **101**, 117, (1985)

H.H. Klahr, P. Bodenheimer: ' Disks, Planetesimals and Planets'. In: *ASP Conference Series* , eds. F.Garzn, C. Eiroa, D. de Winter, T.J. Mahoney, (Astronomical Society of the Pacifc, 2000)

E. Kokubo, S. Ida: Icarus, **131**, 171, (1998)

D.G. Korycansky, J.E. Pringle: MNRAS **272**, 618, (1995)

J.S. Lewis: Science **186**, 440, (1974)

H. Li, S.A. Colgate, B. Wendroff, R. Liska: ApJ **551**, 874, (2001)

C.C. Lin, F.H. Shu: ApJ **140**, 646, 1964

D.N. C. Lin, J. Papaloizou: MNRAS **186**, 799, (1979)

D.N. C. Lin, J. Papaloizou: MNRAS **191**, 37, (1980)

D.N. C. Lin, P. Bodenheimer, D.C. Richardson: Nature, **380**, 606, (1996)

D.N. C. Lin, J.C. B. Papaloizou, C. Terquem, G. Bryden, S. Ida: *Protostars and Planets IV*, (V. Mannings, A.P. Boss, and S.S. Russell, University of Arizona Press, Tucson, 1111, 2000)

J.J. Lissauer: Ann. Rev. Astron. Astrophys. **31**, 129, (1993)

J.J. Lissauer, G.R. Stewart: *Protostars and Planets III*, (E.H. Levy and J. Lunine, University of Arizona Press, Tucson, 1061, 1993)

R.V. E. Lovelace, H. Li, S.A. Colgate, A.F. Nelson: ApJ **513**, 805, (1999)

S.H. Lubow, P. Artymowicz: *Protostars and Planets IV*, (V. Mannings, A. P. Boss, and S.S. Russell, University of Arizona Press, Tucson, 731, 2000)

S.H. Lubow, G.I. Ogilvie: ApJ **504**, 983, (1998)

S.H. Lubow, J.E. Pringle: ApJ **409**, 360, (1993)

R. Lüst: Z. Naturforsch. **7a**, 87, (1952)

D. Lynden-Bell, J.E. Pringle: MNRAS **168**, 603, (1974)

J.M. Makino, T. Taiji, T. Ebisuzaki, D. Sugimoto: ApJ **480**, 432, (1997)

D. Mihalas: *Stellar Atmospheres* (Freeman, San Francisco 1978)
H. Mizuno: *Progr. Theor. Phys.* **64**, 544, (1980)
Y. Nakagawa, M. Sekiya, C. Hayashi: Icarus, **67**, 355, (1986)
A. Nelson, W. Benz, T.V. Ruzmaikina: ApJ **529**, 357, (2000)
A. Nelson, W. Benz, F.C. Adams, D. Arnett, D: ApJ **502**, 342, (1998)
G.I. Ogilvie: MNRAS **297**, 291, (1998)
G.I. Ogilvie, S.H. Lubow: ApJ **515**, 767, (1999)
B.K. Pickett, P. Cassen, R.H. Durisen, R. Link: ApJ **504**, 468, (1998)
B.K. Pickett, P. Cassen, R.H. Durisen, R. Link: ApJ **529**, 1034, (2000a)
B.K. Pickett, R.H. Durisen, P. Cassen, A.C. Mejia: ApJLett **540**, L95, (2000b)
J.B. Pollack, D. Hollenbach, D. Simonelli, S. Beckwith, T. Roush, W. Fong: ApJ **421**, 615, (1994)
J.B. Pollack, O. Hubickyj, P. Bodenheimer, J.J. Lissauer, M. Podolak, Y. Greenzweig: Icarus, **124**, 62, (1996)
V.S. Safronov: Ann. Astrophys. **23**, 979, (1960)
V.S. Safronov: *Evolution of the protoplanetary cloud and the formation of the Earth and planets*, (NASA TT F-67, chapter 4, 1969)
P. Saha, S. Tremaine: *AJ*, **104**, 1633, (1992)
P. Saha, S. Tremaine: AJ, **108**, 1962, (1994)
N.I. Shakura, R.A. Sunyaev: A&A **24**, 337, (1973)
D.P. Sheehan, S.S. Davis, J.N. Cuzzi, G.N. Estberg: Icarus **142**, 238, (1999)
F.H. Shu: *The Physics of Astrophysics. Vol. II: Gas Dynamics* (University Science Books, Mill Valley, CA., 1992)
F.H. Shu , D. Johnstone, D. Hollenbach: Icarus **106**, 92, (1993)
F.H. Shu, H. Shang, T. Lee: Science **271**, 1545, (1996)
F.H. Shu, H. Shang, M. Gounelle, A.E. Glassgold, T. Lee: ApJ **548**, 1029, (2001)
T.F. Stepinski, D.C. Black: A&A, **371**, 250, (2001)
S.W. Stahler: ApJ **274**, 822, (1983)
D.J. Stevenson: Planet. Space Sci. **30**, 755, (1982)
G.R. Stewart, W.M. Kaula: *Icarus* **44**, 154, (1980)
G.R. Stewart, S. Ida: *Icarus* **143**, 28, (2000)
D. Sugimoto, Y. Chikada, J. Makino, T. Ito, T. Ebisuzaki, and M. Umemura: Nature **345**, 33, (1990)
H. Tennekes, J.L. Lumley: *A First Course in Turbulence* (MIT Press, Cambridge, 1972)
E.W. Thommes, M.J. Duncan, H.F. Levison: Nature **402**, 635, (1999)
H.B. Throop, J. Bally, L.W. Esposito, M.J. McCaughrean: Science **292**, 1686, (2001)
A. Toomre: ApJ **139**, 1217, (1964)
D.E. Trilling, W. Benz, T. Guillot, J.I. Lunine, W.B. Hubbard, A. Burrows: ApJ **500**, 428, (1998)
H.J. Volk, F.C. Jones, G.E. Morfill, S. Roser: A&A **85**, 316, (1980)
M. Wadhwa, S.S. Russell: *Protostars and Planets IV*, (V. Mannings, A. P. Boss, and S.S. Russell, University of Arizona Press, Tucson, 995, 2000)

W.R. Ward: Icarus, **126**, 261, (1997a)
W.R. Ward: ApJLett, **482**, L211, (1997b)
W.R. Ward, J.M. Hahn, J: *ApJ* **440**, L25, (1995)
W.R. Ward, J.M. Hahn: *Protostars and Planets IV*, eds. (V. Mannings, A. P. Boss, and S.S. Russell, University of Arizona Press, Tucson, 1135, 2001)
J.T. Wasson: *Meteorites* (W.H. Freeman, New York, 155, (1985)
S.J. Weidenschilling: Astrophys. and Space Sci, **51**, 153, (1977)
S.J. Weidenschilling: MNRAS **180**, 57, (1977)
S.J. Weidenschilling, D. Spaute, D.R. Davis, F. Marzari, K. Ohtsuki: *Icarus* **128**, 429, (1997)
S.J. Weidenschilling, J.N. Cuzzi: *Protostars and Planets III*, eds. (E.H. Levy and J. Lunine, University of Arizona Press, Tucson, 1031, 1993)
G.W. Wetherill, G.R. Stewart: Icarus, **77**, 330, (1989)
G.W. Wetherill, G.R. Stewart: Icarus, **106**, 190, (1993)
G.W. Wetherill: Icarus **88**, 336, (1990)
G.W. Wetherill: Icarus **100**, 307, (1992)
G.W. Wetherill: Geochim. Cosmochim. Acta **58**, 4513, (1994)
G.W. Wetherill: Icarus **119**, 219, (1996)
J. Wisdom, M. Holman: AJ, **102**, 1528, (1991)
A. Wolszczan, D. Frail: Nature **255**, 145, (1992)
D.S. Woolum, P. Cassen: Met. and Planet. Sci. **34**, 897, (1999)
G. Wuchterl, T. Guillot, J.J. Lissauer: *Protostars and Planets IV*, eds. (V. Mannings, A.P. Boss, and S.S. Russell, University of Arizona Press, Tucson, 1081, 2000)
G. Wurm, J. Blum, J.E. Colwell: Icarus **151**, 318, (2001)
C. Yuan, Y. Cheng: ApJ **340**, 216, (1989)
Astrophysical Implications of the Laboratory Study of Presolar Materials. Edited by Thomas J. Bernatowicz and Ernst Zinner. (1997)
Meteorites and the Early Solar System, Ed. J.F. Kerridge and M.S. Matthews, (1988)
Chondrules and the Protoplanetary Disk, Ed. R.H. Hewins, R.H. Jones and E.R.D Scott, (1996)

Index

ABCD detection, 148
abundance profiles, 336
achromatic null, 181
achromatic phase shifts, 189
adiabats, 249
Alfvén velocity, 393
ammonia, 250, 251
Angel's Cross, 206
anisoplanatism, 156
azimuthal drag, 405

Babinet's Principle, 15
bandwidth smearing, 148
baroclinic instabilities, 392
barotropic equilibrium, 262
baseline vector, 144
Bond albedo, 345
Brünt-Väisälä frequency, 395
Bracewell, 206

$CaTiO_3$, 337
caustic, 82
centrifugal force, 260
chondritic meteorites, 442
chondrules, 413
Clausius–Clapeyron, 334
close-in planets, 43
conductive opacity, 289
corotation radius, 398
Coulomb potential, 269
coupling parameter, 269

DARWIN, 181, 207
degenerate gaz, 268

deuterium, 253, 273
diffusion approximation, 280
disk
    accretion luminosity, 380
    angular momentum, 386
    mass conservation, 384
    mass flux, 388
    Rayleigh's criterion, 390
    temperature-weighted mass, 377
    two-layer model, 376
dual-star system, 168
dusty brown dwarfs, 342

eccentricity decay, 46
Eddington approximation, 254
Einstein radius, 78, 83
Einstein ring, 78
energy conservation, 253
EOS
    models, 275
    polarization, 275
    hydrogen, 275
epicyclic frequency, 395

Fraunhofer diffraction, 135
Fresnel scale, 141
Fried parameter, 135
frozen turbulence hypothesis, 139

GAIA, 171
GEST, 93
giant planets
    mass, 244
Grüneisen parameter, 278

grain growth, 340
gravitational moments, 245, 263
gravity field, 244

Hayashi
  evolution track, 347
  forbidden region, 347
helium, 246
  ionization, 271
  sedimentation, 317
Helmoltz free energy, 267
Hill sphere, 60, 417
Hot Jupiters, 42
hot spot, 250
Hufnagel turbulence profile, 157
hydrogen, 246
  metallic, 269
  PPT, 276
Hydrostatic equilibrium, 425
hydrostatic equilibrium, 253, 262

Iodine Absorption cell, 73
irradiated planets, 259
isoplanatic angle, 140

Jeans criterion, 23
Jupiter
  age, 316
  core, 301, 427
  formation, 425
  growth, 428
  red spot, 251
  tropopause, 254

Kelvin–Helmoltz instability, 356
kinematic viscosity, 387
Kolmogorov model, 130

Lane-Emden equation, 256
Legendre polynomials, 261
LiH, 325
Lindblad resonances, 398, 435
local thermodynamic equilibrium, 373
Lorentzian line profile, 284
luminosity, 253

magnetic fields, 246
Marechal approximation, 138
metallicity excess, 49
methane, 250, 251

migration
  type I, 438
  type II, 438
millisecond pulsars, 6
MkIII Interferometer, 153
modal filter, 186
moist convection, 338
monoatomic gas, 259
monochromatic opacity, 280

narrow-angle astrometry, 168
Neptune
  formation, 425
Neutron star disks, 8
null depth, 182, 188
nulling array, 201

OASES, 206
object
  β Pic, 31
  ι Dra, 44
  ε Eri, 30
  16 Cyg B, 50
  47 UMa, 55
  51 Oph, 33
  51 Peg, 75
  55 Cancri, 38
  55 Cancri, 54
  70 Vir b, 44
  B1620−26, 8
  BP Tau, 379
  CM Dra, 121
  Gliese 876, 55, 161
  GM Aur, 26
  HD 209458, 112, 344
  HD 141569 A, 30
  HD 166435, 68
  HD 192263, 68
  HD 4796 A, 30
  HD 80606, 44
  HD 82943, 49
  HD 83443, 51
  HD 89744 b, 44
  HD 12661, 57
  HD 168443, 57
  HD 37124, 57
  HD 38529, 58
  HD 74156, 58
  HD 82943, 58

HH-30, 26
HL Tau, 26
KH 15D, 31
LkCa7, 379
MACHO 97-BLG-41, 94
MACHO 98-BLG-35, 94
OGLE-TR-3, 118
OGLE-TR-56, 118
OGLE #7, 82
Orion Trapezium, 26
PSR B1257+12, 6
UX Orionis, 33
WW Vul, 33
Oort Cloud, 31

particule growth, 407
Pauli principle, 271
perfect gaz, 269
phase diagram, 271
planet
    albedo, 123
    figure, 260
PLANET collaboration, 94
planetary
    rings, 252
    systems, 51
polytropic index, 256
Poynting-Robertson drag, 28
PPI cycle, 320
Prandtl number, 294

Q dissipation factor, 352

Rankine–Hugoniot relations, 272
Rayleigh criterion, 136
Rayleigh scattering, 287
Reynolds number, 294, 412
Roche lobe overflows, 351
Rossby
    number, 294
    waves, 403
Rosseland opacity, 285
runaway growth, 307

satellites, 252
    irregular, 252
saturated atmosphere, 338

saturation pressure, 335
Saturn
    formation, 425
scintillation, 142
secular resonance, 59
SED, 371
SIM, 163, 173, 177
Simultaneous Thorium Technique, 72
solid body rotation, 264
strehl ratio, 137, 138, 195
Stroke mismatch, 155
structure function, 131
superadiabatic temperature gradient, 390

Taylor columns, 294
Taylor–Proudman, 251
temperature gradient, 337
temperature profile, 249
Thermal escape, 351
thermometric conductivity, 289
tidal circularization, 45
tidal dissipation, 46
tidal torque, 432
TiO, 337
tip-tilt, 195
Toomre's criterion, 398
torque density, 436
transit
    depth, 103
    duration, 102
    egress, 103
    ingress, 103
tropopause, 298
turbulence height, 141

Uranus
    formation, 425

visibility, 151
VLBI, 164

Wiener-Khinchin Theorem, 132
winds
    zonal, 358
WKB approximation, 395